Photoreception and Sensory Transduction in Aneural Organisms

NATO ADVANCED STUDY INSTITUTES SERIES

A series of edited volumes comprising multifaceted studies of contemporary scientific issues by some of the best scientific minds in the world, assembled in cooperation with NATO Scientific Affairs Division.

Series A: Life Sciences

Recent Volumes in this Series

The series is published by an international board of publishers in conjunction with NATO Scientific Affairs Division

A Life Sciences	Plenum Publishing Corporation
B Physics	London and New York
C Mathematical and	D. Reidel Publishing Company
Physical Sciences	Dordrecht, Boston and London
D Behavioral and	Sijthoff & Noordhoff International
Social Sciences	Publishers
E Applied Sciences	Alphen aan den Rijn, The Netherlands, and Germantown U.S.A.

Photoreception and Sensory Transduction in Aneural Organisms

Edited by
Francesco Lenci
and
Giuliano Colombetti

Consiglio Nazionale delle Ricerche
Pisa, Italy

PLENUM PRESS • **NEW YORK AND LONDON**
Published in cooperation with NATO Scientific Affairs Division

Library of Congress Cataloging in Publication Data

Nato Advanced Study Institute on the Photoreception and Sensory Transduction in Aneural
   Organisms, Versilia, Italy, 1979.
   Photoreception and sensory transduction in aneural organisms.

   (Nato advanced study institute series: Series A, Life sciences; v. 33)
   "Lectures presented at the NATO Advanced Study Institute on the Photoreception and
Sensory Transduction in Aneural Organisms, held in Versilia, Italy, September 3–14,
1979."
   Includes index.
   1. Photobiology–Congresses. 2. Photoreceptors–Congresses. 3. Senses and sensation–
Congresses. I. Lenci, F. II. Colombetti, Giuliano. III. Title. IV. Series.
QH515.N22 1979   574.19'153   80-12426
ISBN 0-306-40437-0

Lectures presented at the NATO Advanced Study Institute on the
Photoreception and Sensory Transduction in Aneural Organisms, held in
Versilia, Italy, September 3–14, 1979.

© 1980 Plenum Press, New York
A Division of Plenum Publishing Corporation
227 West 17th Street, New York, N.Y. 10011

Printed in the United States of America

PREFACE

This book collects all the lectures presented during the NATO
Advanced Study Institute on "Photoreception and Sensory Transduction
in Aneural Organisms," held in Villa Le Pianore (Versilia, Italy),
September 3-14, 1979.

In order to publish the lectures in the shortest possible time,
we had to make the decision not to include the free communications,
the informal seminars, and the panel discussions, notwithstanding
their very high scientific level and interest. Only the final
panel discussion has been summarized by Prof. W. Haupt (whose
effort we gratefully acknowledge), because it gives a comprehensive
view of the state of the art in this field.

The ASI was intended to be a high-level course, characterized
by an interdisciplinary approach to the problem of photoreception
and photosensory physiology in aneural organisms, bringing together
scientists from different fields and specializations. We hope that
these characteristics are reflected in the content of the book,
which is meant to be both an advanced textbook for researchers and
students entering the field and a critical overview of the problems
of photosensory transduction in aneural organisms. The topics presented range from a phenomenological description of the different
photomotile responses in various microorganisms to a discussion of
the molecular processes involved in the primary events of photoreception as well as in the subsequent steps of the transduction
chain.

We owe a deep debt of gratitude to Maria Antonia Baldocchi
and Graziella Baldeschi, whose organizational and typing expertise
proved invaluable. Thanks are also due to Vincenzo Passarelli,
who helped us in the final preparation of the book.

Finally, we gratefully acknowledge the generous financial
sponsorship of the Scientific Affairs Division of the NATO
(Brussels), which has made this ASI (and this book) possible.

Francesco Lenci and Giuliano Colombetti
C.N.R.-Lab. Studio Proprietà Fisiche
Biomolecole e Cellule
56100 PISA (Italy)

Pisa, November 27, 1979

CONTENTS

CONTENTS

SENSORY TRANSDUCTION IN ANEURAL ORGANISMS

Michael J. Carlile

Dept. of Biochemistry, Imperial College of Science
and Technology
London SW7 2AZ (England)

Transducer. Any device by which variations in one quantity
(e.g. pressure, brightness) are quantitatively converted into variations in another (e.g. voltage, position) - Oxford English
Dictionary (1). It is from this term that transduction, as employed
in sensory physiology, appears to be derived.

Aneural organisms. Microorganisms (bacteria, algae, protozoa,
and fungi) and plants, which unlike animals, lack a nervous system.
There has been relatively little recent research on the sensory
responses of plants and much on microorganisms, and the present
review will discuss the latter only.

Many microorganisms are motile by means of cilia or flagella
or can glide or crawl. Such behaviour is itself valuable for dis-
persing organisms to new sites, but becomes far more useful if
coupled to a sensory system which will permit favourable locations
to be reached and harmful ones avoided. We may hence expect a motile
microorganism to possess a sensory system capable of responding to
those environmental variables of major importance to it and of
modifying motility in an appropriate way. A sensory system can also
be of value to non-motile organisms. Most fungi and some algae show
highly polarized growth in the form of hyphae which spread over the
substratum, sometimes penetrate it, and sometimes rise above to facilitate spore dispersal. Here too, there are sensory systems able
to guide polarized growth in an advantageous direction.

Microorganisms have three outstanding advantages for research
on sensory responses - many species can be obtained in large numbers

and uniform condition for biochemical studies, many are exceptio-
nally favourable material for genetic analysis, and the entire
pathway from stimulus to response occurs within a single cell. No
other material has all these advantages. Sensory responses by higher
plants involve the interaction of many cells, as do those of
nervous system. Studies on sensory responses by individual animal
cells (e.g. receptor cells, leucocytes), are limited by the problems
of obtaining sufficient homogeneous material for biochemical examination and the difficulties of genetic analysis. Hence one of the
great objectives of biological research, the complete description
in molecular terms of a sensory reaction from stimulus to response,
is most likely to be achieved with a microorganism.

      Photoresponses have some important advantages for such research
– for example, the magnitude and duration of the stimulus can be
more precisely defined than is possible with a chemosensory response.
Studies on the sensory responses of microorganisms towards chemicals
and physical agents other than light, are however of interest to
photobiologists, since parallels may be expected between the mechanisms involved in sensory transduction, and part of the pathway
from stimulus to response may be shared between different sensory
systems. This is especially likely to be true for photosensory and
chemosensory systems, if, as seems possible, microbial photosensory
systems may arise by the modification of a pre-existing chemosensory
system (2).

BASIC CONCEPTS AND SEMANTIC PROBLEMS

      The polarized growth of a non-motile organism towards or away
from a stimulus is referred to as a tropism or tropic response.
Unhappily there is no such unanimity with respect to the terms used
to describe the behavioural response of motile organisms, the same
terms being employed by different workers in slightly or wholly
different ways. I have elsewhere tabulated the main types of responses and some alternative terms used (2) and a group of photobiologists
have made recommendations on terminology (3).

      The existence of a sensory response in motile organisms is
usually recognized through the observation that some agent has
brought about an accumulation or aggregation of organisms or alternatively has dispersed them from a locality. The two effects may be
intimately related, with a stimulus of low intensity being attractive
and one of high intensity being repellant. A single expression to
cover both effects is hence needed and the term taxis (pl. taxes)

has been widely adopted. It has also been used in a more restricted
sense (see below). In this paper, it will be used in the broad
sense, and the way other terms are being used will be indicated.

Orthokinesis

Accumulation or dispersal requires that motility be modified
in some way and taxes can be classified according to the nature of
the modification. Orthokinesis (some workers refer instead to
kinesis with a qualification as to the stimulus, e.g. photokinesis)
refers to the speed of movement being determined by the intensity
of a stimulus. Positive orthokinesis (an increase in stimulus
increases speed) will disperse organisms from a locality. For
example, the first effect of illuminating part of a population
of Rhodospirillum rubrum hitherto in darkness, is to expel the
cells from the most intensely illuminated area - see Fig. 1 of
Clayton (4) - Negative orthokinesis, on the other hand, leads to
the accumulation of organisms. This is most easily understood by
considering the extreme case in which any organism coming near the
source of the stimulus ceases moving, which will lead inevitably
to accumulation of cells.

Temporal Sensing - Klinokinetic or Phobic Responses

Accumulation or dispersal can also result from klinokinesis,
in which a stimulus alters the frequency with which an organism
changes direction. Klinokinesis in Escherichia coli was studied
(5) by means of a microscope devised by Berg (6) and capable of
precise tracking of the path of an individual cell in three dimensions. In the absence of a stimulus, the path of a cell consisted
of approximately straight "runs" interspersed at irregular intervals
with "tumbles" which brought about a directional change. Truly
random directional changes would average 90°, but those observed
by Berg and Brown (5) averaged about 60°, i.e. the cells tended to
continue forward. It was found that if the cells were travelling
up a gradient of a chemical known to be an attractant then the
frequency of tumbles was diminished, so favourable conditions are
approached by a biassed random walk. Note that the favourable
stimulus suppresses spontaneous turning; it hence seems inappropriate
to use the expression "phobic response" (see below) for this form
of klinokinesis. Attraction through klinokinesis requires that
temporal comparison of stimulus be made, so that a "memory"

allowing the stimulus intensity to be compared with that experienced
a little earlier is essential. MacNab and Koshland (7) found, with
<u>Salmonella typhimurium</u>, an organism closely related to <u>E. coli</u>, that
a sudden increase in the concentration of an attractant produced by
means of a rapid mixing device suppressed tumbling, a result consistent with the hypothesis of temporal sensing. Unlike Berg and Brown
(5) they found also that a sudden reduction in attractant concentration increased tumbling, the discrepancy is perhaps due to a far
sharper change in stimulus intensity in the <u>S. typhimurium</u> experiments.

    Suppression of tumbles was also found when cells of <u>S. typhi-</u>
<u>murium</u> descend a repellent gradient or experience a sudden diminution
in repellent concentration (8), hence repellents act in a comparable
but opposite way to attractants. It should be noted that the changes
in tumble frequency seen in the rapid mixing experiments were temporary, adaptation to the new level of attractant or repellent soon
occurred and tumble frequency returned to the usual value. Clearly
although a "memory" is needed for klinokinesis, it has to be a short
one, the cells must soon adapt to the changed stimulus intensity
and treat departures from it as a new stimulus. Only in this way
will cells continue to travel along an intensity gradient in an
appropriate manner or remain at the most favourable site when they
reach it. A computer simulation study has confirmed that klinokinesis
with adaptation can bring about aggregation but that in the absence
of adaptation it is ineffective (9).

    <u>Escherichia coli</u> and <u>Salmonella typhimurium</u> are rods about
2 μm long and 0.5 μm diameter. Cells of this size show marked
Brownian motion due to random thermal agitation and in consequence
during an average run of about one second a mean directional change
of about 20° occurs (Table 1 in Ref. 5). Cells therefore, can per-
sist in a given direction for at most a few seconds. Hence, a
biassed random walk with directional changes occurring about once
per second, with their frequency modulated by whether progress was
in the "right" direction as determined by a short memory, seems
the most feasible tactic for such cells. In fact, the klinokinetic
response of <u>E. coli</u>, and <u>S. typhimurium</u>, although random with
respect to stimulus gradient and nearly so with respect to the
previous direction of movement, are a very effective method of
reaching favourable sites.

    Spirilla, such as <u>Rhodospirillum rubrum</u> and <u>Thiospirillum</u>
<u>jenense</u>, and some other bacteria such as <u>Halobacterium halobium</u>,
have bundles of flagella at opposite ends of the cells, and at

interval reverse their swimming direction through reversing the
direction of rotation of their flagella (10). The result is an
approximately 180° change in direction of movement in contrast
to the less marked (average 60°) turn produced by tumbling in
Escherichia coli. In normal sized cells (0.5 μm wide x 2-5 μm
long) of R. rubrum, the resulting path is still however a random
walk (see Fig. 4 in Ref. 11), due presumably to Brownian movement
and perhaps also to other causes such as a frequently asymmetrical
cell shape and possibly imperfect coordination of flagellum activity. When R. rubrum, T. jenense or H. halobium cross a boundary
from light into darkness, reversal in swimming direction occurs,
constituting a classical phobic response (10). Such a response
resembles that of Escherichia coli in that its direction is not
determined by the direction of the stimulus gradient but differs
in that its direction is determined by the previous path of the
organism. The changed direction is thus indirectly related to the
direction of the stimulus gradient and the phobic response is able
to "send the organism in the right direction" through a single
direction change. Such a tactic is likely to be more effective
with large bacteria such as T. jenense (2.5 - 4 μm x 30-40 μm)
than with small ones; it has been found that long forms of R. rubrum
(10-70 μm) show a more precise reversal than normal sized cells
(11). When a stimulus gradient is not steep, a single phobic response will be ineffective, and phototaxis will presumably result
from a biassed random walk in which fewer reversals occur when
the organism is moving in the "right" direction than in the "wrong"
one.

The term taxis is sometimes restricted to responses in which
cells become oriented with respect to the stimulus gradient and
move directly towards or away from the stimulus source instead of
by a biassed random walk. In view of the broad use of taxis it is
probably best to term such responses klinotaxis or topotaxis (see
below). Klinotaxis can occur in an organism capable of a series of
phobic responses each producing a small angle turn oriented in the
same direction with respect to the immediately preceeding path.
Thus as the eukaryotic flagellate Euglena moves in its helical path
periodic shading of the photoreceptor leads to a series of small
turns which cease when the organism is pointing towards the light
and periodic shading of the anteriorly located receptor no longer
occurs (12).

The responses discussed in this section have the common feature
that they are based on temporal comparisons of stimulus intensity.

The paths taken will depend on the morphology of the organism, its
method of locomotion and the steepness of the temporal gradient to
which it is exposed. The latter, in natural situations, will be
determined by the steepness of a spatial gradient and the speed
with which the organism moves. Gunn (13) has recently re-examined
the use of the term klinokinesis and Kennedy (14) has discussed
the significance of responses oriented with respect to the previous
path.

Spatial Sensing: Topotaxis and Tropism

Some organisms are able to carry out spatial sensing - to make
simultaneous comparisons of stimulus intensity at different points
on the body surface and to orient themselves appropriately - Such
orientation by a motile organism is often termed topotaxis; oriented
growth by a stationary organism, as already indicated, is a tropism.
The usefulness of spatial sensing is determined both directly by
the size of the organism and also by the size in relation to speed
of movement (15). A very small organism will only be able to detect
differences in stimulus intensity between front and rear if the
gradient being sampled is very steep, and, moreover, in the case
of chemoreception intensity differences over very small distances
may be due to random fluctuation in the concentration of the chemical. A sensory response, involving reception, transduction and the
action of the motor apparatus takes a finite time, and with a small
organism the simultaneous difference in intensity at different points
on the body surface may be less than the change a receptor experiences due to cell movement during the time taken for a response.
Escherichia coli (length, 2 μm; speed ca. 30 μm/sec) moves several
body lengths per second, and with such small fast organisms more
useful information about shallow gradients can be obtained by
"second-to-second" comparisons during approximately straight runs
than by spatial comparisons. Topotactic sensory responses almost
certainly do not occur in swimming prokaryotes and are probably
infrequent in swimming eukaryotes of microscopic dimensions. Spatial
sensing, however, is likely to be of value to relatively large
micro-organisms that move slowly in relation to body length. These
are gliding prokaryotes (blue-green algae, myxobacteria), gliding
eukaryotes (diatoms) and eukaryotes that crawl, the various amœboid
forms. The amoebae of the slime mould Dictyostelium discoideum show
a direct approach in responding to the attractant acrasins, and
Berg (15) has shown the feasibility of spatial sensing for these

amoebae (diameter 10 μm; velocity about 0.2 μm sec). Mammalian leucocytes, which reach 15-20 μm long can orient precisely in chemotactic gradients (16); it is estimated that a concentration difference of 1% is detectable between "head" and "tail" (17). The large
grex ("pseudoplasmodium") of D. discoideum, consisting of up to 10^5
amoebae, shows phototactic orientation based on side-to-side comparisons of light intensities (18). Tropism of eukaryotic hyphae is
also based on side-to-side comparisons of stimulus intensity, as
for example in the phototropism of the sporangiophores (the giant
hyphae that bear spore-containing sporangia) of fungi (19). Tropism
has not been reported in hyphal prokaryotes; cell diameters are
normally too small (ca. 0.5-1 μm) for useful side-to-side intensity
comparisons.

The spatial sensing involved in topotaxis and tropism has
sometimes been termed "one-instant" sensing in contrast to the
"two-instant" temporal sensing involved in klinokinesis and klinotaxis (20,21).

Adaptation

Adaptation is a very widespread feature of sensory systems –
organisms respond when there is a change in stimulus intensity but
soon became adapted to the new intensity – and cease to respond
in the absence of any further change. Thus Escherichia coli responds
to an increased concentration of an attractant or a reduced concentration of a repellant by smooth swimming but within about a minute
the normal frequency of tumbling is resumed (22). An increased concentration of a repellant or a reduced concentration of an attractant
lead to increase tumbling, but within a few seconds, tumble frequency returns to the normal level. The slow adaptation to a favourable
stimulus which causes smooth swimming and the fast adaptation to an
unfavourable one which causes rapid direction changes constitute
a sound strategy (23). A useful approach in the analysis of sensory
responses is that of a level of excitation, which changes very
rapidly in response to a change in stimulus intensity, and a level
of adaptation, which changes more slowly. This approach has been
applied, explicitly or implicitly, to both bacterial chemotaxis
(25) and the phototropism of the Phycomyces sporangiophore (25,26).
The molecular basis of sensory adaptation in E. coli and Salmonella
typhimurium is now being clarified, and will be discussed in a later
section.

Sensory adaptation, as already indicated, is essential if

klinokinesis is to bring about aggregation. Adaptation has, however, a more general significance in sensory perception. A stimulus of constant intensity conveys no useful information to an organism; only differences in stimulus intensity, experienced for example when an organism moves up a concentration gradient, are meaningful. As a result of adaptation, a new stimulus intensity soon becomes "normal" and only changes from this intensity, such as a return to the earlier value, can be detected. Such adaptation permits a sensory system to operate with a nearly constant efficiency over an enormous range of stimulus intensities. This renders adaptation of value in spatial as well as in temporal sensing - adaptation to changes in light intensities in the <u>Phycomyces sporangiophore</u>, for example, enable phototropism to occur over a 10^9-fold range of light intensities.

The advantages of adaptation are so great that adaptation is likely to be an integral part of most sensory systems. There is however one type of response for which adaptation is unnecessary and perhaps even undesirable. An avoidance response by an organism capable of reversal or a nearly 180° turn as it encounters harmful conditions. Thus zoospores of the fungus <u>Phytophthora</u> are repelled by H^+ ions above a critical concentration (27); there is no repulsion at lower concentrations and adaptation does not occur (Cameron & Carlile, to be published). The phobic response of <u>Halobacterium halobium</u> to darkness also does not show adaptation (10).

Orthokinesis, Klinokinesis and Topotaxis - Distinctions and

Interactions

The sensory responses seen in present day organisms have evolved from less sophisticated responses to meet the needs of a great diversity of species in varied environments. It is hence not surprising that classification of the responses into types which can be precisely defined is not easy. Moreover, a single species may show orthokinesis, klinokinesis and topotaxis and its behaviour will be correspondingly complex.

It is questionable as to whether orthokinesis should be regarded as a sensory response. Negative orthokinesis can act as a rather ineffective means of aggregation and positive orthokinesis can cause repulsion. Temporal sensing can be very effective; in <u>Escherichia coli</u> klinokinesis gives migration speeds in the "right" direction of about one-tenth of the swimming speed of the organism

(23). As suggested above, it may sometimes be convenient to distinguish between klinokinesis and phobic responses, although more information on turning angles is needed before it will be clear as to whether the distinction is generally useful. Temporal sensing can approach spatial sensing in its effectiveness; the klinotaxis of Euglena results in a path almost as direct as that of topotaxis.

Although positive orthokinesis has by itself a weak "repellent" effect, computer simulation studies indicate that it increases the aggregation (9). An organism in the vicinity of an attractant source may first experience very shallow gradients, detectable only by temporal sensing, and on closer approach, steep gradients that permit spatial sensing. It is hence likely that many organisms combine orthokinetic, klinokinetic and topotactic capabilities to give optimal results. Leucocytes, for example, have been shown to respond to attractants by increased speed, diminished angles of turn, and precise orientation (28).

TYPES OF STIMULI

Microorganisms show taxes and tropism to a wide range of stimuli (2). The comments below stress areas of intense interest or aspects likely to be of interest to photobiologists.

Light. Studies of phototaxis have been confined largely to organisms obtaining their energy from light. In such organisms it may prove difficult to distinguish the reactions involved in phototaxis from the effects of photosynthesis, and it seems surprising that more use has not been made of non-photosynthetic organisms. There have been limited studies of the phototaxis of the cellular slime mould Dictyostelium discoideum (18,29), plasmodia of the myxomycete Physarium (30,31) and zoospores of the aquatic fungi Allomyces (32) and Phlyctochytrium (33). Studies on phototropism, however, have been carried out largely on non-photosynthetic organs, the aerial hyphae of fungi, especially the sporangiophores of the Zygomycetes. The photoresponses of the sporangiophore of Phycomyces have been intensively studied, yielding a detailed physical description of the relationship between stimulus and response and more recently a mutational analysis of the transduction path including inputs from receptors for stimuli other than light (19,34).

Information on biochemical aspects of Phycomyces photoresponses remain,however, slight. A carotenoid free mutant has been used in a sophisticated action spectrum study demonstrating optical excitation of the lowest triplet state of riboflavin (35); it is sur-

prising that more use has not been made of albino mutants in
photosensory studies.

Chemicals. Chemoresponses are probably the most widespread of
all microbial sensory reactions, and range from responses to factors as simple and widespread as the hydrogen ion (27) to those as
complex and specific as sex attractants (36,37). The intensively
studied chemotactic responses of Escherichia coli and Salmonella
typhimurium will be considered in more detail later. The chemotaxis of the amoebae of the cellular slime mould Dictyostelium
discoideum (38,39,40) and of mammalian leucocytes (28,41), are also
receiving intensive study. Chemotropism has been less thoroughly
studied than chemotaxis.

Mechanical forces. The most detailed studies on the effects
of contact and the resulting pressure on the cell membrane are
those carried out on the avoidance reaction (actually a collision
reaction) of the ciliate Paramecium (42,43). This organism is
suitable for both genetic and electrophysiological work and impres-
sive progress has been made in the analysis of the steps from stimulus to response. The contact guidance shown by fibroblasts and
other amoeboid cells is probably not really a sensory response but
the result of the interaction of the cell surface with the physical
properties of the substratum, especially shape and adhesiveness (44).

Heat. There has been a recent resumption of interest in thermotaxis, with studies on Escherichia coli (45), Paramecium (46),
plasmodia of the myxomycete Physarum polycephalum (47) and the grex
of Dictyostelium discoideum (48). The topotactic response of the
latter is of very high sensitivity, a gradient of 0.004 mm^{-1} (0.0004
across the grex) being detected and also shows adaptation to changed
temperature, effective temperatures being dependent on growth
temperature. The response in Paramecium is also adaptive, and
both may depend on phase transitions in the plasma membrane lipids
(46,48). The thermoreceptor in E. coli, however, is the chemoreceptor
involved in attraction by serine (45) and presumably is a protein.

Electricity. A distinction has been drawn (3) between tactic
responses to an ionic electric current (galvanotaxis) and to an
electrical field (electrotaxis). Galvanotaxis has received considerable attention in Paramecium (43) and is of general interest in
view of the widespread role of membrane potential changes in
sensory responses.

Gravity. Many microorganisms including algae, e.g. Chlamydomonas (49), protozoa, e.g. Paramecium (50,51), and fungal zoospores,
e.g. Phytophtora (52), show negative geotaxis, i.e. tend to swim

upwards. Gravity differs, however, from the stimuli so far discussed in that it is constant in direction and intensity. It is hence feasible for geotaxis to be produced, in organisms of suitable morphology,by purely physical mechanisms without the intervention of a sensory system. In Tetrahymena the center of propulsive effort is in front of the center of mass, and negative geotaxis has been explained in terms of the interaction of gyrational torque produced in swimming and sedimentation torque produced by gravity, but the precise mechanisms involved in negative geotaxis remain controversial (49,50,51) and may not be the same in different organisms. It seems clear however that physical rather than physiological mechanisms are involved; mutants lacking other sensory responses remain geotactic, and geotaxis can be used in selecting such mutants (43). Negative geotropism of the Phycomyces sporangiophores is brought about through a transduction pathway which is in part common with the photosensory and other pathways controlling growth (19,34). Stretch receptors and perhaps vacuolar displacement are involved (2).

    Magnetism. Magnetotactic bacteria (54) containing chains of magnetite crystals (55) have been reported. Magnetism, like gravity, is at any particular location nearly constant in direction and intensity, and magnetotactic orientation in these microorganisms is likely to be a purely physical reaction rather than one involving a sensory system.

AN INTENSIVELY STUDIED SYSTEM: CHEMOTAXIS OF FLAGELLATE BACTERIA

    Chemotaxis in bacteria has been known for nearly a century. The modern phase of study dates from about 1965 when Adler (56) showed that the most intensively studied of all bacteria, the gram-negative enteric (gut) organism, Escherichia coli, was suitable for the investigation of chemotaxis. Other workers subsequently became involved, and another very closely related enteric bacterium, Salmonella typhimurium, was also studied, yielding, as might be expected, very similar results. More limited studies have been carried out on other flagellate bacteria, especially the gram-positive Bacillus subtilis. The major conclusions from the first decade of these studies were summarized by Adler (57) and Berg (58). Since 1975 the volume of work has increased and numerous reviews have been published on bacterial chemotaxis (59,60), special aspects of bacterial chemotaxis including control mechanisms (25,61), genetics, (62), and adaptation (24,63) and on the related topic of bacterial

flagella (64-68). A massive review of all these areas is that of
MacNab (23). These studies on bacterial chemotaxis have now reached
the stage where complete analysis in molecular terms of a sensory
transduction path from stimulus to response becomes a feasible
objective. In view of the volume of the work and the quantity and
quality of reviewing activity it is here necessary only to summarize the major conclusions. These were obtained, unless otherwise
stated, with E. coli, or S. typhimurium.

As indicated earlier, these bacteria reach favourable locations by means of a biassed random walk, the frequency of random
directional changes (tumbles) being greater if the organism is
going in the "wrong" direction than if it is going in the "right"
one.   A bacterial flagellum is a semi-rigid helix which is
rotated by a "motor" located at the plasma membrane. During normal
smooth swimming the flagella are left-handed helices which are
rotated in a counter-clockwise direction, as viewed along a flagellum towards the cell. Hydrodynamic and mechanical interaction causes
the flagella to form a bundle which as a smoothly rotating "tail"
propels the cell forward. Transient reversal to the clockwise mode
by the flagellum motors causes the flagella to convert to tighter,
right-handed helices through reverse viscous torque. The flagellum
bundle flies apart and tumbling occurs (23,64). The problem of
sensory transduction hence is reduced to the question of how attractants diminish, and repellants increase, the frequency of transient
reversal of the flagellum motor, and how adaptation to changed
stimulus levels occur.

Many attractants and repellents have been demonstrated. Most
of the attractants can be metabolized, but there are some substances
that can be metabolized without causing attraction, and some attractants that are not metabolized. Theories postulating that
attractants or repellents act through gross effects on metabolism
are hence untenable and the evidence strongly indicates a range of
receptors, located in the plasma membrane or the periplasmic space
between membrane and cell wall, and each specific for one or a few
structurally related compounds. A few have been isolated and shown
to be proteins. Genetic studies have demonstrated that some mutations affect a single chemoreceptor, abolishing sensitivity to
one or a few structurally related compounds, and that others pre-
vent a response to a wider range of substances. It has hence been
postulated that there are groups of primary chemoreceptors which
can interact with a secondary receptor that in turn influences the
motor apparatus of the cell. Three such secondary receptors have

so far been found, each interacting with a different group of primary receptors. These secondary receptors have been designated
transducer proteins, or methyl-accepting chemotaxis proteins (MCP's),
since their methylation has an essential role in chemotaxis. It is
postulated that the binding of an attractant or repellent to an
appropriate receptor causes a conformation change first in that
receptor and as a consequence in the appropriate MCP. This is
thought to bring about an immediate change in the level of a hypothetical "tumble regulator". The presence of an attractant will
result in the methylation of a carboxyl group of the appropriate
MCP and a repellent its demethylation. The rates at which these
two processes occur correspond to the slow adaptation of E. coli
to attractants and the fast adaptation to repellents (63). The
consequent changes in the MCP's are thought to restore the tumble
regulator to its normal level. It has been suggested on the basis
of experiments with Bacillus subtilis that the tumble regulator
is Ca^{++}, high levels causing tumbling, and that the MCP's act by
controlling gates admitting exogenous Ca^{++} to the cell (24). Other
ions, however, may also influence the flagellum motor, it has been
shown that in Bacillus subtilis Mg^{++} is required for smooth swimming (70), and postulated that Ca^{++} may act by displacing Mg^{++}
from a binding site (69).

The nature and significance of changes in transmembrane potential in bacterial sensory transduction remains uncertain. Direct
electrophysiological methods of determining membrane potential are
not feasible for bacteria, and use has been made of a tritium-label-
led lipid-soluble permanent cation (71), cyanine dye fluorescence
(72), and absorbance changes of membrane carotenoids (73,74); in
E. coli, Szmelcman and Adler (71) found that both attractants and
repellents caused an initial increase in membrane potential (hyperpolarization) and Armitage and Evans obtained similar results with
Rhodopseudomonas sphaeroides (74). Since attractants and repellents
have opposite effects on the flagellum motor, it seems probable
that the nature of the ion flux caused by attractants and repellents
and responsible for hyperpolarization differs.

Harayama and Iino (73), studying phototaxis of Rhodospirillum
rubrum found that a decrease in light intensity caused a fall in
membrane potential (depolarization) and motility reversal, and an
increase in light intensity hyperpolarization and smooth swimming.
Miller and Koshland (72), with Bacillus subtilis, did not detect
membrane potential changes when attractants were added but found
that an imposed increase in membrane potential caused a period of

smooth swimming and a decrease caused transient tumbling; these
effects parallel those obtained with R. rubrum as a result of
light-intensity changes. Caraway & Krieg (75) found with Spirillum volutans that reversing the polarity of an electrical field or
switching on or off an electrical circuit caused an instantaneous
reversal in swimming direction. The precise coordination of flagellum activity normally occurring between the two ends of these
very long bacteria would seem to require electrical signalling.
Further work is clearly needed to clarify the relationship between
tactic signals, ion fluxes and membrane potential changes. Ion
fluxes can cause membrane potential changes; conversely, a change
in membrane potential can alter permeability to ions. Depending on
the organism and the nature of the stimulus, the answer to the
question "which drives which?" may differ.

The above studies on bacterial chemotaxis demonstrate a number of
converging transduction pathways which control the rotary motor.
To what extent are these pathways utilized in sensory responses to
other stimuli? Thermosensory transduction in E. coli utilizes the
chemosensory transducing pathway for L-serine and the thermoreceptor
and L-serine chemoreceptors are probably the same protein (45).
In photosynthetic bacteria and Halobacterium halobium changes in
light intensity will directly influence electron flow or proton
gradients across the membrane; Hildebrand has suggested how these
could influence Ca^{++} level and flagellum activity (10). Work on
bacterial chemotaxis and phototaxis is likely to prove complementary in elucidating the final stages in the transduction pathways
from stimulus to response in bacteria, especially adaptation mechanisms and the way the rotary motor is controlled.

Studies on the sensory responses of other prokaryotes lag far
behind those on flagellate bacteria, an inevitable result of slower
progress in achieving pure culture and elucidating their genetics
and biochemistry. Some progress is being made with the study of
the motility and chemotaxis of spirochaetes (76), organisms propel-
led by the action of flagella located inside the cell wall. The
chamotactic and phototactic behaviour of gliding prokaryotes,
myxobacteria and cyanobacteria (blue-green algae), is a fascinating
topic, but the elucidation of their sensory transduction pathways
will be delayed until the basis of gliding movement is better
understood (77).

SENSORY TRANSDUCTION IN EUKARYOTIC MICROORGANISMS

Modes of Motility

Gliding movements in eukaryotes, as in prokaryotes, remain
ill-understood (77). It is possible that they are essentially
similar to those of prokaryotes, and were present in a common
ancestor; on the other hand, eukaryotic gliding may have evolved
independently and be based on different mechanisms. So far, progress towards understanding sensory transduction in gliding eukaryotes is limited.

A feature of all eukaryotes is an ability to perform intracellular movements. Chromosome movements during nuclear division
are universal, and protoplasmic streaming and exocytosis and
endocytosis very common. These activities involve actin microfilaments or tubulin microtubules or both, either in force generation
or in an essential, complementary skeletal role.

The capacity of the eukaryotic cell for intracellular movement
appears to have formed a basis for the evolution of means
of moving the whole cell. Amoeboid movement involves protoplasmic
streaming (78,79) and adhesion to the substratum (79,80) and may
have evolved from protoplasmic streaming more than once. As in
protoplasmic streaming, actin-myosin interaction is involved in
force generation.

The flagella and cilia of eukaryotes are closely similar organelles, and differ completely in structure and mode of action from
bacterial flagella; thus force is generated within the eukaryotic
flagellum by tubulin-dynein interaction and the flagellum is
bounded by the plasma membrane (81). It is clear that prokaryotic
and eukaryotic flagella must have evolved independently.

Polarized growth of eukaryote hyphae has some of the attribu-
tes of cell motility. Thus it involves movement of the materials
required for growth from a zone that may extend for several mm
behind the growing point (82), orientation in response to stimuli
can occur, and protoplasm can be moved forward within the hypha by
protoplasmic streaming, leaving a vacuolate or empty region behind.

Intensively Studied Systems

A "eukaryotic E. coli" has not so far emerged. Instead, sensory
systems in a wide range of microorganisms have been studied, and
more intensive work carried out on a few having experimentally

useful features.

The most extensive studies on the behaviour of an amoeboid
microorganism are those on the various chemotactic responses of
the cellular slime mould Dictyostelium discoideum (38,39,40,83,84).
The very effective response to the chemotactic factor involved in
morphogenetic aggregation (acrasin = 3',5'-cyclic AMP), permits
a range of ingenious experiments, and the organism is also suitable for genetic and biochemical investigation. More limited work
has been carried out on phototaxis (18,29) and thermotaxis (48) in
D. discoideum. Less extensive studies have been carried out on
the large multinucleate plasmodia of the myxomycete Physarum polycephalum (83) which shows chemotaxis (85), phototaxis (30,31) and
thermotaxis (47). The organism can be produced in large amounts
for biochemical study, and electrophysiological work is feasible
(86). Mammalian leucocytes are components of organisms with nervous
systems but, treated as if they were microorganisms, they are the
subjects of an increasing amount of work on chemotaxis (28).

The photoresponses (21,87-88) and chemoresponses (36, 37, 89)
of a wide variety of flagellate organisms – algae, protozoa, and
fungal zoospores and gametes – have been studied. Considerable work
has been done on the photoresponses of Euglena (12,21,87,88).
Another promising organism is Chlamydomonas, which shows chemoresponses (89) as well as photoresponses, and is a suitable subject
for genetical work. Extensive studies have been made on the electro-
physiology of behavioural mutants of the ciliate Paramecium tetrau-
relia (42,43,90); work has so far centered on mechanical stimulation although other behavioural responses are now being studied
including chemotaxis (42,91). Paramecium caudatum, although less
suitable for genetic work, is being used for electrophysiological
studies. Further work on Paramecium bursaria, which contains photosynthetic endosymbionts and shows phototaxis (92) would be of interest.

Chemotropic (93) and phototropic (94) studies have been carried
out on a wide variety of fungal hyphae, with a major effort being
devoted to the sporangiophore of Phycomyces (19,34). Unfortunately
biochemical information on this intensively studied system remains
sparse.

GENERAL FEATURES OF MICROBIAL SENSORY TRANSDUCTION

A comparison of bacterial chemotaxis with other intensively
studied microbial sensory systems, suggests that there are common

principles and mechanisms underlying the observed diversity. I
here comment on successive steps in sensory transduction pathways.

1) Primary receptors. A diversity of receptors reflects the
diversity of stimuli. These receptors are commonly located in or
near the plasma membrane. This facilitates speedy sensing of the
environment and swift transmission of information to the motor
organelles. In bacteria and amoeboid cells it seems likely that the
receptors are widely distributed in the cell membrane, but in ciliates and flagellates it is possible that receptors are largely
located in the plasma membrane of the motor organelles; the evidence
for this is strongest for mechanosensory responses (90,96).Some primary
mary receptors have been shown to be protein molecules - binding
proteins for chemoreceptors and proteins - with chromophore groups
for photoreception. The stimulus has in some instances been shown
to produce a conformational change in the protein.

2) Secondary receptors. Stimuli received from primary receptors of several different types may be processed by a secondary
receptor. In E. coli, three such secondary receptors (transducer
proteins) deal with chemical stimuli received by numerous primary
receptors. In Physarum polycephalum the pattern of competition
between attractants (85) suggests a similar system. In Phycomyces
evidence from behavioural mutants indicates the convergence of
stimuli of different types through a common processing system
(19,34). In E. coli, carboxyl methylation of the transducer proteins
has been shown to have a key role in the chemotactic response and
adaptation to attractants and repellents. Mammalian leucocytes have
also been found to respond to an attractant by carboxyl methylation
(95).

3) Changes in transmembrane potential. The most extensive
studies in the significance of electrical activity in microbial
behavioural responses have been carried out with Paramecium (42,
43,90), an organism in which measurements of resting potential and
the transient action potentials produced by stimuli can readily be
carried out. Depolarization of the cell membrane in Paramecium, which
results from a collision or some repellents (91,97), causes an
influx of calcium ions, reverse beating of cilia, and reversed
swimming direction (90). Hyperpolarization, which can result from
some attractants (91), suppresses ciliary reversal and causes
smoother swimming. Chemotactic agents operating through an orthokinetic mechanism act differently - attractants strongly depolarize
and repellents strongly hyperpolarize (91). Information on the
electrical activity associated with behaviour in other microorganisms
is limited, but should be forthcoming as methods of measuring

membrane potential with slow-response and fast-response dye indi-
cations improve (98).

4) Ionic control of motor activity. There is increasing evidence of a key role for calcium ions in controlling motor activity.
When internal Ca^{++} activity rises above $10^{-6}M$, tumbling of bacteria
(69) and reverse beating of cilia of Paramecium and flagella of
Chlamydomonas (99) occurs. In Paramecium depolarization causes an
inward Ca^{++} current, but subsequently a rectifying K^+ current (90);
calcium entry also leads to the inactivation of the calcium channel
(100). Calcium is also involved in the control of actomyosin activity (101), and hence of protoplasmic streaming and amoeboid movement. A major role for calcium ions in controlling motor activity
seems clear, and further work on this topic, and on the interaction
of calcium with other ions, especially magnesium and potassium,
desirable.

A pathway, signal → primary receptor → secondary receptor →
→ membrane potential changes → changed internal calcium ion concentrations → changed motor activity → changed path → tactic
response, has been sketched. This omits negative feedback involved
in adaptation, and the probability that some types of stimuli,
chemical or physical, may bypass much of the path.

REFERENCES

1. "The Shorter Oxford English Dictionary on Historical Principles"
      3rd edition. Addenda, Clarendon Press, Oxford (1973).
2. M. J. Carlile, in "Primitive Sensory and Communication Systems:
      The Taxes and Tropism of Microorganisms and Cells", M. J.
      Carlile, ed., Academic Press, London (1975).
3. B. Diehn, M. Feinleib, W. Haupt, E. Hildebrand, F. Lenci,
      E. Nultsch, Photochem. Photobiol. 26:559 (1977).
4. R. K. Clayton, Archiv. für Mikrobiol. 27:344 (1957).
5. H. C. Berg, & D. A. Brown, Nature 239:500 (1972).
6. H. C. Berg, Rev. Sci. Instrum. 42:868 (1971).
7. R. M. MacNab, & D. E. Koshland, Proc. Natl. Acad. Sci. USA,
      69:2509 (1972).
8. N. Tsang, R. MacNab, & D. E. Koshland, Science 181:60 (1973).
9. J. F. Rohlf, & D. Davenport, J. Theor. Biol. 23:400 (1969).
10. E. Hildebrand, in "Taxis and Behaviour: Elementary Sensory
      Systems in Biology", G. L. Hazelbauer, ed., Chapman & Hall,
      London (1978).
11. A. G. Lee, & J. T. R. Fitzsimmons, J. Gen.Microbiol. 93:346

(1976).

12. B. Diehn, Science 181:1009 (1973).

13. D. L. Gunn, Anim. Behav. 23:409 (1975).

14. J. S. Kennedy, Physiol. Entomol. J. 3:91 (1978).

15. H. C. Berg, & E. M. Purcell, Biophys. J. 20:193 (1977)

16. S. H. Zigmond, Nature 249:450 (1974).

17. S. H. Zigmond, J. Cell. Biol. 73:606 (1977).

18. K. L. Poff, & W. F. Loomis, Exp. Cell. Res. 82:236 (1973).

19. K. W. Foster, Ann. Rev. Biophys. Bioeng. 6:419 (1977).

20. M. A. Feinleib, Photochem. Photobiol. 21:351 (1975).

21. F. Lenci, & G. Colombetti, Ann. Rev. Biophys. Bioeng. 7:341
    (1978).

22. H. C. Berg, & P. N. Tedesco, Proc. Natl. Acad. Sci. USA
    72:3235 (1975).

23. R. M. MacNab, C.R.T. Crit. Revs. in Biochem. 5:291 (1978).

24. M. F. Goy, & M. S. Springer, in "Taxis and Behavior: Elemen-
    tary Sensory Systems in Biology", C. L. Hazelbauer, ed.,
    Chapman & Hall, London (1978).

25. D. E. Koshland, Science 196:1055 (1977).

26. J. R. Medina, & E. Cerdà-Olmedo, J. Theor. Biol. 69:709 (1977).

27. R. N. Allen, & J. D. Harvey, J. Gen. Microbiol. 84:28 (1974).

28. P. C. Wilkinson, in "Taxis and Behavior: Elementary Sensory
    Systems in Biology", G. L. Hazelbauer, ed., Chapman & Hall,
    London (1978).

29. D. P. Häder, & K. L. Poff, Photochem. Photobiol. 29:1157
    (1979).

30. J. Bialczyk, & L. Rakoczy, Bull. Acad. Polon. Sci. Sér. Sci.
    Biol. 23:571 (1975).

31. M. Hato, T. Hueda, K. Kurihara, & Y. Kobatake, Cell Struct.
    and Function 1:269 (1976).

32. J. A. Robertson, Arch. Mikrobiol. 85:259 (1972).

33. F. Y. Kazama, J. Gen. Microbiol. 71:555 (1972).

34. E. Cerdà-Olmedo, Ann. Rev. Microbiol. 31:535 (1977).

35. M. Delbrück, A. Katzir, & D. Presti, Proc. Natl. Acad. Sci.
    USA 73:1969 (1976).

36. G. Kochert, Ann. Rev. Plant Physiol. 29:461 (1978).

37. M. Levandovsky, & D. C. R. Hauser, Int. Rev. Cytol 53:145
    (1978).

38. M. Darmon, & P. Brachet, in "Taxis and Behavior: Elementary
    Systems in Biology", G. L. Hazelbauer, ed., Chapman & Hall,
    London (1978).

39. P. C. Newell, in "Microbial Interactions" J. L. Reissig, ed.,
    Chapman & Hall, London (1977).
40. J. T. Bonner, Mycologia 69:443 (1977).
41. S. H. Zigmond, J. Cell Biol. 77:269 (1978).
42. D. L. Nelson, & C. Kung, in "Taxis and Behavior: Elementary
    Systems in Cell Biology", G. L. Hazelbauer, ed., Chapman &
    Hall, London (1978).
43. C. Kung, S. Y. Chan, Y. Satow, J. Van Houten, & H. Hansma,
    Science 188:898 (1975).
44. G. A. Dunn, & T. Ebendal, Zoon 6:65 (1978).
45. K. Maeda, & Y. Imae, Proc. Natl. Acad. Sci. USA 76:91 (1979).
46. T. Hennessey, & D. L. Nelson, J. Gen. Microbiol. 112:337
    (1979).
47. W. W. Tso, & T. E. Mansour, Behav. Biol. 14:499 (1975).
48. K. L. Poff, & M. Skokut, Proc. Natl. Acad. Sci. USA 74:2007
    (1977).
49. B. Bean, J. Protozool. 24:349 (1977).
50. A. Grebecki, & G. Nowakowska, Acta Protozool. 16:351 (1977).
51. G. Nowakowska, & A. Grebecki, Acta Protozool. 16:359 (1977).
52. J. N. Cameron, & M. J. Carlile, J. Gen. Microbiol. 98:599
    (1977).
53. H. Winet, & T. L. Jahn, J. Theor. Biol. 46:449 (1974).
54. R. Blakemore, Science 190:377 (1975).
55. R. B. Frankel, R. P. Blakemore, & R. S. Wolfe, Science
    203:1355 (1979).
56. J. Adler, Cold Spring Harbor Symp. Quant. Biol. 30:289 (1965).
57. J. Adler, Ann. Rev. Biochem. 44:341 (1975).
58. H. C. Berg, Ann. Rev. Biophys. Bioeng. 4:119 (1975).
59. G. L. Hazelbauer, & J. S. Parkinson, in "Microbial Interac-
    tions", J. L. Reissig, ed., Chapman & Hall, London (1977).
60. R. M. MacNab, in "Encyclopedia of Plant Physiology, New Series
    Vol. 7: Physiology of Movements", W. Haupt, & M. E. Fein-
    leib, eds., Springer Verlag, Heidelberg, (1979).
61. D. E. Koshland, in "Symp. Soc. Gen. Microbiol. Vol. 27: Micro-
    bial Energetics", B. A. Haddock,and W. A. Hamilton, eds.,
    Cambridge University Press, Cambridge (1977).
62. J. S. Parkinson, Ann. Rev. Genet. 11:397 (1977).
63. M. S. Springer, M. F. Goy, & J. Adler, Nature 280:279 (1979).
64. R. M. MacNab, in "Encyclopedia of Plant Physiology: New Series,
    Vol. 7: Physiology of Movements", W. Haupt, & M. E. Fein-
    leib, eds., Springer Verlag, Heidelberg, (1979).

65. R. M. MacNab, Trends in Biochem. Sci. N10:N13 (Jan. 1979).

66. T. Iino, Ann. Rev. Genet. 11:161 (1977).

67. M. Silverman, & M. I. Simon, Ann. Rev. Microbiol.31:397 (1977).

68. M. Simon, M. Silverman, P. Matsumura, H. Ridgeway, Y. Komeda, M. Hilman, Symp. Soc. Gen. Microbiol. 28 "Relations between Structure and Function in the Prokaryotic Cell", R. Y. Stanier, H. J. Rogers, & B. J. Ward, eds., Cambridge University Press, Cambridge (1978).

69. G. W. Ordal, Nature 270:66 (1977).

70. G. W. Ordal, J. Bact. 126:706 (1976).

71. S. Szmelcman, & J. Adler, Proc. Natl. Acad. Sci. USA 73:4387 (1976).

72. J. B. Miller, & D. E. Koshland, Proc. Natl. Acad. Sci.USA 74:4752 (1977).

73. S. Harayama, & T. Iino, J. Bact. 131:34 (1977).

74. J. P. Armitage, & M. C. W. Evans, FEBS Letters 102:243 (1979).

75. B. H. Caraway, & R. N. Krieg, Canad. J. Microbiol. 18:1749 (1972).

76. E. Canale-Parola, Ann. Rev. Microbiol. 32:69 (1978).

77. L. N. Halfen, in "Encyclopedia of Plant Physiology, New Series, Vol. 7: Physiology of Movements", W. Haupt, M. E. Feinleib, eds., Springer Verlag, Heidelberg (1979).

78. R. D. Allen, & N. S. Allen, Ann. Rev. Biophys.Bioeng. 7:469 (1978).

79. M. Abercrombie, G. A. Dunn, & J. P. Heath, in "Cell and Tissue Interactions", J. W. Lash, M. M. Burger, eds., Raven Press, New York (1977).

80. D. A. Ress, C. W. Lloyd, & D. Thom, Nature 267:124 (1977).

81. P. Satir, & G. K. Ojakian, in "Encyclopedia of Plant Physiology, New Series, Vol. 7: Physiology of Movements", W. Haupt, & M. E. Feinleib, eds., Springer Verlag, Heidelberg (1979).

82. A. P. Trinci, Sci. Prog. 65:75 (1978).

83. K. L. Poff, & B. D. Whitaker, in "Encyclopedia of Plant Physiology, New Series, Vol. 7: Physiology of Movements", W. Haupt, &. M. E. Feinleib, eds., Springer Verlag, Heidelberg, (1979).

84. W. F. Loomis, Devel. Biol. 70:1 (1979).

85. D. J. C. Knowles, & M. J. Carlile, J. Gen. Microbiol. 108:17 (1978).

reaction, although not all the absorbed photons follow the same pathway. The "quantum yield" Φ gives the fraction of photons which are effective in giving the considered phenomenon.

The wavelength range of interest can be empirically limited at 285 nm (the short-wavelength minimum of sunlight that reaches the earth's surface when total skylight is considered (4)) and 1000 nm in the infrared region. In this range energy changes by a factor of 3.5, being respectively $7 \cdot 10^{-19}$ and $2 \cdot 10^{-19}$ J for the single photon (conversion factor $E [J] = 1986 \cdot 10^{-19}/\lambda [nm]$; for one einstein ($6.022 \cdot 10^{23}$ photons) the energy is between 100.3 and 28.6 Kcal/mol (conversion factor $E [Kcal/mol] = 28590/\lambda [nm]$.

As an introduction to the study of the light stimuli, we will examine the general properties of light and its interaction with matter. The presentation is descriptive, with reference to texts of specific and general interest. Consideration of how light is generated (sources and modifying devices) and how it can be detected and measured will hopefully add material to discuss limits and possibilities of experiments with light.

CHARACTERIZATION OF THE LIGHT BEAM

In order to transfer information, signals must have variety. For the optical stimulus, the fulfillment of this condition is due to wave properties of light, which are associated not only to the entire beam, but also with each photon. Light is a transverse electromagnetic wave, i.e. a series of rapid alterations of electric and magnetic fields perpendicular to each other, and both transverse to the direction of propagation. Such a wave can be considered as the superposition of purely sinusoidal waves, each of fixed frequency. The energy per quantum, E, is related to the frequency (ν_o, measured in Hz or s^{-1}), to the wavelength (λ_o, measured in nm) and to the wavenumber ($\bar{\nu}_o$, measured in cm^{-1}, with the unit Kaiser, K) by the equation:

(1) $E = h\nu = hc/\lambda_o = hc\bar{\nu}_o$

where h is the Planck constant ($6.63 \cdot 10^{-34}$ J·s) and c the vacuum speed of light ($3 \cdot 10^8$ m·s^{-1}). Also λ_o is measured in the vacuum; in a medium of index of refraction n, the wavelength is λ_o/n and the light speed is $v = c/n$. The units given are those typically used in each case, although they are not consistent.

The basic parameter for the characterization of a light beam
is wavelength, because it allows a direct experimental determination. Wavenumber, which is directly proportional to energy, would
be a better wave parameter, and also frequency, as it remains unchanged as the ray passes through different media. Accordingly, a
beam of light composed of only one type of photons is said to be
monochromatic (i.e. of the same colour).

We will directly observe that a perfectly monochromatic radiation cannot transport information. A light signal must have some
parameter which varies with time (f.i. amplitude, phase, polarization) and this change alters the monochromaticity. Nevertheless
this observation does not apply to biological work, because the
receiving molecular systems have absorption bands, and not lines.
As a consequence a beam of light showing $\Delta\lambda/\lambda = 1\cdot10^{-3}$ is considered monochromatic. As an example, a beam of light with $\lambda_o = 500$
nm and $\Delta\lambda = 0.5$ nm, has a frequency of $\nu = 6\cdot10^{14}$ Hz and a band
width $\Delta\nu = 6\ 10^{11}$ Hz: a change at the frequency of 1 M Hz, that
introduces a band width of $1\cdot10^6$ Hz, does not substantially modify
the chromaticity.

Polarization is another radiation parameter to be considered,
and it can be thought of in terms of the orientation of either the
electric or the magnetic field vector (5). Usually the electric
field vector (E) is taken to specify the cases of linear, circular,
and elliptical polarization. To visualize them, we use a single
steadily radiating oscillator at a distant point on the negative
Z axis of a rectangular coordinate system. If we continuously record
the instantaneous electric field present at the origin, it will lie
in the X-Y plane. The more general expression of the electric field
would be

$$E_x = V_x \cos (2\pi\nu t + \delta_x)$$

(2)

$$E_y = V_y \cos (2\pi\nu t + \delta_y)$$

Different polarization can be described as special combination of
amplitudes (V) and phases (δ). If the horizontal component E_x is
always zero, the light is linearly polarized in the vertical direction. A more general example of linear polarization is when the two
components E_x and E_y are exactly in phase with one another, but
their amplitude V_x and V_y differ. When the two amplitudes are equal

but exactly a quarter of cycle out of phase with one another, $\delta_y = \delta_x + \pi/2$, the light is circularly polarized. As we look back to the source, the electric vector appears to trace a circle in the clockwise direction, and the polarization is called right circular (left circular if $\delta_y = \delta_x - \pi/2$). In the absence of special relationships between amplitudes and phases, the light is elliptically polarized. When the two components E_x and E_y have unequal amplitudes and are out of phase by an angle $\gamma = \delta_y - \delta_x$, with $|\gamma| \leq \pi$, the polarization is called right (left) elliptical if γ is positive (negative). When the phase between E_x and E_y varies at random, the light is unpolarized. Polarization can be changed by means of birefringent optical systems.

A linearly polarized ray can be considered as a resultant of two equal vectors corresponding to a right circularly polarized wave and a left circularly polarized wave. When the ray is refracted through a non absorbing asymmetric medium, the two waves travel at different speeds. The emerging ray is still linearly polarized, but the plane of polarization is rotated (optical activity). In the spectral region in which optically active absorption bands are present, the two components are also differentially absorbed; the resulting light is elliptically polarized (circular dichroism) (6).

Another property of the light beam can be envisaged by considering more than one source, along with the possibility of interference and diffraction. The main point is the phase difference between the waves; for simplicity we consider only linear polarization. Two point sources of radiation are perfectly coherent if they send out wavetrains having a phase difference which is independent of time. In the case of two real sources, the phase difference between the two waves varies both point by point in space and in time. Spatial incoherence is due to the fact that the wavefront does not emanate from an area of the source in which all points are in phase; spatial coherence is determined by the Young interference experiment (7). Temporal incoherence results from the finite length of pulses of radiation produced by the sources; temporal coherence is related to chromaticity, the coherence time t_c being given by:

(3) $t_c = 1/2\pi\Delta\nu$

t_c can be directly measured by the Michelson interferometer (7).

The nearest approach to a source of coherent radiation is the laser.

The beam characterization has to take into account also the number of photons which are present. Two parallel descriptions can be built, the one founded on photometric quantities (energy) the other based on photon-flux quantities.

The most complete characterization for incoherent unpolarized radiation is given by spectral radiance (L_λ), since it represents the distribution of the energy (or photon) flux with respect to all the variables: position (area normal to the beam, A cos θ), direction (solid angle Ω) and wavelength:

(4) $L_\lambda (x,y,\theta,\psi,\lambda) = d^3\phi (x,y,\theta,\psi,\lambda) /A \cos\theta \, d\Omega \, d\lambda$

where $L_\lambda (x,y,\theta,\psi,\lambda)$ $\left[W \cdot m^{-2} \cdot sr^{-1} \cdot nm^{-1}\right]$ is the spectral radiance at the point x,y in the direction θ,ψ and at the wavelength λ; $d^3\phi (x,y,\theta,\psi,\lambda)[W]$ is the radiance flux through the surface element $dA = dx \cdot dy \left[m^2\right]$ about the point x,y ,within the element of solid angle $d\Omega = \sin\theta \, d\theta \, d\psi \, \left[sr\right]$ in the direction θ,ψ and within the elementary wavelength interval dλ $\left[nm\right]$ about the wavelength λ.

The quotient of the radiance divided by the square of the index of refraction is an invariant quantity as a beam undergoes any combination of reflections or refractions in accordance with the laws of geometrical optics. In this case all spectral quantities are expressed in λ_o, the vacuum wavelength. The invariant or basic spectral-wavelength-radiance, in terms of the local wavelength $\lambda(n)$, in a medium of varying refractive index n, is L/n^3. Accordingly, in going from a medium of index of refraction n_1 to a medium of index n_2, the radiance $L_{\lambda 1}$ becomes (8)

(5) $L_{\lambda 2} = (n_2/n_1)^3 \, L_{\lambda 1}$

Provided that the distribution L_λ is known in enough detail, by integration it can be transformed in simpler distributions.

A general agreement on an operational terminology for the various distributions (9) is needed also for the interdisciplinary works, as Smith and Tyler (10) observe in the discussion of the transmission of solar radiation into natural waters. Here we will examine only one example, which is frequently met in photobiology, i.e. the flux distribution with respect to position only, in the form of flux per unit area as a function of position. Two possibilities arise. To distinguish them, Nicodemus (11), refers to one

as a "direct surface distribution" and to the other as "omnidirec-
tional - surface distribution". In photobiology the former is called irradiance or photon-flux density (exitance when light emerges
from the surface) and the latter (radiant or photon) fluence rate.
Fluence rate in the quantity needed when effects are to be compared,
due to widely different irradiation geometries. As an example,
Rupert (12) observes that for the interior illumination from a
uniformly emitting hemisphere (approximately equivalent to skylight
on a uniformly overcast day) fluence rate would be twice the energy
per unit area falling on a horizontal surface. In the case of parallel beams on a normal surface, the distinction between the two
distributions is only conceptual and in practice they would differ
by only 2 % even at the focus of a uniformly illuminated lens of
relative aperture (focal length/diameter) f/2 (12).

We must recall that there are special photometric units in
the SI system (13), which have been used in photobiology in the
past, and at present are used especially in illuminating engineer-
ing. These units are valid for the visible range of wavelength.
In this case, the spectral distributions of radiation power is
multiplied by an arbitrary human visual response function, and integrated over the wavelength limits of the human response function,
with the results expressed in lumens. The light adapted eye shows
the maximum efficiency at 556 nm; for this light we observe the
minimum value of the equivalent in watts for the lumen

$$(6) \quad 1 \text{ lumen } (\lambda = 556 \text{ nm}) = 1.47 \cdot 10^{-3} \text{ watts}$$

At either lower or higher values of λ the equivalent is obtained
by dividing this figure by the proper relative luminous efficiency
(which is less than 1). Only for incandescent light sources can
these units of physiological photometry be related to the true
spectral intensity distribution (14).

The advent of lasers has increased light beam resolution in
time and space, so providing interesting parameters for irradiation of biological samples. Pulsed dye lasers have been made with
pulses ranging from about 1 μs in duration from a flash lamp-pumped
dye laser down to less than 1 ps for a mode-locked laser. A well
collimated beam can be focused to a sub-millimeter spatial region.
The focal spot diameter is given approximatively by the product of
the divergence angle and the focal length of the lens. High ir-
radiances can also be obtained with unfocused beams by using pulsed
lasers with laser beam amplifiers.

MODIFICATION OF THE LIGHT BEAM

Interaction of light and matter affects the properties of the light beam. For the sake of simplicity, we will consider two cases, classically distinguished as those of "resonant" and "non resonant" light. Correspondently we will use two descriptions of matter.

The "non resonant" case will be treated here by replacing the atom (quantum system) by an electric oscillator: this will preserve all the essential features of the light-matter interaction. When the electromagnetic wave of frequency ν passes an electron oscillator with a natural frequency of vibration ν_0, the periodic electric field induces a vibration on the oscillator. The amplitude and phase of the motion depend on the relative value of ν and ν_0.

In a dielectric, where ν is much smaller than ν_0, the bound electrons in the dipoles are set into motion with a small amplitude and in phase with the driving electric force of the light. Since the result of an accelerated electric charge is production of electromagnetic radiation, there will be emission of light (scattered light, weak, and with the same frequency ν of the incident light).

This is the case when each molecule reradiates independently of the other, as in a gas. When the average distance between the oscillators is small compared with the wavelength, and they are more or less regularly arranged (as in liquids and solids), the emitted waves interfere with one another. In the bulk, there is constructive interference in one direction and destructive interference in all other directions: by this mechanism, a refracted wave is constructed. Near the surface of the material, there is a thin layer of oscillators (about as deep as half a wavelength), for which the back radiation is not completely cancelled by interference: backward radiation of these oscillators add up to a "reflected" wave. Also the exit boundary from the medium gives rise to reflection.

A rough surface of a metal diffuses light (diffuse reflectance), and a smooth surface is an ideal mirror. This depends on the fact that for the conduction electrons of the metal $\nu \gg \nu_0$ and, under light, the electrons vibrate in opposite phase to incident light. Under these conditions a refracted light wave cannot be propagated, if the density of electrons and the amplitude of their vibration is above a certain limit (plasma frequency, ν_p), and all energy of

the incoming light must go into the reflected wave. Since ν_p is
frequently in the UV region, some metals, as silver, allow considerable transmission in this region.

The classic phenomenon of the resonance of light occurs when
$\nu = \nu_0$. It is easier here to change the description of matter, and
discuss the atoms as having discrete energy levels. When the atom
is in the lower level, by the action of resonant light it undergoes a transition to an upper level (excited state) by absorbing
the energy of the incident photon (absorption of light). From this
state, it tends to decay spontaneously with a series of different
mechanisms, some of which are radiative (fluorescence and phospho-
rescence; the emitted frequencies usually differ from that of the
incident light).

When the atom is found by the resonant light in the upper
level, it can decay to the lower level, by giving its energy change
to the incident light wave (induced emission; in this case the emitted light cannot be distinguished from the incident light). For
matter in thermal equilibrium, the number of atoms in each level is
uniquely determined by the temperature of the system, and the number of atoms in a lower level is larger than that in an upper one.
We have absorption of light and so matter will attenuate an incident radiation. However there are systems that, though in a steady
state, are not in thermal equilibrium. If there is population inversion, (i.e. the average numbers of atoms in the upper level is
larger than the number in the lower level) the incident resonant
light can be amplified as it passes through the medium. The "Light
Amplification by Stimulated Emission of Radiation" is the underlying principle of a light source of unique properties, the laser.

So much for a general presentation of the various processes.
Since we are interested both in noticing the modifications of the
light beams as they interact with the different media they pass
through, as well as in describing some relevant aspects of light
technology, it appears interesting to add details in reporting the
single isolated processes.

Scattering

When a photon strikes a layer of finite path ℓ, containing N
particles per unit volume, the probability P that the photon comes
in interaction is given by

(7) $P = \sigma_s N\ell$

where σ_s has the dimension of a section (cm^2) and is named cross section.

Isolated, non-absorbing $(\nu << \nu_0)$ and optically isotropic particles, with radius significantly minor than the wavelength of the incident light $(r \le 0.1 \lambda)$, show Rayleigh scattering of light, according to

(8) $\sigma_s \propto (n - 1)^2/\lambda^4$

The light scattered normally to incident radiation is polarized, with the electric field orthogonal to the plane of the incident and scattered light.

When scattering particles are large $(r > 25 \lambda)$ ordinary geometrical optics applies; but when the size is between this and the limit for Rayleigh scattering, and the refractive index is considerably different from the surrounding medium, the effect is called Mie scattering. At small particle size (relative to λ) Mie scattering occurs in all directions relative to the incident light path, although the scattering is never uniform as in Rayleigh scattering of unpolarized light. As size increases, Mie scattering is directed forward to an increasing extent. Nearly all scattered light is directed forward when the radius of the particles is about the size of wavelength. Mie scattering depends in a complicated way on $x = 2\pi r/\lambda$ and n values, so that the transmitted radiation has a different spectral distribution than the incident radiation. Numerical solutions of the Mie relations can be obtained in special cases. When x and nx are below 0.8, the Mie formulae can be expressed by series expansion (15)

(9) $\sigma_s = \pi r^2 (8x^2/3) \left[(n^2-1)/(n^2+1) \right]^2 \cdot$

$\cdot \{1+1.2\left[(n^2-1)/(n^2+1)\right]x^2 + \ldots\}$

and the radiance ratio between forward and backward scattering is

(10) $R = 1+(4/15) (n^2+4) (n^2+2) x^2/(2n^2 + 3)$

In the case of scattering by non-absorbing molecules, the bulk of scattered light is at the frequency of the incident beam; however, a small portion of the scattered energy is found at definite frequencies above or below that of the incident light (Raman effect). The agency modulating the frequency of the incident light turns out to be the normal frequencies of the sample, and Raman effect in scattering is a sensitive tool for spectroscopy (16).

The scattering effect is applied in technology, f.i. in the sharp-cut glass filters for visible light. The mechanism depends on scattering crystals formed within the glass mass through a post-production thermal treatment. Lower wavelength are scattered, whereas higher wavelengths are unaffected, so the visible region is isolated from the near UV region. It is possible also to produce gaseous host filters, blue sky coloration and red sunsets being the most popularized results of a gaseous red filter.

Reflection and refraction

The law of reflection states that the angle of reflection is equal to the angle of incidence, the incident ray, reflected ray, and normal all being located in the same plane. Although reflection can be described as a special case of scattering, the geometry of the layer that gives rise to it cancels the $1/\lambda^4$ effect on the spectral distribution. As a result, the reflected wave has the same composition of the impinging light.

On a micro-rough surface, light is multiply reflected at the surface, and hence emerges in various directions (diffuse reflectance). An ideally diffusing surface obeys Lambert's law

$$(11) \quad I_\theta \propto \cos \theta$$

where I_θ is the radiant exitance in the direction which makes an angle θ with the normal to the surface.

Passing from medium of refraction index n_1 to a medium of refraction index n_2, the direction of the light beam is changed by following the law

$$(12) \quad n_2/n_1 = \sin \theta_1 / \sin \theta_2$$

where θ_1 is the angle between the normal to refracting surface and the incident light (angle of incidence) and θ_2 is the angle between

the normal and the reflected ray, the two rays and normals being
located in the same plane.

In both reflected and refracted light, the polarization is
affected: if the impinging light was not polarized, depending on
the incidence angle, the reflected and refracted light beams are
polarized at different extent. The maximum polarization of the
light occurs when the reflected and refracted rays are at 90° from
one another. Electric field vibrations parallel to the plane of
incidence are eliminated from the reflected beam, but they appear
in the refracted and transmitted beam.

Refraction is at the base of lenses, pieces of optical glass
with surfaces so curved to converge or diverge the transmitted rays
of an object, thus forming a real or virtual image of that object.
When $n_1 > n_2$, the refracted light is bent away from the normal: at
a certain critical angle (for glass-air interface is 42°) no refracted light emerges in medium 2: instead there is total internal
reflection within medium 1. This observation is at the base of fiber
optics, by which light can be conducted through paths which are
far from the straight line (17).

Absorption (18)

When a photon strikes an infinitesimal layer $d\ell$, which contains
N particles (atoms, molecules) per unit volume, the probability dP
that the photon is absorbed is given by

(13) $dP = \sigma_a \ N \ d\ell$

where σ_a (cm^2) is the cross section.

Usually the cross section is function of the radiation frequency ν, and a line of natural linewidth $\Delta\nu$ has a Lorentian shape
given by

(14) $\sigma(\nu) = \sigma_a \ g(\nu) = \sigma_a \ \Delta\nu/2\pi \left[(\nu-\nu_0)^2 + (\Delta\nu/2)^2 \right]$

where $\int g(\nu) \ d\nu = 1$ and $\sigma(\nu_0) = 2 \ \sigma_a/\pi\Delta\nu$; $\sigma(\nu)$ is a differential
cross section $(cm^2 \ s)$. Equation 14 holds for all the lines in which
$\Delta\nu$ is due exclusively to the finite life time τ of the involved
electronic levels by following

(15) $\Delta\nu \cdot \tau = 1/2\pi$

Very often the lines are broadened, due to the Doppler effect, molecular interactions, etc.; in these cases

(16) $\Delta\nu \cdot \tau > 1/2\pi$

As an example for a transition $\lambda = 500$ nm and $\tau = 1$ ns, $\Delta\nu = 160$ MHz

$$\Delta\nu/\nu \sim \Delta\lambda/\lambda = 27 \cdot 10^{-8}, \quad \text{and so} \quad \Delta\lambda = 1.35 \cdot 10^{-4} \text{ nm}$$

Usually in the visible region, the absorption bands of molecular systems are much broader, typically $\Delta\lambda \sim 10$ nm. This is explained by assuming that the band results by superimposing many lines, and some levels show $\tau \sim 10^{-12}$ s. The obtained absorption curve is not a Lorentian curve.

By integrating equation (13), and substituting P with the ratio between the incident irradiance I_0 (incident watts per square meter) and the exitance I after the passage through an absorbing layer of finite length ℓ, we obtain

(17) $I = I_0 e (- \sigma_A N\ell)$

For sufficiently weak absorbers and sufficiently small changes of irradiation as frequently met in photobiology, a linear approximation of equation (17) can be used (19). When the medium is a solution, it is more convenient to make use of concentration of the chromophores (the light absorbing molecules), and common logarithms. As the log (I_0/I) is called absorbance (A), the relationship (17) becomes

(18) $A = \varepsilon c \ell$

where ε is the molar absorption coefficient $\left[\text{liter mol}^{-1} \text{ cm}^{-1}\right]$, measuring c in mol liter^{-1} and ℓ in $\left[\text{cm}\right]$. As a consequence

(19) $\sigma_a = 3.82 \ 10^{-21} \varepsilon$

Whereas ε is frequently used for the characterization of solutions,

σ_a is used in kinetic studies by photobiologists. The photosensitivity σ (which is the ratio of the number of events to the number of photons incident per cm^2 (20)) is related to the quantum
yield by

(20) $\sigma = \sigma_a \cdot \Phi$

As a measure of transition probability, ε is related to the oscillator strengh, f, the effective fraction of an electron in a molecule set in oscillation by the radiation field,

(21) $f = 4.32 \cdot 10^{-9} \; F \int_{\nu_1}^{\nu_2} \varepsilon \, d \, \nu$

where ν_1 and ν_2 are the limit frequencies of the band, and F, near
unity, is a correction for the refractive index of the sorrounding
medium.

Equation (17) is valid when the light is monochromatic and
collimated, the chromophores act independently of each other, and
different processes (reflection, emission, non-linear effects) are
absent. For very high irradiance ε is no longer independent of the
irradiance (saturation effect).

We must note here that absorption is a directional phenomenon,
and the probability per unit time that excitation of an absorbing
molecule will occur is proportional to the squared amplitude of
the electric vector of radiation field, to which it is subjected
(21).

Absorption filters have been developed for the control of
spectral energy. The relative transmission is obtained by simple
ionic absorption. Host materials are plastic or glass; also liquid
filters are used for special applications: an important drawback
of these filters is solarization, that is shift in their transmission spectra after molecular photodamage. Absorption filters are
rather insensitive to temperature changes, and the beam incidence
angle variations are mainly changes in filter thickness.

Polarization by selective absorption occurs in some crystal-
lized substances because they are not isotropic, i.e. some thermal
and electrical properties depend on direction relative to crystal
axes. Crystals of quinine iodosulphate absorb energy from electric
field variation in a particular direction. Microcrystals of this
material within a cellulose film have their normally random orientations changed to one preferential direction in the manufacturing

process, which involves stretching of the film. The greenish material produced in large thin sheets (Polaroid) is used as a polarizer, notwithstanding its intrinsic colour and the fact that the polarization is not complete. One interesting property is the fact that the angle of incidence of light is quite non-important.

Dispersion

The refracted wave travels with a velocity v that differs from the ordinary light velocity c. The ratio c/v is the refractive index n of the medium. This definition is for one frequency, so that we must consider the variation of the value of n with frequency. In "normal dispersion" n decreases with decrease of frequency; at frequencies in the neighbourhood of ν_0, the resonant absorption frequency, the refractive index n decreases with increase of frequency (anomalous dispersion; glasses and water are so behaving in the ultraviolet and in the infrared regions).

Here we must note that, in a rigorous treatment, dispersion and absorption are to be taken together, because the two phenomena are connected, and bound by equations (dispersion relations) that have general validity (22). In particular, due to the anomalous dispersion, in some regions of frequencies $n > 1$ and $v > c$. We observe that v is a phase velocity, that cannot carry signals: in fact no monochromatic wave can convey information, a signal requiring modulation of some radiation parameter. This modulation travels with a velocity less or equal to the light velocity in vacuum, in accordance with the relativity theory (group velocity).

The normal dispersion is at the basis of the separation of the radiation into its constituent components by passage through a prism. The same property is an important hindrance in the manufacture of lenses because it causes "chromatic aberration": focusing of different frequencies to different places in space. Chromatic aberration is also present in the eye.

Interference and diffraction

The propagation of anyone beam of radiation is unaffected by the presence of other beams. If a beam of light crosses a second beam, superposition will occur in the region of intersection but the one beam will subsequently proceed as it would have done if the other had not been present. In the region where superposition oc-

curs, two or more progressive waves produce electric field vectors
in the same region around any point: the resultant displacement is
the vector sum of the separate displacements due to individual
waves.

Coherent light beams from the same source, that reach the same
locus in space by different routes, interfere. They add, if they
are in phase (constructive interference), and substract if they
are out of phase (destructive interference). The different paths
can be obtained by division of wavefront, when a single beam is
divided by two apertures, or by division of amplitude, by means of
a partial reflection on a thin plate.

Diffraction is concerned with the interactions between light
waves, because an obstacle is placed between the source and the
screen, f.i. the passage through an aperture. The phenomenon is explained by the Huygens principle. It is assumed that each point of
a wavefront may be considered as a center of secondary waves. The
envelope of the secondary waves forms a new wavefront. If part of
the original wavefront is blocked by an obstacle, the system of
secondary waves is incomplete. Diffraction phenomena occur giving
rise to the spatial energy distribution in which there is alternating presence or absence of radiation (diffraction pattern).

Thin layer interference is used to produce high performance
mirrors, which have selective properties of reflection/transmis-
sion with respect to wavelength, with the result of narrowing the
spectral distribution of the reflected light. Being based on die-
lectrics, they are important when electrical conductivity of metal
mirrors causes problems. Also the eyes of some marine animals which
live in deep sea, are based on these mirrors (23).

Interference filters are used in isolating spectral regions:
there are broad band filters (half height width 50 nm, transmit-
tance 70 %), narrow band and super-narrow band filters, which ar-
rive to less than 1 nm half width,30-60 % transmission. The spectral region immediately surrounding the passband is rejected,
shorter and longer wavelengths are eliminated by absorption filters. Interference filters reversibly shift the passband to longer
wavelength with increasing temperature, with a coefficient of about
0.15 Å per degree centigrade change. They are normally designed
for normal incidence of radiation, and non-normal incidence results
in a shift of the passband towards shorter wavelengths. In the visible region, the approximate amount of shift can be determined from
the equation

$$(22) \quad \lambda_\alpha = \lambda \cdot (n^2 - \sin^2 \alpha)^{\frac{1}{2}}/n$$

where α is the angle of incidence of the radiation, λ_α is the central wavelength at the angle of incidence and n is the effective refraction index of the filter.

Interference is used to create dispersion. Ruled gratings, which are plates ruled in precisely spaced lines, or holographic gratings, prepared by printing interference patterns on an optically flat glass, are used in collimated light to isolate narrow regions of wavelengths (grating monochromators).

SOURCES OF LIGHT   (24)

The blackbody

Used as a standard for absolute determination of radiation power, is based on the Planck formula of emission. It can be approached by a radiator which has the form of a small hollow cylinder of pure fused thoria, immersed in platinum maintained at its freezing point (2,046 °K).

Incandescent lamps

A filament (generally of tungsten) is enclosed in a hermetically sealed bulb of glass or quartz, that is either filled with an inert gas, or operated at a very low pressure. The filament is heated to incandescence by an electric current. Lamps are available with electric power ratings from a fraction of watt to 10 Kwatt, yield $\leqslant$ 20 %. Efficiency of energy conversion to light increases with filament operating temperature, whereas life time decreases: conventional lamps may decline as much as 50 % in light output during life, primarily due to bulb darkening from evaporated tungsten deposit. Special lamps are added of iodine or bromine compounds, have small bulbs and operate at high temperatures. The halogens serve to combine with tungsten deposited on the bulbs, and the compounds are decomposed on the hot filament, carrying back the tungsten over it. The spectral distribution is essentially the same of a blackbody radiating at the temperature T_c (color temperature).

Low pressure discharge lamps

The lamps are used mainly to have some monochromatic lines; they are employed as a standard for the set-up of instrumentation, and for irradiating with monochromatic light. Their power density is low, and the yield is quite high. For example mercury lamps operate at a pressure of a few millimiters of Ar containing 10^{-3} mmHg. Electrons from the hot filament, and mercury ions produced by them excite mercury atoms by collision. Low-lying metastable triplet states are created, which fall to ground state by emitting at 254 nm. This light covers about 85 % of the total energy emitted, whereas the longer wavelengths represent transitions from high energy states down to the metastable state (germicidal lamp, with quartz tube). If the tube is made of glass, and its inside wall is coated with phosphors, such as calcium halo-phosphate, the 254 nm radiation is absorbed, and visible light is emitted in a smooth continuum. The spectral distribution depends on temperature of operation; the lamps must be aged, and warmed up for 10 min before constant operation conditions (fluorescent lamps).

High pressure discharge lamps (25)

These lamps are used when high irradiance is required. They emit a continuum, with a spectral distribution that in special cases almost resemble that of the blackbody. Optical radiation is generated by an electric discharge through an ionized gas. Carbon arcs were the first practical electrical light sources, and are used for high irradiance. In present day lamps, the discharge takes place between tungsten electrodes, contained within a transparent bulb. The properties of light are principally determined by the elements and compounds that fill the bulbs. Special starting devices are necessary, as well as current regulating auxiliary devices. The high pressure xenon lamp is the most useful high intensity source for the visible region, because it emits a continuum spectrum, similar to daylight. It is a short arc lamp, and special lamps can be built as virtually point sources. They reach rapidly the operating conditions. The high pressure mercury lamp is used for the near-UV and visible regions. Due to the high pressure (up to 250 atm), the 254 line is absorbed. The output is a continuous with broadened mercury lines. The density of mercury vapors, and consequently the irradiance, depends on temperature: approximately

15 min are required to obtain constancy of luminous flux.

Lasers (26)

At present many types of lasers can be bought, with wavelengths from the U.V. to the far I.R. region. These sources are coherent, but only in few cases the coherence properties of light are exploited in the experiment (f.i. quasi-elastic scattering). Normally lasers are used because of their outstanding properties: directionality, monochromaticity and radiance. With these sources non-linear effects are easily obtained (saturation, two-photon effects (27), generation of second harmonics (28).

Some examples can be reported, by dividing them for convenience in fixed frequency lasers and tunable lasers (both pulsed or c.w., i.e. continuous wave operating).

Fixed frequency lasers. The first laser, and up to day the most used, is c.w. helium-neon laser with $\lambda = 633$ nm and typical power $\simeq 1$ mW; it is lower priced, easy to handle, expecially used for optical alignments.

The ruby laser, $\lambda = 694$ nm, usually used in the Q-switched mode, gives $\simeq 30$ ns pulses, with typical energy of 1 J/pulse, repetition rate $\simeq 1$ pulse/s.

The argon laser is used in the visible region with one or more spectroscopic lines, $\lambda = 457-528$ nm, typical power 1 W c.w.

The nitrogen laser works with pulses of few ns, energy per pulse $\simeq 1$ mJ, repetition rate 50 pulses per s.

Tunable lasers require a luminous source as a pump. The dye lasers are very common. They emit stimulated light in the fluorescence band of the dye. They usually work with pulsed light, yield $\simeq 10\%$; continuous wave operation can be achieved with certain dyes, if the solution flows rapidly enough and a c.w. laser is used as the pump.

LIGHT DETECTORS (24)

Radiation sensors can be generally divided in three classes: photoemissive, semiconductor and thermal detectors.

Photoemissive and Semiconductor Detectors (29)

The process of absorption of radiation directly produces a measurable effect, e.g. generation of photoelectrons in photomul-

tipliers or charge carrier pairs in a semiconductor, so that these
detectors have a rapid response. Photomultiplier tubes are useful
over the wavelength range of about 140 nm to 1.2 μm. The short
wavelength limit is due to transmission of the window material. No
single photomultiplier tube is optimum for the full wavelength range,
although the multialcali and GaAs cathode can be used from about
160-910 nm. A major advantage of the photomultiplier tube over solid state devices is the ability of the dynode system to amplify
current without adding a major amount of noise to the signal.
Practical gains of 10^6 make the detection of single photon events
possible. Another advantage is that they can have large photosensitive surfaces (several cm^2). The major disadvantage is the
requirement of well regulated voltages of between 350 and 2,000
volts. The most recent advantages are on special cathodes. At present photomultiplier tubes are commercially available with risetime
of less than one nanosecond, quantum yield of about 0.2.
Semiconductor photodiodes cover a range of light detectors from
the simple solar cells and photovoltaic photodiodes, to sensitive
and accurate photoconductive photodiodes. The two groups also differ in frequency responses, the latter being necessary when high
speed light pulses or high frequency modulation of a light beam
are to be detected. Quantum yield 0.4, risetime 1 ns.

Thermal Detectors

     The absorbed radiant energy is converted first to heat, which
in turn produces a measurable effect. These detectors are capable
of uniform response over a broad wavelength range, limited only
by the ability of the sensor to absorb the radiation.
Examples are the thermistors, conductors with a resistance that
is markedly dependent on temperature; usually they are mounted
as a branch of a bridge, and so attain considerable sensitivity.
The thermopile is composed usually of 4 to 12 thermocouples in
series, the alternate junctions of which, covered by carbon black,
are exposed to light, and the others shielded from light and kept
at constant temperature. The voltage produced is very low and
must be measured by a sensitive microvoltmeter. Calibration
with standard sources is essential for absolute radiation determinations. Typical risetime 1 ns.
Pyroelectric detectors are based on the temperature dependence
of the electric polarization in certain materials, of which ferroelectric crystals are the most common. At temperature below its

Curie temperature, a ferroelectric crystal has a spontaneous
electric polarization. By proper preparation this polarization
can be aligned in the same direction along the polar axis for the
entire crystal. Heat resulting from radiation absorption produces
alterations in the crystal lattice spacing, and affects the value
of the electric polarization. A polarization charge is developed
on the surface of the crystal normal to the polarization axis.
The polarization charge is balanced when electrodes are applied to
these surfaces and connected through external circuit. The current
generated in the circuit is proportional to the rate of change of
the crystal temperature. This allows the pyroelectric detectors
to operate with much faster time response than those of the other
thermal detectors, the advantage of a flat response over a wide
region of wavelengths being retained.

ACKNOWLEDGEMENTS

    Support of C.N.R. on the Cooperative Research Project Italy-
Japan is acknowledged for this work.

REFERENCES

1. J. P. Hailman, in "Optical Signals: Animals Communication and
      Light", Indiana Univ. Press, Bloomington and London (1977).
2. R. K. Clayton, in "Light and Living Matter: The Biological
      Part" McGraw-Hill, New York, N.Y. (1971).
3. D. F. Walls, Nature 280:451 (1979).
4. W. H. Klein, quoted by W. Shropshire, Jr., in "The Science of
      Photobiology", K. C. Smith, ed., Plenum Press, New York,
      N. Y. (1977).
5. W. A. Shurcliff, "Polarized Light: Production and Use" Harvard
      Univ. Press, Cambridge, Mass. (1962).
6. L. Velluz, M. Legrand, & M. Grosjean, "Optical Circular
      Dichroism, Principles, Measurements and Applications
      Verlag Chemie, Weinheim (1965).
7. D. W. Tenquist, R. M. Whittle & J. Yarwood, "University
      Optics", Iliffe Books, London (1969).
8. F. E. Nicodemus & H. J. Kostkowski, in "Self-Study Manual on
      Optical Radiation Measurements", F. E. Nicodemus, ed., Nat.
      Bur. Stand., U.S., Tech. Note 910-1 (1976).
9. C. S. Rupert, & R. Latarjet, Photochem.Photobiol.28:3 (1978).

10. R. C. Smith & J. E. Tyler, in "Photochemical and Photobio-
     logical Reviews", K. C. Smith, ed., Plenum Press, New York,
     N. Y. (1976).
11. F. E. Nicodemus, in "Self-Study Manual on Optical Radiation
     Measurements", F. E. Nicodemus, ed., Nat. Bur. Stand., U.S.,
     Tech. Note 910-2 (1978).
12. C. S. Rupert, Photochem. Photobiol. 20:203 (1974).
13. C. H. Page & P. Vigoreux, "The International System of Units
     (SI)", Nat. Bur. Stand. U.S.,Spec. Publ. 330 (1974).
14. H. H. Selinger & W. D. McElroy, "Light: Physical and Biologi-
     cal Action", Academic Press, New York (1965).
15. H. C. Van de Hulst, "Light Scattering by Small Particles",
     Wiley, New York, N. Y. (1957).
16. M. C. Tobin, "Laser Raman Spectroscopy", Wiley, New York, N.Y.
     (1971).
17. W. B. Allan, "Fiber Optics, Theory and Practice", Plenum
     Press, London (1973).
18. W. West, ed., "Chemical Application of Spectroscopy", in
     "Technique of Organic Chemistry", A. Weissberger, ed.,
     Wiley-Interscience, New York, N. Y. (1968).
19. R. K. Clayton, "Light and Living Matter: The Physical Part",
     McGraw-Hill, New York (1970).
20. J. Jagger, in "The Science of Photobiology", K. C. Smith, ed.,
     Plenum Press, New York, N.Y. (1977).
21. W. Kauzman, "Quantum Chemistry", Academic Press, New York,
     N. Y. (1957).
22. M. Born & E. Wolf, "Principles of Optics", Fifth edition,
     Pergamon Press, Oxford (1975).
23. M. F. Land, Progr.Biophys. Molec. Biology 24:75 (1972).
24. J. C. Calvert & J. N. Pitts, "Photochemistry", J. Wiley,
     New York, N.Y. (1965).
25. F. E. Carlson & C. N. Clark, in "Applied Optics and Optical
     Engineering", R. Kingslake, ed., Academic Press, New York,
     N. Y. (1965).
26. C. A. Sacchi & O. Svelto, in "Analytical Laser Spectroscopy",
     N. Omenetto,ed.,in"Chemical Analysis"P. J. Elvin & J. D.
     Winefordner, eds., J. Wiley, New York, N.Y. (1978).
27. M. W. Berns, Biophys. J. 16:973 (1976).
28. A. Yariv, "Introduction to Optical Electronics", Holt Rinehart
     & Wilston, New York, N.Y. (1971).
29. A. H. Sommer, "Photoemissive Materials, Preparation, Properties
     and Uses", Wiley, New York, N.Y. (1968).

PHOTOMOTILE RESPONSES IN FLAGELLATES

Mary Ella Feinleib

Biology Department, Tufts University
Medford, Ma. 02155 (USA)

The photomovement of flagellates provides us with excellent model systems for investigating sensory-transduction pathways. The responses to light are rapid and dramatic, and they show a fascinating degree of diversity. (See recent reviews by Forward, Hand, Feinleib, Lenci and Colombetti, Häder, Nultsch and Häder, Diehn; 1-7). After summarizing some general features of flagellate photoresponses, this chapter will examine the behavior of three flagellates that show diverse photomovement "strategies": Euglena, Volvox and Chalmydomonas.

GENERAL FEATURES OF FLAGELLATE PHOTOMOVEMENT

Certain features are common to almost all photoresponsive flagellates. (a) The majority of flagellates that show photomovement are photosynthetic; these movements offer an organism the selective advantage of positioning itself in an optimally illuminated region of its environment.

(b) Unlike the photosynthetic bacteria, however, (Nultsch, this volume), flagellated algae have different photoreceptors for photosynthesis and photomovement. In general, flagellate photomovements are "blue light responses", with action-spectra maxima between ca. 420 nm and 520 nm, and with some activity in the near-ultraviolet range. Red light is not normally effective as a stimulus. The photoreceptor pigments are presumaly flavoproteins or caroteno-

proteins. Exception to the blue-light "rule" are the flagellates
Prorocentrum (8) and Cryptomonas (9), both of which are maximally
responsive to yellow light.

(c) In a flagellate, unlike a bacterium, the photoreceptor
pigment for photomovement is thought to be localized in a special
photoreceptor organelle. In most cases, however, this organelle
has not yet been identified.

(d) Most photoresponsive flagellates also have a characteristic structure that is thought to play an accessory role in photomovement: the stigma. Under the light microscope, the stigma, appears as a reddish spot; it is composed of osmiophilic granules
that contain carotenoids. The position of this structure in the
cell tends to be characteristic of the algal class (10).

(e) Flagellates tend to accumulate in a lighted area (sometimes called a "light trap") at low and moderate intensities, and
to disperse from the lighted area at high intensity. According to
the terminology recently recommended for motile microorganisms (11),
this kind of phenomenon is referred to simply as "photoaccumulation"
or "photodispersal", regardless of the kind of movement that led to
that accumulation of dispersal.

Flagellates display several kinds of photomovement, any of
which, theoretically, could lead to photoaccumulation or photodispersal (11).

(1) In photokinesis, the steady-state rate of activity is controlled by the absolute magnitude of the light intensity[+]. As long
as light intensity is maintained at the same level, the rate of
activity continues to be the same: no adaptation occurs. In order
for photoaccumulation to occur by photokinesis, organisms should
move more slowly in the light than in the dark; i.e., they should
demonstrate "negative photokinesis" (12). In general, however, the
opposite is true. Flagellates that show photokinesis tend to move
faster in the light than in the dark (except at very high intensity),
and some are motile only in the light. Photokinesis per se is probably not used by any flagellate as a major photomovement "strategy".

[+] According to Diehn et al. (11), if the linear velocity of movement is the activity controlled by stimulus intensity, the phenomenon
is called "orthokinesis". The term "kinesis", used without prefix,
is taken to denote orthokinesis.

It should also be noted that, unlike the other kinds of photomovement in flagellates, photokinesis may not be a "blue-light response". Red light has been reported to be effective, presumably
via the photosynthetic system (5,13,14).

(2) In a photophobic response, a change in light intensity--usually
a sudden change--elicits a transient change in the activity of the
organism. In many organisms, the photophobic response includes a
cessation of forward movement (stop response), followed by a change
in direction of movement. This change in direction is not oriented
with respect to the direction of the stimulus. Instead, the basic
pattern of activity change is stereotyped: direction of movement
is fixed with respect to the morphology of the organism and is independent of the direction of the nature of the stimulus.
    A photophobic response may be elicited by a "step-up" (in-
crease) in light intensity or by a "step-down" (decrease) in intensity (11). Although the terms "step-up" and "step-down" refer to
the stimulus, they are also used, for convenience, to describe the
response. For example, if an organism changes direction whenever
it crosses a border from a lighted zone into a dark zone, it may
be said to show a "step-down photophobic response". It is easy to
visualize how such step-down responses can lead to photoaccumulation: if every organism in a population changes direction upon entering the dark, but none shows a direction change upon entering
the light, the organisms will accumulate in the lighted area.
    Unlike photokinesis, photophobic responses show adaptation.
After completing its photophobic activities, which typically last
a matter of seconds, the organism resumes its normal movement.
This is true even if the light intensity persists at its new level.
    Flagellates show highly diverse photophobic-response patterns.
Nonetheless, all photophobic responses demand that the same kind
of information be acquired about the stimulus; namely, the organism
must be able to detect a temporal change in light intensity (dI/dt).

(3) The characteristic photobehavior of most flagellates is phototaxis: movement that is oriented with respect to the direction of
the stimulus light. Typically, a flagellate steers toward the light
source (shows positive phototaxis) at low or moderate light intensity, and steers away from the source (shows negative phototaxis)
at high intensity (15). (Phototaxis can lead to accumulation in a
light trap by an indirect route: if a stimulus beam traverses part

of an algal suspension, organisms outside the beam may become oriented toward light scattered from organisms in the beam).

Phototactic orientation demands that the system have a sophisticated capability; namely, it must be able to detect the direction
of light. In order to perceive light direction, the organism must
detect a gradient in light absorption; this gradient may be spatial
or temporal.

An organism may use a one-instant mechanism (16) to detect a
spatial gradient in light absorption; i.e., it may compare the
light absorbed in two regions of the organism at one instant in
time (essentially instantaneously). A simple hypothetical example
of a one-instant mechanism is shown in Figure 1A. Two photoreceptor organelles are separated by screening pigments. If a stimulus
light comes from the left, the photoreceptor organelle on the left
side of the cell absorbs more light than the photoreceptor organelle
on the right side. To account for positive phototaxis on this model,
one postulates that the cell (or organism) always turns toward the
side where maximum light absorption occurs--in this case, toward
the left side. We do not know whether any flagellate uses a one-
instant mechanism for phototactic orientation, but such a mechanism almost certainly occurs in some gliding algae (5).

A rapidly-moving flagellate may well make use of an alternative
mechanism for perceiving light direction; namely, it may use a two-
instant mechanism (16) to detect a temporal gradient in light absorption. The essential feature of this mechanism is that the organism compares the light absorbed in one photoreceptive region at
two instants in time; i.e., it takes successive light-absorption
"readings" at different positions with respect to the light source.
This kind of mechanism has also been referred to as "movement-modu-
lated" (7). Rather different versions of a two-instant mechanism
are found in the flagellates Euglena (17-19,4,7) and Volvox (20-
23), and will be described below.

Both basic mechanisms for perception of light direction (one-
and two-instant) require establishment of a light-absorption gradient. Figure 1 illustrates three ways in which such a gradient could
be set up in a hypothetical cell (shown in cross section). The
stimulus light comes from the left. The sketches on the left side
depict establishment of a spatial gradient, as needed in a one-
instant mechanism for perceiving light direction; the sketches on
the right side, in each case, depict the corresponding scheme for
establishing a temporal gradient, as needed in a two-instant mechanism. In each case on the right side, two successive positions of

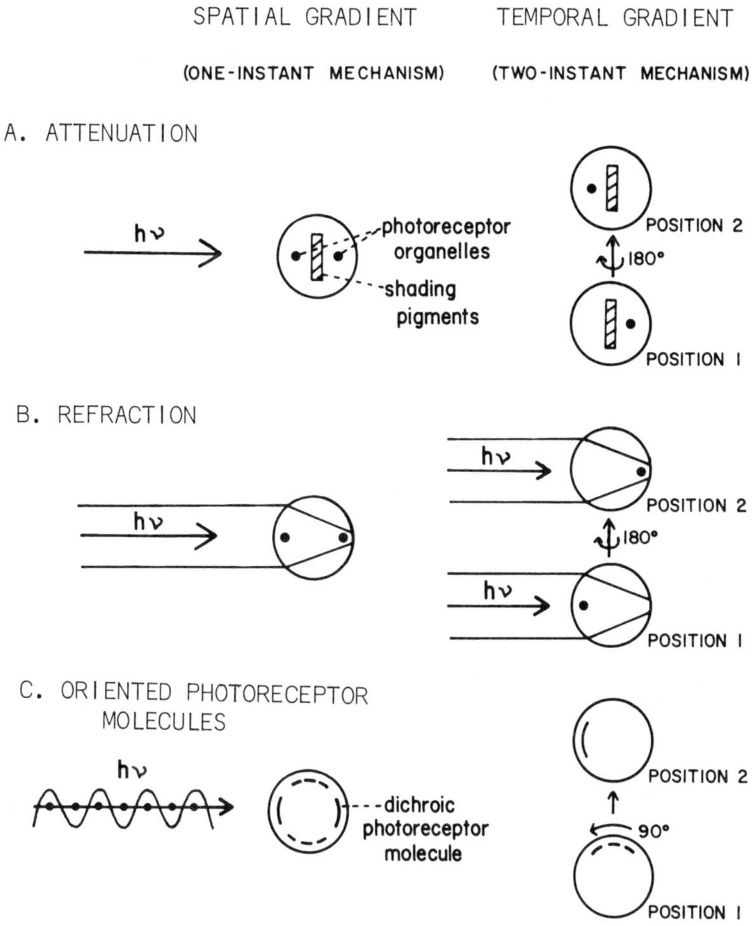

FIGURE 1. Schematic illustration of three fundamental ways in which a light-absorption gradient could be established across a hypothetical cell (shown in cross-section):   A. Attenuation;   B. Refraction;   C. Oriented (dichroic) photoreceptor molecules.
In each case, the sketch on the left side depicts establishment of a spatial gradient, as occurs in a one-instant mechanism, and the sketches on the right side depict the corresponding scheme for establishing a temporal gradient, as occurs in a two-instant mechanism. For each case on the right, as the cell moves from position 1 to position 2, it also rotates as indicated. See text for further details.

the cell are shown: as the cell moves forward from position 1 to
position 2, it also rotates as indicated. Note that, in cases of a
one-instant mechanism, (a minimum of) two photoreceptive points
are involved; in cases of a two-instant mechanism, one photoreceptive point is sufficient.

Case A. Attenuation

A light-absorption gradient can be established by shading of
photoreceptor molecules via absorption by other molecules of the
same pigment or by molecules of some other pigment. Alternatively,
light may be attenuated via scattering by cellular organelles. A
two-instant mechanism involving shading pigments is thought to occur in Euglena (see below).

Case B. Refraction

The cell acts as a converging lens, with the result that the
photoreceptor molecules on the side further from the light source
receive more light than those on the side nearer to the source. A
lens effects is known to occur in the fungus Phycomyces (Hertel,
this volume), but is less likely to be found in a microscopic flagellate.

Case C. Oriented (dichroic) photoreceptor molecules

If the photoreceptor molecules are dichroic, they will absorb
different amounts of light depending on the orientation of their
transition moments with respect to the light source. For simplicity,
the electrical vector of the unpolarized light beam is shown here
as having only two components: one vibrating in the y-axis (wavy
line) and the other vibrating in the z-axis (dots). On the left,
four photoreceptor molecules are shown schematically: the bars represent the molecular transition moments for absorption, aligned
parallel to the cell surface. The molecules at the left and right
sides of this cell (indicated by solid bars) have their transition
moments oriented parallel to one of the planes of polarization of
the stimulus beam, and thus absorb light, whereas the molecules at
top and bottom (dashed bars) are unable to absorb. Thus, there is
an absorption gradient between left/ right and top/bottom. This
case is different from A and B in that the organism can detect
only the axis of stimulus propagation (left vs. right), but not

the direction of propagation (toward left vs. toward right[(+)].
Figure 1C depicts the analogous case for a two-instant mechanism.
Again, the solid bar represents a photoreceptor molecule in an
orientation allowing it to absorb light; the dashed bar represents
a molecule unable to absorb.

A one-instant mechanism involving dichroic receptors is known
to operate in chloroplast orientation in the filamentous alga
Mougeotia (Haupt, this volume). Recently, a two-instant mechanism
of this type has been proposed for phototactic orientation in
Euglena (24), as an alternative to the commonly-accepted periodic-
shading hypothesis (see below).

EUGLENA

Photomovement has been studied more intensively in Euglena
than in any other flagellate. There is now very good evidence that
the photoreceptor pigment for photomovement is a flavin, and that
it resides in the paraflagellar body, a swelling at the base of
the emergent flagellum (25-29). An important clue to the flavin nature of the Euglena photoreceptor is the presence, in various photomovement action spectra, of a peak at ca. 370 nm (30,31,28,7).
By contrast, in other flagellates for which UV activity has been
explored, no such peak is evident; e.g., in Volvox (23), Chlamydomonas (32), Gyrodinium (33,34), and Gymnodinium (35). The stigma in
Euglena is independent of the chloroplasts and lies adjacent to the
paraflagellar body (36,37,10). This structure is thought to play
an accessory role in phototactic orientation.

During normal forward movement, the emergent flagellum of
Euglena trails backward along the so-called "ventral" side of the
cell (Figure 2A). The flagellum beats in a helical, asymmetric pattern, causing the cell to rotate about its longitudinal axis and
to describe a helical swimming path (19,4,7). This kind of swimming path is typical of unicellular flagellates. Euglena traces a
series of cones as it moves along, with the front end of the cell

---

(+)
    Note that this mechanism would work equally well, in theory,
with only two photoreceptor molecules, and is shown here with four
merely to illustrate the point about axis vs. direction of propagation.

at the base of the cone and the rear at the vertex of the cone (19).
The same side of the cell is always directed toward the axis of the
helix--the usual situation in helical swimming paths. In Euglena,
the ventral side is directed toward the axis, and the dorsal side
(where the stigma is located) is directed outward.

Euglena is capable of positive and negative phototaxis, but
its fundamental photomovement is the photophobic response. These
responses have been analyzed by Diehn et al.(38), and by Barghigiani et al. (31). Below a certain threshold intensity, a step-down
signal elicits the photophobic response. (The value of that threshold depends on a variety of conditions; 39,31,4,7). The first observable event, the so-called "elementary motor response" (19), is
a sharp movement of the flagellum in the ventral direction, bringing
it to lie at 90° to the long axis of the cell (Figure 2A). This
change in direction of the flagellar beat causes the cell to turn
toward its dorsal side. Depending on the magnitude of the signal,
the cell may execute only a partial turn about its lateral axis,
or it may undergo several complete 360° turns (31).

This photophobic response constitutes the basis of   phototactic orientation according to the commonly-accepted periodic-
shading hypothesis (17-19,4,7); see Figure 2. In this model, the
stigma is believed to act as a shading body (essentially as in the
basic scheme shown earlier in Figure 1A, right side). As the cell
swims forward, it rotates on its longitudinal axis. If a stimulus
beam is introduced perpendicular to that axis (from the left in
Figure 2), the stigma periodically comes between the light source
and the photoreceptor organelle (the paraflagellar body), modulating
the signal to the photoreceptor.

To account for positive phototaxis, this model postulates that
whenever the photoreceptor experiences a step-down signal, the cell
displays the photophobic response described above. The step-down
signal is produced when the stigma comes into the light path, as
in position 2 of Figure 2A. The cell responds by turning toward
its dorsal side. The result is shown in Figure 2B: the axis of the
swimming path swings over slightly in the direction of the stimulus
beam; i.e., there is a slight broadening of the "swimming cone" in
that direction (19). The cell continues to make these small, stepwise adjustments in its swimming path (one adjustment per rotation)
until it is heading toward the light source. Once the cell is heading toward the light, there are no further modulations of the signal;
the photoreceptor is continuously illuminated. Thus, in positive

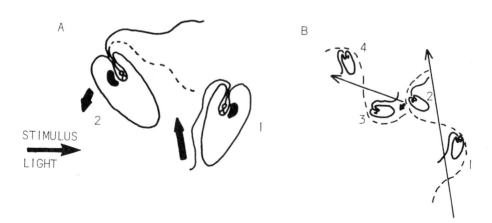

FIGURE 2A. Two successive positions of Euglena during swimming. Thickening at flagellar base: paraflagellar body. Adjacent dark patch: stigma. In position 2, dashed-line flagellum: orientation before (photophobic) response; solid-line flagellum: reorientation during response; arrow toward left: direction of turn response.

FIGURE 2B. 1,2: Axis of swimming path before response, with two cell positions. 3,4: Axis of path after response, with two further cell positions.

phototaxis, the step-down photophobic response acts as a feedback mechanism, maximizing stimulation of the photoreceptor (7). (A similar mechanism is thought to underlie negative phototaxis, but in this case the photophobic response is elicited by a step-up signal).

The main feature of this two-instant model is that phototactic orientation results from a series of photophobic responses. This has the intriguing implication that the capabilities demanded for perception of light direction in phototaxis are the same as those demanded for photophobic response--namely, the system must be able to detect a temporal change in light absorption.

A puzzling question about photomovement in Euglena has been raised by Diehn's report (7), that, in his experience, native E. gracilis is rarely phototactic. Thus, photoaccumulation and dis-

persal would be attributed directly to photophobic responses. If
Euglena does not use phototaxis as its principal photomovement
"strategy", this organism would seem to be an exception among the
flagellates.

VOLVOX

　　Phototaxis in the colonial flagellate Volvox occurs by an
interesting version of a two-instant mechanism (20-23), very different from that of Euglena. The Volvox colony is a hollow ball,
several hundred μm in diameter, made up of hundreds (sometimes
thousands) of interconnected Chlamydomonas-like cells. Each cell
has two flagella and one cup-shaped chloroplast. There is also a
stigma in each cell, but the stigmata of the anterior cells are
larger than those of the posterior cells. The photoreceptor for
photomovement has not been identified, but it has been suggested
to be a carotenoprotein and to be located in the region of plasmalemma just peripheral to the stigma (23). There is no paraflagellar body.
　　Volvox shows a photophobic stop response as well as positive
and negative phototaxis. The stop response is intimately associated
with the orientation mechanism, but in a subtle way. Hand and Haupt
(22) studied flagellar activity in Volvox by obtaining photomicro-
graphs of colonies immersed in a suspension of microscopic poly-
styrene spheres, and by observing the pattern of movement of these
particles. When Volvox is exposed to a single high-intensity flash
of white light (2s) from below, the entire colony shows a stop response: all the flagella stop beating for several seconds. (The
duration of the stop depends on the magnitude of the step-up signal). This cessation of movement is the fundamental photophobic response of Volvox; unlike the case of Euglena, there is no direction-
change inherent in the response pattern. Nonetheless, the stop response of Volvox does lead to phototactic orientation.
　　Orientation results from an asymmetric stop response, via a
mechanism known, appropriately, as the "rowboat effect" (20,23).
The basic concept of this mechanism is simple. In a Volvox colony
under unilateral illumination, the cells on the lighted side experience a step-up signal, which induces them to stop beating their
flagella. Meanwhile, the cells on the side away from the light
source are shaded by the rest of the colony; they therefore receive
no signal and continue to beat their flagella. The result is a turn

toward the light source[(+)]. The shading is attributed to combined
absorption by chloroplasts and stigmata (23). In Volvox, unlike
Euglena, it is thus a step-up photophobic response that leads to
positive phototaxis. (The same is true in Gyrodinium; 40).

In visualizing the phototaxis of Volvox, it is important to
note that the colony rotates about its anteroposterior axis during
forward movement. In any given cell, a stop response is elicited
when that cell crosses the border from the dark hemisphere into
the lighted one. The stop lasts on the order of half the period of
colony rotation (20), by which time that cell is on the dark side
again. As each new semicircular "band" of cells is brought into
the light, those cells undergo a stop and then a subsequent recovery in the dark.

The orientation process in Volvox is complicated further by
two curious features.

(a) The cells in the anterior hemisphere are more responsive
to light than those in the posterior hemisphere (22). When a stimulus beam is presented from the front of the colony, the center of
the stop-response area is at the anterior pole, as expected. Surprisingly, when the stimulus is presented from the side, the stop
response still occurs mainly in the vicinity of the anterior pole;
unless the step up in intensity is very large, the posterior cells
continue to beat their flagella. Nonetheless, the center of stop-
response area is shifted slightly in the direction of the stimulus-
by 20-30° from the anterior pole. This small shift in stop-response
distribution does result in reorientation toward the light, but the
turn is a gradual one.

(b) Even when Volvox has become oriented, it does not swim
directly toward the light. The colony moves in a straight line,
but (viewed from behind) it always veers to the left of the light
source (21). This eccentric behavior derives simply from the physical properties of the colony. Because the flagella of the individual cells beat in a fixed, oblique plane, the colony rotates about
its antero-posterior axis in a constant sense. Viewed from behind,

---

[(+)] For this mechanism to work, there must be no signal transmission from the illuminated to the non-illuminated portion of the
colony; i.e., the stop response must be localized. Huth (20) showed
that this is indeed the case.

Volvox rotates counterclockwise. As a result, the spherical colony should follow a curved path in the direction of rotation. This effect is analogous to a well-known phenomenon in baseball: if one puts a spin on a ball, it will curve in the direction of the spin ! Indeed, in the absence of stimulation, Volvox follows a curved path to the left (21). Presentation of an orienting stimulus (photic or galvanic) counteracts this tendency somewhat, tending to straighten out the curve. The final slanted path is the resultant of these opposing tendencies. This phenomenon of deviation to the left was used by Schletz (23) in determining and interpreting action spectra (Colombetti and Lenci, this volume).

CHLAMYDOMONAS

Chlamydomonas, the organism used for photomovement work in our laboratory, provides a good case study of problems to be solved. The photoreceptor for photomovement has not been identified. We do not know the mechanism by which Chlamydomonas perceives light direction, nor the possible role of the stigma in that mechanism, but recent work has begun to untangle these problems (see below).

Although much smaller than Euglena or Volvox, Chlamydomonas offers some attractive features. It shows a quick and clear phototaxis, and a striking photophobic response. This organism also lends itself well to genetic analysis, and an intriguing array of photomovement mutants has already been isolated (41-45).

Chlamydomonas swims with a kind of "breaststroke", while rotating on its longitudinal axis. During phototactic orientation, Chlamydomonas steers quite precisely toward or away from the light source. The higher the stimulus intensity, the more direct is the swimming path, both in positive and in negative phototaxis (15).

The photophobic response is best observed in cells kept in a background of red light and exposed to a single high-intensity flash (of blue or white light). This stimulus elicits a dramatic stop response. Analyzed in slow motion on videotape, the stop response reveals a complex pattern (46). The cell stops within 1/20s after a flash, and remains stationary for ca. 1/10s. Then it backs up by about one cell length. Next, the cell may stop again, or move backward or sideways. In any event, within 1 s following the flash, the cell resumes forward motion, usually is a new direction. Reversal of flagellar beat during the backward-swimming phase of this response is under Ca^{2+} control (47,48). The stop response of Chlamydomonas

is evidently a step-up response, since it often occurs at the onset
of continuous illumination (15), but not when the light is turned
off.

Recent studies in our laboratory have centered on two related
problems: the mechanism of perceiving light direction and the role
of the stigma in photomovement, particularly in phototaxis. Our
most fruitful approach to the direction-perception problem has been
to pose the question: can the cell become oriented in response to a
single light flash, too short to allow comparison of light absorption at two different cell positions ? If most of the cells became
oriented following such a short flash, a one-instant mechanism
would be most likely.

In studies using the "population system" of Feinleib and Curry
(49), a single flash of blue light, lasting 6 μs, elicits a small
but definite net movement of a cell population, toward the stimulus
source at low intensity, and away from it at high intensity (16).
Evidently, some cells do become oriented in response to a flash.

In order to establish the proportion of cells responding to
a flash, studies have been conducted on single cells, using the
video-microscope system shown in Figure 3 (50,46). A high-resolution
video camera is mounted over a microscope. The algal suspension is
placed on the microscope stage in a clear plastic chamber and the
cell movements are monitored from below with a dim red light (phototactically inactive). The image is viewed on a video monitor and
is simultaneously recorded on video tape. The stimulus lights enter
through the side walls of the chamber, at right angles to the mon-
itoring beam. This arrangements of beams permits us to record oriented movement (phototaxis) toward or away from the stimulus source.

The cells are first exposed to a continuous blue (442 nm)
laser light of high intensity. Under the conditions of these experiments (1 W·m^{-2}, cells kept in dark 4h prior to testing) this
stimulus elicits strong negative phototaxis in all cells. After
10s, a single 6 μs flash from a stroboscopic lamp (used with a
blue filter) is superimposed on the laser beam, at 90° to that
beam. The laser stimulus continues for 10s after the flash. The
data are analyzed by turning the recorder reel manually, and examining the tape by single frames. An acetate sheet is placed over
the face of the monitor, the tape is stopped at the desired frame,
and the position of an individual cell is traced into the acetate.
A similar method has recently been used to study photophobic response in Euglena (31).

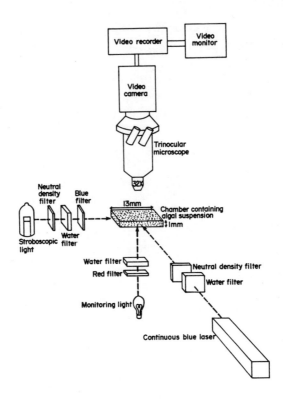

FIGURE 3. Schematic diagram of videoscope system, with stimulus lights for single-flash experiments

      About half the cells studied (182 of 392) responded to the
flash as shown in Figure 4. Cell A shows a transient deflection
away from the flash source ("negative turn response"); cell B shows
a deflection toward the source ("positive turn response"). We have
several lines of evidence that a turn response represents the initial stages of phototactic orientation. Most importantly, a majority of the cells showing a turn response (70%) turn away from the
light source (as would be expected under conditions eliciting negative phototaxis). This predominance of negative turns is statisti-
cally significant at the 0.01 confidence level.
      The fact that only about half the cells show a turn response
to a flash suggests the interesting possibility that the only ones
to respond are those in which the photoreceptor happens to be in a
"correct" position with respect to the light source at the instant
of the flash. If so, the cell might be comparing light absorbed at

two instants: before and during the flash (or, possibly, during and after the flash). According to this interpretation, the data are consistent with a two-instant mechanism for detection of light direction.

In order to resolve this problem, we should know the precise orientation of the cell about its longitudinal axis at the time of the flash. If a two-instant mechanism operates, we can make several predictions: (a) The position of the cell about its axis during a flash should be related to the subsequent occurrence of a turn response. (b) When the direction of a continuous stimulus is switched; e.g., by 90°, the position of the cell about its axis at the time of the switch should be related to the lag time for the turn into the new stimulus direction. (c) Initially, however, when the light-direction is switched, the cell should turn in a

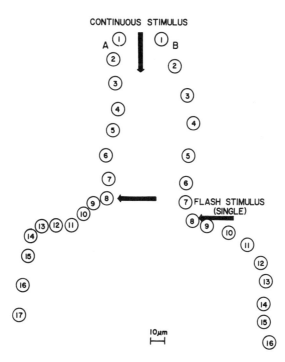

FIGURE 4A (left). Tracing of negative-turn response elicited by a single flash. Cell positions are shown at 0.1s intervals and are numbered consecutively in direction of movement.

FIGURE 4B (right). Same as 4A, but positive turn response

fixed direction with respect to its morphology, regardless of the
direction of the stimulus (cf. Euglena, above).

We have found a promising technique for keeping track of the
position of the cell about its axis. Between crossed polarizers,
the stigma appears as a bright spot against the darker background
of the cell. The stigma is an ideal marker for the purposes of this
work. Since it is asymmetrically located in the cell[+], it provides a convenient label for following rotation. Moreover, the position of the stigma is of specific interest because of its suspected role in photomovement. If the stigma is involved in perception of light direction, its position at the onset of stimulation
should be related to the pattern of subsequent response.

Evidence that the stigma is indeed involved in perception of
light direction comes from our recent work with so-called "eyeless"
mutants (51). About twenty-five years ago, Hartshorne (41) reported
isolation of a stigmaless ("eyeless") mutant of Chlamydomonas,
scored by observing the absence of stigmata under the light microscope. He noted that these cells were phototactic, but that their
response was slower and less well defined than that of the wild
type (WT). In his studies, movement of a cell population was observed simply with the naked eye.

Solely on the basis of Hartshorne's observations, it has since
been assumed that the stigma of Chlamydomonas aids in phototaxis,
but is not essential for it. This assumption has only recently
begun to be tested, using videomicroscope analysis of photomovement
in individual cells (51,52). Several strains of "eyeless" (ey)
mutants have been obtained from R.Smyth (53).

A morphological difference between ey mutants and WT cells can
readily be seen with a light microscope (400x), using transmitted
light. The contrast is even more vivid when cells are viewed between
crossed polarizers. No stigmata are visible in mutant cells from
cultures in early through mid-exponential phase (in non-synchronous
batch cultures). In contrast with earlier observations, however
(41,53), we find that the ey mutants are not completely "eyeless".
As the culture "ages", approaching stationary phase, some stigmata
do become visible. We suspect that degree of "eyelessness" may be
related to stage of the cell cycle. This hypothesis is being tested

---

[+] The stigma lies just under the chloroplast envelope, about
halfway between the anterior and posterior ends of the cell.

by using synchronous cultures and examining their stigmata with
the electron microscope at various stages of the cycle.

A preliminary EM study of our mutants, performed by H.Laibold
(Botan. Inst. Univ. of Erlangen, Germany), reveals a reduced stigma
with abnormal morphology. The stigma granules are fewer than in WT
and are highly variable in size.

Photomovement of eye mutants is being studied using the video-
microscope system shown in Figure 3; in these experiments, however,
there is a single continuous stimulus entering from the side: a
blue-green light from an argon laser (488 nm). Videotapes are analyzed as described above.

The ey mutants show both positive and negative phototaxis,
but they differ markedly from WT in the straightness (directness)
of their swimming paths. Figure 5, A and B are tracings of swimming tracks typical, respectively, of WT and mutant ey 627, in red
light only. Cell position is shown every 1/3s for 2s. The swimming
paths are fairly straight, but there is no evidence of "preferred"
direction. Figure 5, C and D show swimming paths for WT and mutant
ey 627, during phototactic response to the continuous blue-green
stimulus. Tracing was begun 40s after the onset of illumination.
The WT cells swim almost parallel with the stimulus beam, heading
either toward or away from the light source. By contrast, the swimming paths of the mutant are extremely erratic.

Straightness of path is analyzed using a program for TI-59
calculator, recently developed by K.L.Poff (personal communication)
and based on methods of Batschelet (54,55). Since for the purposes
of this analysis, we are interested in the direction rather than
the rate of travel, each segment of the swimming path is treated
as a unit vector; i.e., all segments are weighted equally, regardless of the actual distance travelled. For each cell, we calculate
the empirical mean vector; i.e., the average direction taken by the
cell, and the length, r, of that mean vector. The value r is an
index of the straightness of path; it is, in essence, a measure of
the clustering of the positions of the cell about its mean direction
of travel. The maximum value, r=1, means that the cell positions
fall on one line; i.e., the cell has a perfectly straight path. A
low r value, by contrast, indicates a path with many changes in
direction. (For an explanation of these methods, see the Batschelet
references cited above).

Table 1 summarizes r values for WT and mutant ey 627 in a typical experiment. In red light, both WT and mutant have fairly

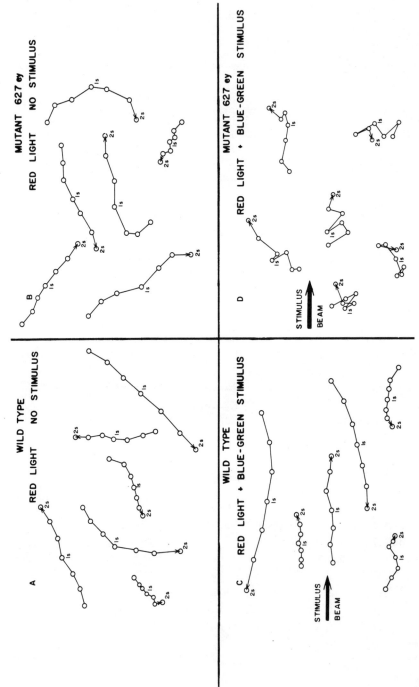

FIGURE 5. Tracings of swimming paths from videomicroscopy. Cell positions are shown every 1/3s for 2s. Arrows indicate direction of movement. A. Wild type, red light only; B. "Eyeless" mutant 627, red light only; C. Wild type, 40–42s after introduction of stimulus; D. Mutant, 40–42s after introduction of stimulus

straight swimming paths, with median r values > 0.9. In the presence of the stimulus, the median r value for WT remains essentially unchanged and the range of values becomes narrower: all paths
are practically straight. On the other hand, the median r value
for the mutant decreases sharply and the range expands: the paths
are highly irregular and they vary from cell to cell. The stimulus
light obviously has a peculiar effect on these mutants.

Our next question was: do the mutant cells head in a preferred
direction, despite their irregular paths ? The relative number of
cells heading in various directions can readily be illustrated by
plotting mean vectors on a polar-wedge histogram. Figure 6, A and
B show the distribution of directions for WT and mutant ey 627,
respectively, in red light. There is no clear indication of a preferred direction. Figure 6, C and D are histograms for WT and mutant,
40s after onset of stimulation. The laser beam enters at 0°. WT cells
show a strong bimodal orientation under the conditions of this experiment (1 W·m^{-2} at 488 nm, cells kept in dark 15 min prior to
testing). Half the cells swim at 0 $\pm$ 30° toward the light source
and half swim at 180 $\pm$ 30° away from it.

TABLE 1. SUMMARY OF r VALUE

	Wild Type		Mutant ey 627	
	Median r	Range of r	Median r	Range of r
Red light	0.96	0.63 - 1.00	0.93	0.69 - 0.99
Red light + Blue-green stimulus	0.97	0.88 - 1.00	0.57	0.26 - 0.99
n = 40 for each cell type and under each set of light conditions				

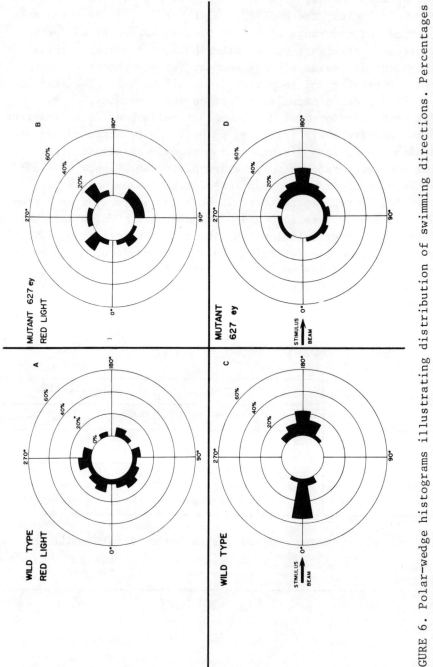

FIGURE 6. Polar-wedge histograms illustrating distribution of swimming directions. Percentages of cells with mean vectors in various directions are plotted on a circle at 20° intervals. A. Wild-type, red light only; B. "Eyeless" mutant 627, red light only; C. Wild type, during unilateral illumination (at 0°); D. Mutant, during unilateral illumination (at 0°)

Figure 6D answers the question just raised: clearly, the mutant cells also show a preferred direction. In this experiment, almost half the cells travel away from the light source. The remaining cells appear to be randomly oriented. The occurrence of negative phototaxis in the ey mutant under conditions eliciting positive and negative response in WT may mean that reversal from positive to negative occurs at lower intensity in the mutant than in the WT (15). In preliminary experiments, mutant ey 627 has shown clear positive phototaxis at lower intensities.

Although their phototactic behavior is so different, cells of WT and ey mutants all show a distinct stop response upon introduction of the stimulus light.

In conclusion, our data indicate that the stigma is not necessary for the step-up photophobic response in Chlamydomonas, but that it is important for normal phototactic orientation. Thus, the stigma appears to be involved in perception of light direction. A corollary to this conclusion is that stop response and phototactic steering are, at least in part, separable phenomena. Additional evidence for this comes from earlier work (56), in which cells were studied by video-microscopy during reorientation from one continuous light source to another: the turn into the new stimulus direction can occur without an obvious stop response. We plan to test our conclusions by comparing photomovements of WT cells with those of ey mutants at various stages of eyelessness; it will be especially interesting to compare the initial orientation movements upon introduction of a stimulus.

SUMMARY

Photomovement "strategies" of three flagellates have been reviewed in this chapter, with special attention to the mechanism by which light direction is perceived. All three organisms: Euglena, Volvox, and Chlamydomonas, show a photophobic response as well as positive and negative phototaxis, but they have diverse photophobic-response patterns.

In the photophobic behavior of Euglena, a turn about the lateral axis results in a change in direction, randomly oriented with respect to stimulus direction. In Volvox, the photophobic response does not involve a change in direction; here, a stop response is the critical feature. Yet, as we have seen, both of these photophobic "strategies" can form the basis of a two-instant mechanism

for phototactic orientation, albeit in very different ways. In
Chlamydomonas, the photophobic pattern comprises an initial stop
response, followed by a reversal in direction of movement, and then
another stop or change in direction. The photophobic response may
not be an essential part of phototactic orientation in this organism. The mechanism by which Chlamydomonas perceives light direction
is currently being investigated.

REFERENCES

1. R. Forward, Jr., in "Photochemical and Photobiological Reviews",
      K. Smith, ed., Plenum Press, New York (1976).
2. W. G. Hand, in "New Topics in Photobiology", K. Smith, ed.,
      Plenum Press, New York (1977).
3. M. E. Feinleib, Photochem. Photobiol. 27:849 (1978).
4. F. Lenci & G. Colombetti, Ann. Rev. Biophys. Bioeng. 7:341
      (1978).
5. D. P. Häder, in "Encyclopedia of Plant Physiology", New Series,
      vol. 7, Physiology of Movements, W. Haupt & M. E. Feinleib,
      eds., Springer Verlag, Berlin (1979).
6. W. Nultsch & D. P. Häder, Photochem. Photobiol. 29:423 (1979).
7. B. Diehn, in "Handbook of Sensory Physiology", H. Autrum, ed.,
      Springer Verlag, Berlin (1979).
8. P. Halldal, Physiol. Plantarum 11:118 (1958).
9. M. Watanabe & M. Furuya, Plant Cell Physiol. 15:413 (1974).
10. J. D. Dodge, The Fine Structure of Algal Cells", Academic
      Press, New York (1973).
11. B. Diehn, M. E. Feinleinb, W. Haupt, E. Hildebrand, F. Lenci
      & W. Nultsch, Photochem. Photobiol. 26:559 (1977).
12. W. Nultsch, in "Primitive Sensory and Communication Systems",
      M. J. Carlile, ed., Academic Press, New York (1975).
13. J. J. Wolken & E. Shin, J. Protozool. 5:39 (1958).
14. C. Ascoli, in "Biophysics of Photoreceptor and Photobehavior
      of Microroganisms", G. Colombetti, ed., Lito Felici, Pisa
      (1975).
15. M. E. Feinleib & G. M. Curry, Physiol. Plantarum 25:346 (1971).
16. M. E. Feinleib, Photochem. Photobiol. 21:351 (1975).
17. S. O. Mast, "Light and the Behavior of Organisms", J. Wiley
      and Sons Inc., New York (1911).
18. B. Diehn, Science 181:1009 (1973).
19. A. Checcucci, Naturwiss. 63:412 (1976).

20. K. Huth, Z. Pflanzenphysiol. 62:436 (1970).
21. K. Huth, Z. Pflanzenphysiol. 63:344 (1970).
22. W. G. Hand & W. Haupt, J. Protozool. 18:361 (1971).
23. K. Schletz, Z. Pflanzenphysiol. 77:189 (1976).
24. C. Creutz & B. Diehn, J. Protozool. 23:552 (1976).
25. P. A. Benedetti & A. Checcucci, Plant Sci. Letters 4:47 (1975).
26. P. A. Benedetti & F. Lenci, Photochem. Photobiol. 26:315 (1977).
27. P. A. Kivic & M. Vesk, Cytobiologie 10:88 (1974).
28. A. Checcucci, G. Colombetti, R. Ferrara & F. Lenci, Photochem. Photobiol. 23:51 (1976).
29. B. Diehn & B. Kint, Physiol. Chem. Phys. 2:483 (1970).
30. B. Diehn, Biochim. Biophys. Acta 177:136 (1969).
31. C. Barghigiani, G. Colombetti, B. Franchini & F. Lenci, Photochem. Photobiol. 29:1015 (1979).
32. W. Nultsch, G. Throm & I. von Rimscha, Arch. Mikrobiol. 80:351 (1971).
33. W. G. Hand, R. B. Forward, Jr. & D. Davenport, Biol. Bull. 133:150 (1967).
34. R. B. Forward, Jr., Planta 111:167 (1973).
35. R. B. Forward, Jr., J. Protozool. 21:312 (1974).
36. P. L. Walne & H. J. Arnott, Planta 77:325 (1967).
37. P. L. Walne, in "Contributions in Phycology", B. C. Parker & R. M. Brown, Jr., eds., Allen Press, Lawrence, Kansas (1971).
38. B. Diehn, J. R. Fonseca & T. L. Jahn, J. Protozool. 24:492 (1975).
39. B. Diehn, Exp. Cell. Res. 56:375 (1969).
40. W. G. Hand & J. A. Schmidt, J. Protozool. 22:494 (1975).
41. J. N. Hartshorne, New Phytol. 52:292 (1953).
42. R. D. Smyth & W. T. Ebersold, Genetics 64:S62 (1970).
43. G. A. Hudock & M. O. Hudock, J. Protozool. 20:139 (1973).
44. G. A. Hudock & H. Rosen, in "The Genetics of Algae", R. A. Lewin, ed., U. Calif. Press, Berkeley-Los Angeles (1976).
45. R. Hirschberg & R. Stavis, J. Bacteriol. 129:803 (1977).
46. J. S. Boscov & M. E. Feinleib, Photochem. Photobiol. 30:499 (1979).
47. J. S. Hyams & G. G. Borisy, J. Cell Biol. 67:186a (1975).
48. J. A. Schmidt & R. Eckert, Nature 262:713 (1976).
49. M. E. H. Feinleib & G. M. Curry, Physiol. Plantarum 20:1083 (1967).
50. J. S. Boscov, M. S. Thesis, Tufts Univ., Medford, Ma. U.S.A. (1974).

51. N. M. L. Morel & M. E. Feinleib, 7th Ann. Mtng. Amer. Soc. Photobiol., Abstr. 115-116 (1979).
52. M. E. Feinleib, in "Research in Photobiology", A. Castellani, ed., Plenum Press, London (1977).
53. R. D. Smyth, G. W. Martinek & W. T. Ebersold, J. Bacteriol. 124:1615 (1975).
54. E. Batschelet,"Recent Statistical Methods for Orientation Data", AIBS Monograph, Washington, D. C. (1965).
55. E. Batschelet, in "Symposium on Animal Orientation and Navigation", NASA Sci. & Tech. Info. Office, Washington, D. C., Doc: NAS 1.21262 (1972).
56. M. E. Feinleib, in "Biophysics of Photoreceptor and Photobehavior of Microroganisms", G. Colombetti, ed., Lito Felici, Pisa (1975).

PHOTOMOTILE RESPONSES IN GLIDING ORGANISMS AND BACTERIA

Wilhelm Nultsch

Fachbereich Biologie Lahnberge

D-3550 Marburg (W. Germany)

1. INTRODUCTION

Gliding microorganisms displaying photomovement belong to the following groups: blue-green algae, diatoms, desmids, red algae, and myxomycetes (Table I). The speed of these gliding movements, compared with that of flagellates (100-250 μm/s), is relatively low (Table II). This is a disadvantage, if the duration of the experiments is concerned. On the other hand, the behaviour of single cells can easily be studied, reaction and presentation times can be measured and movement tracks can be drawn, without using high speed cinematography and other expensive equipment.

Among the bacteria, mainly photosynthetic bacteria and halobacteria were investigated in the past. However, light induced motor responses occur also in colorless bacteria, as recently shown by MacNab and Koshland (1), and Taylor and Koshland (2) for the tumbling response of Salmonella typhimurium. The speed of all these flagellated forms (gliding bacteria have not yet been investigated) is relatively high and in some cases of the same order of magnitude as in flagellates.

2. GLIDING MOVEMENT

Although at first view gliding movement seems to look rather uniform, actually considerable differences exist in the various groups.

TABLE I

GLIDING   ORGANISMS   INVESTIGATED

Group	Genus/Species
Cyanophyceae	Phormidium autumnale
(blue-green algae)	Phormidium uncinatum
	Anabaena variabilis
	Oscillatoria tenuis
	Cylindrospermum licheniforme
	Pseudanabaena catenata
	Spirulina subsalsa
	Nostoc muscorum, N. calcicola
Bacillariophyceae	Navicula radiosa, N. buderi
(diatoms)	Nitzschia communis
	Amphora spec.
	Suriella spec.
	Peurosigma spec.
Chlorophyceae	Micrasterias denticulata
(desmids)	Closterium spec.
	Cosmarium spec.
	Pleurotaenium spec.
Rhodophyceae	Porphyridium cruentum
(red algae)	
Myxomycetes	Dictyostelium discoideum
(slime molds)	

TABLE II

SPEED OF MOVEMENT OF SOME GLIDING ORGANISMS

	Speed ($\mu m\ s^{-1}$)	
	Average rate	Maximum rate
Cyanophyceae		
Phormidium autumnale	2.2	2.8
Anabaena variabilis	0.3	0.5
Bacillariophyceae		
Nitzschia palea	8-10	14
Pinnularia nobilis	2-4	6
Navicula peregrina	10	18
Chlorophyceae		
Micrasterias denticulata	0.5	1
Rhodophyceae		
Phorphyridium cruentum	0.035	0.05
Myxomycetes		
Dictyostelium discoideum	–	0.1

Blue-green algae: Gliding movement in members of the Oscillatoriaceae such as Phormidium is relatively smooth and steady with
the trichomes rotating around their longitudinal axis. Sometimes
they move in one direction for one hour or more, sometimes they
change the direction frequently. Hormogonia and motile trichomes
of the Nostocaceae, such as Anabaena, glide without rotation (3).

Diatoms display an alternating "forward" and "backward"
movement. If no external factors act unilaterally, the direction
of movement is random in cells which lack morphological polarity.
Cells of the Navicula-type move in more or less straight paths,
whereas in the Nitzschia-type the paths are curved or circular (4).

Desmids: Micrasterias cells display an oscillating movement
(5,6) resulting in meander traces of mucilage. Under diffuse light,
the orientation is random, and the amplitudes of the oscillations
are comparatively large. The behaviour of Cosmarium cells is similar.

Red algae: The movement of Porphyridium cruentum is more or
less irregular and random.

Slime molds: The pseudoplasmodia as well as the single amoebae
of Dictyostelium discoideum display an amoeboid movement, which
cannot be regarded as a "gliding" movement in the strict sense.
Nevertheless, they may be included here.

3. MOVEMENT MECHANISMS

Besides the amoeboid and bacterial flagellar movement, the
mechanisms of which are of course completely different from
gliding movement, not even the mechanisms of gliding movements are
the same in the various groups.

Blue-green algae: Halfen and Castenholz (7,8) and Halfen (9)
have suggested that gliding of the blue-green algae Oscillatoria
princeps is the result of unidirectional waves, travelling on the
cell surface, which act against a solid substrate and are produced
by a lateral deformation of fibrils 5-8 nm wide. These fibrils are
arranged in a parallel array and are obliquely aligned creating a
helix with a pitch of about 60° on the surface of the trichomes.

Diatoms are encapsulated in silica shells. The so-called raphe,
i.e. a fissure in the silica layer, appears to function by trans-
mitting the motive force from the protoplast to the solid substrate.
The latter one is a prerequisite for movement. The mechanisms of
the motive force transmission itself, however, is not yet known,
although several hypotheses have been proposed. In some of them

also extramembraneous protein fibrils play an important role (10).

Desmids: In desmids the movement is caused by an unidirectional excretion of mucilage through terminal pores, as mentioned above (5,6).

Red algae: Tracks of excreted slime can also be demonstrated in Porphyridium cruentum. Since changes in the shape of the cells during movement have been observed by Geitler (11) and Vischer (12), active changes of the cell shape might be also involved.

Slime molds: For amoebae, investigated so far, cytoplasmic contractility has been established as the source of mechanical work that brings about cytoplasmic streaming and cell locomotion. Endogenous ATP serves as the energy source (13).

Bacteria: In the bacterial flagellum several filaments are winded up like a rope. Each filament consists of flagellin monomers with a molecular weight of about 40,000 Dalton. As shown by DePamphilis and Adler (14), the flagellum is inserted in the cell wall and in the cytoplasmic membrane with rings of the basal apparatus. Since the flagellin filaments neither show enzymatic activity nor are able to bend actively, the flagellar beating seems to be brought about by rotation of the hook with the aid of the rings of the basal apparatus (15-17).

4. REACTION TYPES

In most of the microorganisms which display photomovement three different types occur: photokinesis, phototaxis and photophobic responses. Most of the bacteria lack phototaxis. In general, I follow the terminology of Diehn et al. (18). For reasons discussed later, however, phototaxis is not only regarded as a result of a response, but as a true response brought about by a mechanism of its own.

Photokinesis describes the effect of light on the linear velocity of movement, which is either increased (positive) or decreased (negative). Under certain conditions photokinetic effects can result also in starting or stopping of movement.

Photophobic responses are single transient motor responses, each caused by a temporal change in irradiance, dI/dt, either by a decrease (step-down) or by an increase (step-up). When a spatial gradient of irradiance is established, any organism traversing this gradient perceives a temporal change in irradiance, since with movement the spatial gradient is transformed into a temporal one. Thus, photophobic responses can result either in photoaccumu-

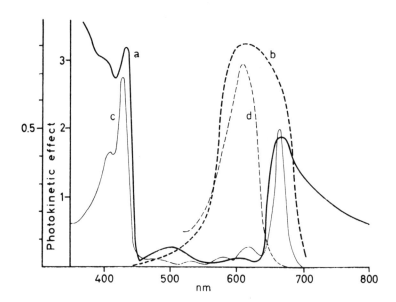

FIGURE 1. Photokinetic action spectra of Phormidium ambiguum (a)
and Anabaena variabilis (b). For comparison the absorption spectra
of chlorophyll a (c) and C-phycocyanin (d) are given. Abscissa:
wavelength in nm; ordinate: photokinetic effect in relative units
and absorbance (after 21, modified).

lations or in photodispersal.

   Phototaxis denotes an oriented movement with respect to the
light direction, resulting in a migration either toward the light
source (positive) or away from it (negative).

5. PHOTOKINESIS

   In most of the microorganisms in which photokinesis occurs
light has a positive photokinetic effect, i.e. movement is accelerated by light related to its speed in the dark, or movement is
even initiated by light in organisms which are motionless in the
dark. Since the photokinetic irradiance-response curves are typical optimum curves (19), the photokinetic effect decreases above
the optimum and finally reaches zero or sometimes, in organisms
which are motile also in the dark, becomes negative, i.e. the speed
is decreased below the dark movement level.

   Photokinetic action spectra were measured with the following

microorganisms: the purple bacterium Rhodospirillum rubrum (20), several blue-green algae (19,21), the diatom Nitzschia communis (22) and
the red alga Porphyridium cruentum (23). All action spectra indicate
that the photokinetically active light is absorbed by the photosynthetic pigments, as shown in Fig.1 for the blue-green algae Phormidium
ambiguum and Anabaena variabilis and in Fig. 2 for the red alga
Porphyridium cruentum. Studies with inhibitors and uncoupler carried
out with Rhodospirillum (20) and with blue-green algae (19,21,24-26)
revealed that the positive photokinetic effect is the result of an
additional energy supply from photosynthetic phosphorylation to the
motor apparatus. If the latter one is energetically saturated, no
photokinesis occurs.

Obviously, both cyclic and non-cyclic (or pseudocyclic) photophosphorylation can serve as energy sources of photokinesis, as the
action spectra for Phormidium ambiguum and Anabaena variabilis
(Fig. 1) show. In Phormidium ambiguum light absorbed by chlorophyll
a is highly active, while the spectral ranges absorbed by carotenoids and biliproteins are not as active as the light absorption
would suggest. This indicates that photokinesis in this species
is coupled with cyclic photophosphorylation by PS I. The action
spectrum of Anabaena variabilis, however, is quite different. Its
maximum coincides with the absorption maximum of C-phycocyanin.
Red light absorbed by chlorophyll a is also active, although no
distinct peak or shoulder can be detected, while blue-light
absorbed by the Soret band is ineffective. Thus, in Anabaena the
photokinetic effect must be ascribed to a phosphorylation process
coupled with the electron transport via both photosystems I and II.
Accordingly, photokinesis of Anabaena is much more sensitive to
DCMU than photokinesis of Phormidium (21). There are, however,
also mixed types in which both photophosphorylation processes supply
energy to the locomotor apparatus (19,27).

6. PHOTOTAXIS

Phototactic orientation can be brought about either by a
true steering or by changing the autonomous rhythm of non-oriented
backward and forward movement, in which the orientation with
respect to light direction is random, but movement in one direction,
either toward the light source or away from it, is preferred.
In the latter type, unilateral illumination has no effect in
organisms which are oriented more or less perpendicularly to the
light beam. However, since the movement of the organisms is never
completely straight, all individuals of a population come, sooner

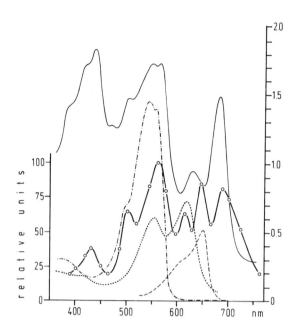

FIGURE 2. Photokinetic action spectrum (open circles, solid line), and in vivo absorption spectrum (fine solid line) of Porphyridium cruentum, and absorption spectra of B-phycoerythrin (dashed dotted line), R-phycocyanin (dotted line) and allo-phycocyanin (dashed line). Abscissa: wavelength in nm; ordinate: photokinetic effect in relative units and absorbance, respectively (after 23).

or later, for a short or a long time, into a position more or less parallel to the direction of light, so that light can act on the autonomous rhythm of reversal.

Among the non-flagellated organisms investigated so far, the steering mechanism is realized in the blue-green algae of the family Nostocaceae, such as Anabaena variabilis and Cylindrospermum licheniforme, in the red alga Porphyridium cruentum, and in the desmid Micrasterias denticulata. The steering mechanism results in a movement of single trichomes as well as of masses of algae toward the light source or away from it.

Phototaxis by means of changes of the autonomous rhythm of reversal occurs in blue-green algae of the family Oscillatoriaceae and in diatoms. The algae do not move strictly toward the light source, but the spreading areas shift toward the light source or away from it the more, the stronger the phototactic effect is.

Most of the phototactic action spectra measured in the past have one feature in common: only visible light of shorter wavelengths (up to 550 nm) and, in some species, near UV is active. Maximal activity in flagellated forms is often between 470 and 500 nm, whereas in the diatom Nitzschia communis light of shorter wavelenghts and especially UV is more active. Therefore, phototaxis has been regarded very often as a typical blue light response, comparable to the phototropism. The effectiveness of radiation of shorter wavelengths points to yellow pigments as photoreceptors. Carotenoids and flavins are discussed as possible photoreceptor pigments.

Most recently, the phototactic action spectrum of Porphyridium cruentum has been measured by Nultsch and Schuchart (unpublished).

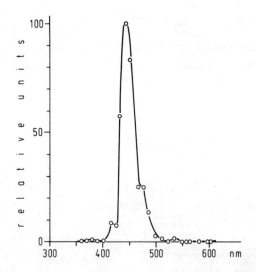

FIGURE 3. Phototactic action spectrum of Porphyridium cruentum. Abscissa: wavelength in nm; ordinate: phototactic effect in relative units (Nultsch and Schuchart, unpublished).

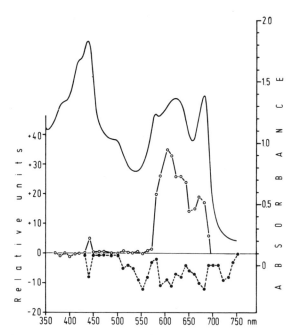

FIGURE 4. Action spectra of positive (open circles, solid line) and negative (closed circles, dashed line) and in vivo absorption spectrum (fine solid line) of Anabaena variabilis. Abscissa: wavelength in nm; ordinate: phototactic effect in relative units and absorbance (after 31).

It displays a maximum at 443 nm and shoulders at 416 and around 470 nm (Fig. 3). UV is not completely ineffective, but no significant peak around 370 nm exists. Consequently, this spectrum favours carotenoids rather than flavins.

It must be emphasized, however, that there are some phototactically reacting organisms the action spectra of which cannot, or at least, not exclusively, be interpreted on the basis of the carotenoid/flavin concepts, since they are considerably extended to longer wavelengths. The phototactic action spectra of Phormidium autumnale and Ph. uncinatum show three main peaks (28,29). The broad one between 350 and 400 nm might be interpreted as to be indicative for flavins, the second one at 490 nm with a shoulder at 460 nm points to carotenoids rather than flavins, but the third one at 560 nm coincides with the absorption maximum of C-phycoerythrin. C-phycocyanin participation is indicated by a small but significant peak at 615 nm. Although the phycobiliproteins are part

of the photosynthetic apparatus, correlation between phototaxis
and photosynthesis can be excluded, since red light above 640 nm
is completely ineffective.

     The participation of photosynthetic pigments in photoreception
of photomotile response have been demonstrated also with other
organisms, e.g. with Micrasterias denticulata (6), with Cosmarium
margaritiferum and C. botrytis (30), and by Nultsch et al. (31)
with the strain B 1403-10 of Anabaena variabilis. In Anabaena
(Fig. 4), participation of both C-phycocyanin and chlorophyll a,
is indicated by peaks at 440, 615 and 670 nm. The same is true for
negative phototaxis. In addition, however, radiation between 500
and 560 nm which is inactive at irradiance levels causing positive
phototaxis at other wavelengths, induces clear negative responses
at higher photon fluence rates. These different action spectra
for positive and negative phototaxis point to different photoreception pigments. The phototaction action spectrum of the desmid
Cosmarium botrytis (Fig. 5) indicates the participation of chlorophyll a. Only with Anabaena inhibitor experiments have been carried
out. They have shown that the trichomes orient themselves perfectly

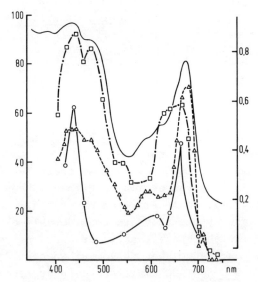

FIGURE 5. Action spectra of phototaxis (circles), photokinesis
(squares) and photophobic response (triangles) and in vivo
absorption spectrum (solid line) of Cosmarium. Abscissa: wavelength in nm; ordinate: relative response and absorbance (modified
after 30).

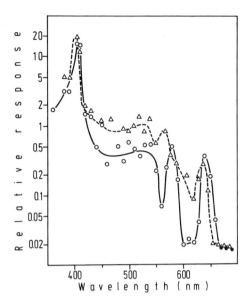

FIGURE 6. Action spectra of photoaccumulation (circles) and photodispersal (triangles) of Dictyostelium amoebae. Abscissa: wavelength in nm; ordinate: relative response (modified after 33, 34).

well even in presence of DCMU and DBMIB at concentrations at which
photosynthetic O_2 production is almost completely inhibited.
Consequently, correlations between phototaxis and photosynthesis
can be excluded even in this case.

     The phototactic action spectrum of colorless pseudoplasmodia
of Dictyostelium discoideum which shows maxima at around 430 and
560 nm also cannot be interpreted by assuming carotenoids or flavins
as the only photoreceptor pigments. Between 520 and 600 nm this
spectrum agrees fairly well with the action spectrum of absorbance
changes at 411 nm. The isolated and purified pigments show strong
absorption at 430 nm and two broad bands between 530 and 590 nm
(32). Recently, the action spectra of photoaccumulation and photodispersal of Dictyostelium amoebae have been measured by Häder
and Poff (33,34). Since these authors have shown that photoaccumulation and photodispersal in this organism are the result of positive and negative phototaxis, respectively, the action spectra of
both the reactions which turned out to be essentially identical
(Fig. 6) can be regarded as action spectra of positive and negative

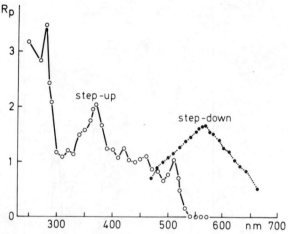

FIGURE 7. Action spectra of the step-up and the step-down photophobic response of <u>Halobacterium halobium.</u> Abscissa: wavelength
in nm; ordinate: relative response (after 14).

phototaxis of single amoebae. Both spectra show a main maximum at
405 nm and a broad band in the blue extending to longer wavelengths.
The action spectrum of photoaccumulation shows two further sharp
peaks at 570 and 640 nm, the one of photodispersal a secondary
maximum at 635-640 nm. Both action spectra are different from
that of pseudoplasmodia phototaxis and do not coincide with any
pigment known to be present in Dictyostelium.

7. PHOTOPHOBIC RESPONSE

Since the classical work of Engelmann (35) we know that in
purple bacteria the photophobically active light is absorbed by the
photosynthetic pigments. This was confirmed later by Manten (36),
Thomas (37), Duysens (38), and Clayton (39). These findings led to
the conclusion that in purple bacteria the photophobic responses
are due to a sudden change in the rate of photosynthesis.

More recently, the photophobic responses of <u>Halobacterium halo-
bium</u> have been investigated by Hildebrand and Dencher (40,41). They
measured action spectra (Fig. 7) and found that the step-down response occurring at wavelengths between 460 and 650 nm is controlled
by a photosystem called PS 565, the photoreceptor pigment of which
seems to be bacteriorhodopsin. Step-up photophobic responses which
occur in UV, violet and blue light are controlled by PS 370, the

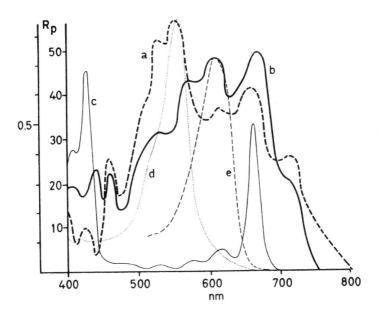

FIGURE 8. Photophobic action spectra of Phormidium uncinatum (a)
and Phormidium ambiguum (b). Absorption spectra of chlorophyll a
(c), C-phycoerythrin (d) and C-phycocyanin (e). Abscissa: wavelength in nm; ordinate: relative response and absorbance (after
44,45).

photoreceptor pigment of which is presumed to be a retinylidene
protein, probably a precursor of bacteriorhodopsin (42). Action
spectra measurements with the aid of a population method (43) gave
similar results.
      Action spectra studied with blue-green algae of the genus
Phormidium (44,45) have shown that in general the spectral ranges
absorbed by the biliproteins are effective (Fig. 8). Red light
absorbed by chlorophyll a is active, too, while the effectiveness
of blue light is far out of proportion to its absorption by the
Soret band. These action spectra agree essentially with the photosynthetic ones measured by Duysens (38), Haxo (46,47) and Nultsch
and Richter (48). The coupling between photosynthesis and photophobic responses has been confirmed by studying the effects of
photosynthesis inhibitors, uncouplers, and artificial redox systems
on the photophobic response of Phormidium uncinatum (24-27, 49-52).
      Attempts were made to elucidate the role of PS I and II in the
photophobic response of Phormidium uncinatum with the aid of a dual

wavelength technique. In this experimental system the photophobic
reaction depends qualitatively and quantitatively on whether the
background and the trap wavelengths are absorbed by the same or
by different photosystems. If a background wavelength of 563 nm
mainly exciting PS II was used and the trap wavelength was varied,
a maximum of accumulation in the light trap around 700 nm was
observed. Below 440 and above 700 nm migrations out of the light
field occurred. If a background wavelength of 723 nm exciting only
PS I was used, maximum activity was observed in the range between
500 and 600 nm (53). With some reservations the spectrum measured
with a constant background of 723 nm can be regarded as an action
spectrum of PS II, whereas a constant background wavelength of 563
nm roughly yields an action spectrum of PS I, in which even the
dispersal from the light field must be regarded as indicating strong
photophobic activity. This was shown by Häder (54) using a constant
trap wavelength of 578 nm and varying the background wavelength.
Since at this wavelength no migrations out of the light field occur,
maxima were found at 434 nm in the blue and at 647, 687 and 715 nm
in the red, but a depression between 500 and 600 nm absorbed by PS
II. This is a quasi "positive" version of the action spectrum of
PS I. Conversely, if a constant trap of 723 nm is used and the
background is varied, again an action spectrum of PS II is obtained
(Fig. 9).

     Since photophobic action spectra of flagellates measured so
far do not show red light activity, the photophobic responses of
eukaryotes often have been regarded as typical blue light responses.
However, investigations of other groups of eukaryotes, such as
diatoms, desmids and red algae, have shown that this is not true.
On the contrary, in most of these organisms correlations between
photosynthesis and photophobic responses can be judged from action
spectra.

     The first action spectrum of photophobic responses in diatoms
was measured by Nultsch (22) with Nitzschia communis. It shows
activity of UV, violet and blue light. Wenderoth (55) studying
other forms found that in diatoms two photophobic reaction/types
exist: a blue light type and a photosynthetic type. An action
spectrum of the latter one was measured with single cells of Navi-
cula peregrina. Its shape between 370 and 560 nm resembles that
of Nitzschia communis, pointing to chlorophyll a (and c?) and carotenoids as photoreceptor pigments. The red maximum at 669 nm coincides with the in vivo absorption maximum of chlorophyll a. Since
Wenderoth was able to induce photophobic responses also in single

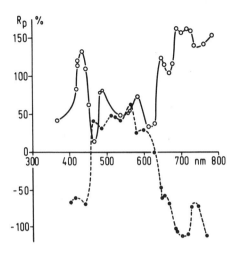

FIGURE 9. Photophobic action spectra of _Phormidium uncinatum_
obtained by variation of background wavelength under constant
trap wavelength of 578 nm (circles, solid line) and 723 nm (dots,
dashed line). Abscissa: wavelength in nm; ordinate: phobic response
in relative units (after 54).

cells of _Navicula peregrina_ by partial irradiation with red light,
there is no doubt that the accumulations in red light fields are
the results of true photophobic responses, although DCMU inhibits
only at concentrations $> 10^{-5}$ mol.

Since single cell studies revealed that photoaccumulations
of _Cosmarium cucumis_ are results of photophobic responses (30),
the photoaccumulation action spectrum can be regarded as the action
spectrum of the photophobic response (Fig. 5). It shows maxima
around 443 and 669 nm and a shoulder at 470 nm. Although the effects
of inhibitors have not yet been studied, correlations between photophobic responses and photosynthesis seem to be obvious.

Porphyridium cells show true photophobic responses (56).
However, photoaccumulations can also result from phototactic reactions toward the light scattered laterally from the organisms
within the light trap. Moreover, also photokinetic effects can
influence the density of photoaccumulations. Therefore, Schuchart
(56) measured the action spectrum of photoaccumulation using a
blue background irradiation (Fig. 10). Since the background light
is photokinetically and phototactically effective, interference

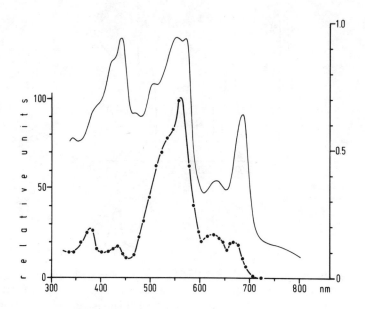

FIGURE 10. Photophobic action spectrum (dots and heavy solid line) and in vivo absorption spectrum (fine line) of <u>Porphyridium cruentum</u>. Abscissa: wavelength in nm; ordinates: relative response and absorbance (after 56).

with phototaxis and photokinesis is excluded. The main maximum at 560 nm coincides with the absorption maximum of B-phycoerythrin. Smaller but significant peaks around 440 and 670 nm are indicative for chlorophyll a, another one at 627 nm for R-phycocyanin. This action spectrum resembles the PS II spectrum of <u>Porphyridium cruentum</u> (57). Since the photophobic responses are inhibited by DCMU, correlations between photophobic responses and photosynthesis seems to be well established even in this organism. Thus, the "photosynthetic" type of photophobic response control is widespread also among eukaryotes.

8. CONCLUSIONS

The results reported here clearly demonstrate that three different reaction types of photomovement exist in microorganisms: photokinesis, phototaxis and photophobic response. This is unequivocal in those cases in which the action spectra are completely different, as shown the first time for the blue-green algae of

the genus Phormidium (28,29,44,45). It is also true for the eukaryote Porphyridium, in which the whole visible range is more or less
photokinetically active (Fig. 2), whereas phototaxis is a true
blue light response, restricted to the spectral range between 350
and 520 nm with a main maximum at 443 nm (Fig. 3). Finally, the
photophobic response has a main maximum at 560 nm, indicating
B-phycoerythrin as main photoreceptor pigment (Fig. 10). Consequently, in these forms, phototaxis cannot be the result of either
photokinetic or photophobic responses, but must be regarded as a
reaction type of its own.

    Further evidence comes from the phototactic investigations with
Anabaena. In this organism the phototactic orientation is certainly
not the result of a series of successive photophobic responses.
Moreover, it can also not be the result of different photokinetic
effects on both sides of the trichome, for the following reasons:
At lower photon fluence rates, where photokinesis is positive,
the trichomes bend toward the light source, although the irradiated side perceives more photons per time and area unit and should
therefore move faster than the shaded side. At higher photon fluence
rates, when the photokinetic effect becomes zero or negative, the
filaments bend away from the light source, although now the shaded
side should move faster, if the bending were the result of different photokinetic effects.

    In some organisms, these three reaction types have similar
action spectra, as in the desmid Cosmarium (Fig. 5). This indicates that the stimulus light is perceived by the same photoreceptor
pigments. However, it does not necessarily mean that these reactions are triggered by the same reaction chain, and that one of
them is the result of another one. In order to decide whether or
not photomotile responses are coupled with each other, it is
necessary to know their whole reaction chains.

REFERENCES

1. R. MacNab, & D. E. Koshland Jr., J. Mol. Biol. 84:399 (1974).
2. B. L. Taylor, & D. F. Koshland, J. Bacteriol. 123:557 (1975).
3. G. Drews, & W. Nultsch, in "Handbuch der Pflanzenphysiologie",
    W. Ruhland, ed., Springer Verlag, Berlin-Göttingen-Heidelberg
    (1962).
4. W. Nultsch, Arch. Protistenk. 101:1 (1956).
5. W. Neuscheler, Z. Pflanzenphysiol. 57:49 (1967a).
6. W. Neuscheler, Z. Pflanzenphysiol. 57:151 (1967b)

7. L. N. Halfen, & R. W. Castenholz, Nature 225:1163 (1970).

8. L. N. Halfen, & R. W. Castenholz, J. Phycol. 7:133 (1971).

9. L. N. Halfen, J. Phycol. 9:248 (1973).

10. R. W. Drum, & J. T. Hopkins, Protoplasma 62:1 (1966).

11. L. Geitler, Öst. Bot. Z. 100:672 (1953).

12. W. Vischer, Ber. Schweiz. Bot. Ges. 65:459 (1955)

13. N. S. Allen, & R. D. Allen, Ann. Rev. Biophys. Bioeng. 7:479
    (1978).

14. M. L. DePamphilis, & J. Adler, J. Bacteriol. 105:383 (1971).

15. H. C. Berg, Nature 249:77 (1974).

16. H. C. Berg, Nature 254:389 (1975).

17. P. Läuger, Nature 268:360 (1977).

18. B. Diehn, M. Feinleib, W. Haupt, E. Hildebrand, F. Lenci,
    W. Nultsch, Photochem. Photobiol. 26:559 (1977).

19. W. Nultsch, "Abhandl. Marburger Gelehrtengesellschaft" W. F.
    Verlag, München (1972).

20. G. Throm, Arch. Protistenk. 110:313 (1968).

21. W. Nultsch, & W. Hellmann, Arch. Microbiol. 82:76 (1972).

22. W. Nultsch, Photochem. Photobiol. 14:705 (1971).

23. W. Nultsch, H. Schuchart, & M. Dillenburger, Arch. Microbiol.
    122:207 (1979).

24. W. Nultsch, Z. Pflanzenphysiol. 56:1 (1967).

25. W. Nultsch, Photochem. Photobiol. 10:119 (1969).

26. W. Nultsch, & J. Jeeji-Bai, Z. Pflanzenphysiol. 54:84 (1966).

27. W. Nultsch, in "Primary Molecular Events in Photobiology"
    A. Checcucci, & R. A. Weale, eds., Elsevier Scientific
    Amsterdam, London, New York (1973).

28. W. Nultsch, Planta 56:632 (1961).

29. W. Nultsch, Ber. dt. Bot. Ges. 56:632 (1961).

30. K. Wenderoth, & D. P. Häder, Planta 145:1 (1979).

31. W. Nultsch, H. Schuchart, & M. Höhl, Arch. Microbiol. 122:85
    (1979).

32. K. L. Poff, W. F. Loomis, Jr., & W. L. Butler, J. Biol. Chem.
    249:2164 (1974).

33. D. P. Häder, & K. L. Poff, Photochem. Photobiol. 29:1157
    (1979).

34. D. P. Häder, & K. L. Poff, Experim. Mycol. (in press) (1979).

35. Th. W. Engelmann, Pflügers Arch. Ges. Physiol. 30:95 (1883).

36. A. Manten, Thesis, Utrecht (1948).

37. J. B. Thomas, Biochim. Biophys. Acta 5:186 (1950).

38. L. N. M. Duysens, Thesis, Utrecht (1952).

39. R. K. Clayton, Arch. Mikrobiol. 19:107 (1953).

40. E. Hildebrand & N. Dencher, Ber. Dt. Bot. Ges. 87:93 (1974).
41. E. Hildebrand & N. Dencher, Nature 257:46 (1975).
42. E. Hildebrand Biophys. Struct. Mechanism 3:69 (1977).
43. W. Nultsch & M. Häder, Ber. Dt. Bot. Ges. 91:441 (1978).
44. W. Nultsch, Planta 57:613 (1962).
45. W. Nultsch, Planta 58:647 (1962).
46. F. T. Haxo, in "Handbuch der Pflanzenphysiologie", W. Ruhland,
    ed., Springer, Berlin-Göttingen-New York (1960).
47. F. T. Haxo, in "Comparative Biochemistry of Photoreactive
    Systems", M. B. Allen, ed., Academic Press, New York, London
    (1960).
48. W. Nultsch & G. Richter, Arch. Mikrobiol. 47:207 (1963).
49. W. Nultsch, Photochem. Photobiol. 4:613 (1965).
50. W. Nultsch, Arch. Mikrobiol. 55:187 (1966).
51. W. Nultsch, Arch. Mikrobiol. 63:292 (1968).
52. W. Nultsch, in "Behaviour of Micro-organisms" A. Perez-Miravete,
    ed., Plenum Press, London, New York,(1973).
53. W. Nultsch & D. P. Häder, Ber. Dt. Bot. Ges. 87:83 (1974).
54. D. P. Häder, Arch. Mikrobiol. 96:255 (1974).
55. K. Wenderoth, Thesis, Marburg (1975).
56. H. Schuchart, Thesis, Marburg (1978).
57. A. C. Ley & W. L. Butler, Photosynthetic Organelles, Special
    Issue of Plant & Cell. Physiol. 33:46 (1977).

PHOTOTROPISM OF LOWER PLANTS

Rainer Hertel
Institut für Biologie
III Universität Freiburg
D-78-Freiburg
Federal Republic of Germany

1. INTRODUCTION AND SKETCH OF REACTION CHAIN

Certain growing plant organs bend towards the light when exposed to asymmetric illumination. The phenomenology of this - positive-phototropism in lower plants will be our main topic. The implicit aim, however, is an analysis leading towards the molecular understanding of the reaction chain. It is likely that on one hand the initial part of the phototropic reaction chain is common to several other photobiological processes (photomorphoses) and that on the other hand, the bending reaction may be common to other tropistic responses such as geotropism.

Negative phototropism and diaphototropism will not be dealt with; examples are rare and completeness is not required since at least two excellent recent treatises on phototropism exist (1,2; see there for further references). Polarotropism, however, will briefly be mentioned where the growth direction is oriented by the plane of light polarisation rather than by the direction of illumination.

Phototropic bending is found in growing single cells, in coenocytic systems such as the sporangiophore of Phycomyces and multicellular organs, such as grass coleoptiles. Among lower plants the multicellular type is described only briefly for some mushrooms (e.g. Coprinus, see in 3). We therefore can concentrate on the single cell type, especially since the coenocytes can be treated practically like single cells with regard to tropism: one growing tube with a cell wall surrounding plasmalemma, protoplast plus vacuole.

In order to ask meaningful questions a preliminary answer shall be given in advance: A model chain of information flow from the photoreceptor pigment to the asymmetric growth output is shown in Fig. 1A. The idea that a unilateral "proportional" stimulation of an enzyme involved in growth could account for the observed phototropic systems is much too simple; at least signal amplification and adaptation or sensitivity changes have to be included. Genetic analysis in Phycomyces (4) indicates that we may expect 5-10 proteins in the transduction chain. Furthermore we may presuppose that the differential response is not necessarily proportional to the light gradient and that most of the elements are localized in the plasmalemma, the photoreceptor as well as the growth output.

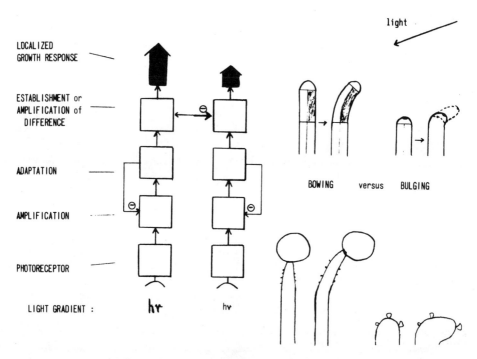

FIGURE 1A (left). Scheme of information flow on two sides of a growing and phototropically sensitive cell. Photoreceptor molecules are localized at the plasmalemma where also the other boxes may be localized. The growth output is expressed in the wall.

B (right). Two types of tropistic growth: bowing-lens effect-versus bulging; Phycomyces and Dryopteris, modified after (1)

Cell elongation in the cases described here consists of an irreversible increase in volume, concomitant with plastic stretching of the walls. In case of single cells or coenocytes, phototropism is based on a change in the spatial distribution of localized wall synthesis or stretching. Since so little is known about the final growth output, we may already deal here briefly with the biochemical possibilities: synthases or hydrolases of wall structures, e.g. of chitin, may be locally activated as a consequence of illumination, or else, vesicles, e.g. from golgi, may be locally attracted in response to stimulation.

The geometric distribution and restribution of growth in phototropism can occur in basically two modes: bowing and bulging (Fig. 1B). Phycomyces sporangiophores - most clearly in stage IV - grow or elongate a zone below the apex. If one flank elongates more, the tube bends away from the side of faster growth. In Dryopteris protonema the growing point at the very tip bulges out and is shifted to one side after asymmetric illumination. The two types can be distinguished by use of markers on the cell surface and must conceptually be kept apart, although in some cases (stage I of Phycomyces and Pilobolus, see 1,3 and Chara rhizoid, see below) the situation is not completely clear.

2. EXAMPLES OF PHOTOTROPIC SYSTEMS

2.1 Sporangiophores of Phycomyces

The positive phototropism of these large specialized fungal cells (more precisely coenocytes) and their blue light controlled growth reactions have been investigated in detail (1,2,4,5,6). Phycomyces blakesleeanus is a mould belonging to the Mucorales. A mycelium grows from spores, and subsequently giant aerial hyphae, the sporangiophores, develop. The sporangiophores are straight, fast growing upward, passing through several stages of development. In the last stage (IVb) a mature sporangium sits on top of the sporangiophore, and a steady growth rate of 3mm/h is maintained over a long time while phototropic experiments can be performed. During this time the sporangium rotates clockwise about 10°/min. Elongation is confined to a small sensitive growing zone, 0.2-2.5 mm below the sporangium. Stretch and twist occur only in this zone, the lower parts of which are continuously converted into insensitive wall material, not elongating any more. The lower part is stiff, the growing zone of the sporangiophore however is flexible but kept straight by the ca. 2 atm. turgor pressure.

Sporangiophores are geotropic, autochemotropic and they respond
to applied stretch with changes in growth rate (5). Furthermore,
they respond to changes in light intensity and, of course, to unilateral illumination-phototropism. The light-growth reaction is
considered to be related to phototropism. A short pulse or a step-
up in light intensity causes a transient increase in growth rate
reaching a maximum at about 5 min after the first stimulus. The
growth rate drops subsequently to or – in case of a pulse – below
normal rate for some min. The light-growth reaction adapts to different levels of light. This light and dark adaptation has been
extensively investigated (1,2,5,6).

Blue light and near UV, when unilaterally impinging on sporangiophores, cause positive phototropism. Since the growing zone is
subapically localized, curvature is caused by faster growth on the
distal side (Fig. 2A). Partial illumination did yield bending away
from the site of illumination. This implies a positive light-growth
reaction responsible for the tropism and – indirectly – a lens effect in order to obtain more light at the distal side. The statement
that there is more light at the distal side is not precise enough.
A deeper analysis how the focusing advantage is expressed in growth
cannot yet be given (see in 1).

The fluence-response curve covers almost a range of 10^9 units
of light intensity, over 10^6 intensity units it is almost constant
(5,6). Generally it should be pointed out that reciprocity holds
only over short intervals, e.g. ca. 1 min of illumination with
Phycomyces. Linearity is hardly ever observed, and the fluence-
response-curves in many, also unicellular, systems are even more
complicated than with Phycomyces: a plateau or a minimum have often
been observed (two reactions; see in 2).

In Phycomyces the phototropic reaction has been correlated with
the light-growth reaction for several reasons (1,5,6). The action
spectra are very similar and the direction of bending under different optical conditions at different wavelengths are in line
with the assumption that a localized light-growth reaction is the
cause for tropism when the illumination is asymmetric. This simple
hypothesis does run, however, into serious difficulties when adaptation is considered (see below).

2.2 Sporangiophores of Pilobolus

The related mould genus Pilobolus has also been investigated,

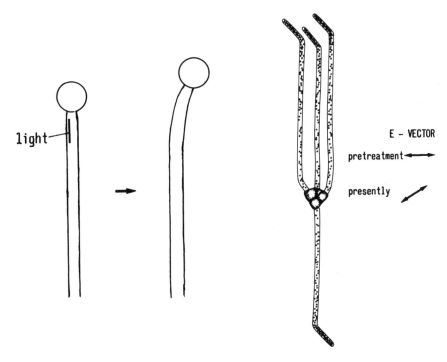

FIGURE 2A (left). Laser light or microbeam (coming from above the plane of paper) hitting left flank of Phycomyces in the sporangio-phore's growing zone; curvature is produced to the right (see in 1,2).

 B (right). Polarotropism of Sphaerocarpus chloronemata germinating from gonospores. Modified after (7)

partly because of its spectacular spore-shooting mechanism (1). The phototropism of its mature sporangiophores is very similar to Phycomyces. Growth and sensitivity is found underneath a lens-like, subsporangial swelling (2,3). This mature stage displays in all cases positive growth reactions, positive phototropism by means of a light gradient due to the lens.

The young sporangiophores, however, may present a more compli-cated pattern with species differences. In Pilobolus longipes the growth rate, as shown by microbeam illumination, is reduced by light, while being increased in the mature stage. Bending, however, is

towards the light, presumably by carotenoid screening; in caro-
tenoidless young sporangiophores, however, tropism is negative and
can be reversed to positive by immersion in oil.

In another species, <u>Pilobolus kleinii</u>, the local light growth
reaction is positive in all cases. A lens effect in the tip region
is invoked to account for positive phototropism (2,3). Theoretically this response should be due to subapical bowing. A phototropic
mechanism based on displacement of the bunging tip - which has been
described (3) - is difficult to rationalize.

2.3 Chloronema of Dryopteris

The phototropism and polarotropism of tips of fern protonemata
is only briefly treated here since this system is mainly relevant
for the elsewhere discussed orientation of photoreceptors (1; see
Haupt, this A.S.I.).

The Dryopteris chloronema are forced to grow horizontally. If
linearly polarized light is given from above the bunging tip is dis-
placed so that the chloronema grows perpendicularly to the E-vector.
The bend is very sharp and the action spectrum indicates that phytochrome is involved in addition to a blue absorbing pigment. The
growing tip is moving to the place of maximum light absorption as
shown with partial illumination. P_r must be oriented parallel to the
surface while P_{fr} surprisingly had to be orientated radially in order to account for the observed action dichroism of the far-red pho-
toreversion (see 1).

2.4 Liverwort and Moss Chloronema

The filamentous chloronema of the Liverwort <u>Sphaerocarpus don-
nellii</u> also displays phototropism and polarotropism (Fig. 2B) similarly to <u>Dryopteris</u>, but the action spectra do not show any involvement of phytochrome (7). Only their blue light photoreceptor,
highly oriented in a dichroitic structure, close to the cell surface, is active. Interestingly, with <u>Dryopteris</u> (8) as well as
<u>Sphaerocarpus</u> the fluence-response curves show different slopes at
different wavelengths in the UV-blue region (see below).

The phototropism of the moss protonema <u>Physcomitrium</u> (9) may
display a phytochrome action spectrum without blue light involvement.

2.5 Algal Systems

The coenocytic alga, Vaucheria geminata, reacts by shifting
its growing tip laterally towards the light source. Only short wavelength light is effective (10). The green alga Mougeotia displays
a bending type photo- and polarotropism (see in 1).

2.6 Relation to Polarity Induction

The polarity induction in spores and also in the Fucus zygote
may have many steps in common with the phototropic reaction chain.
This is especially true since there is a gradual change from the
rhizoid induction at the darkest pole and the negative phototropism
in the same system (see 11). The photoreceptor and the amplification
and stabilization of small differences may use the same mechanism
in both polarity induction and phototropism.

2.7 Comparison of Geotropism of a Single Algal Cell

The rhizoid of Chara is a light-insensitive but highly gravity-
sensitive system (review 12). Its geotropic growth is a "mixture"
of tip bulging and of bending by asymmetric incorporation of Golgi-
vesicles. At the lower side of a horizontally exposed rhizoid
$BaSO_4$-statoliths exert pressure. At this flank, directly behind
the tip, less vesicle incorporation and wall growth can be observed
resulting in downward bending.

3. PHOTORECEPTOR PIGMENT AND LIGHT GRADIENT

3.1 Role of Flavins in Phototropism and their Localization

The photoreceptor pigment for the blue- and UV-induced phototropic reactions in Phycomyces must be a flavin. The universality
of such blue light responses is remarkable and the action spectra
have suggested a flavoprotein or a carotenoid (7). Using mutants
or inhibitors of carotenoid biosynthesis, many cases have been reported where carotenoid deficiencies did not decrease the sensitivity to blue light (see 13). In one example a "multiplemutant" of
Phycomyces containing less than 0.1% of the carotenoids of wild-
type was shown to be phototropically fully active (14). Since this
is below the theoretically required minimal level of photoreceptor

(5), carotenoids are ruled out. Flavins are the most likely candidates for the photoreceptor (for further evidence see 13).

The following dilemma arises: On the one hand the photoreceptor flavin should be a membrane bound flavoprotein. The fine structure of the action spectrum indicates that the pigment active in
tropism must be bound in a hydrophobic environment. Studies with
polarized light (see below) also implicate a dichroic, oriented
photoreceptor localized at or near plasmalemma. On the other hand
we have found (15) than in corn and in Phycomyces the amount of
tightly membrane-bound, non-washable, non-mitochondrial flavoprotein was very low, much lower than the 3×10^{-7} M theoretically required in Phycomyces phototropism (see 5). The amount of free riboflavin or "FX", a riboflavin analog, is very high in grass
coleoptiles and Phycomyces (see in 15).

3.2 Candidate for Photoreceptor

The dilemma may be solved by a new type of reversibly bound
flavin (15). Saturable and reversible in vitro binding of riboflavin
was found to occur to sites associates with subcellular, sedi-
mentable membrane vesicles from maize and Cucurbita and Phycomyces.
The K_D was ca 6 μM, the pH optimum was 6 and the number of binding
sites amounted to 0.1-0.5 μM on a fresh weight basis. When the reducing agent dithionite was present, riboflavin binding was increased - the K_D was 2.5 μM, and the pH optimum above 8. A saturation curve is shown in Fig.3A documenting also specificity: FMN
and FAD bound less tightly to these sites than riboflavin. The ribo-
flavin-binding sites were localized on vesicles derived from plasmalemma and endoplasmic reticulum by analysing sucrose and metri-
zamide density gradients and marker enzymes. This binding site
together with its reversibly bound ligand could represent the blue
light photoreceptor "cryptochrome". Localization and amount are in
agreement with physiological requirements. Its relation to the
light-induced absorbance changes in b-cytochrome remains to be investigated (see in 15).

If the primary act in phototropism was a photo-reduction of
the bound flavin and if a substrate-driven shuttle $Fl^{\cdot} \rightleftharpoons Fl^{\cdot\cdot}$
was the "action" at the photoreceptor then a comproportionation of
a bound $Fl^{\cdot\cdot}$ + free Fl_{ox} could yield 2 $Fl^{\cdot}$ and thus an autocatalytic
amplification mechanism. The large reservoir of free flavin which
is found in many plant and fungal cells, around 5 μM (15), could

have this photobiological function.

The proposed autocatalytic mechanism could be relatively slow
and may account for the long latency periods found for light reactions after very small stimuli, e.g. in Phycomyces (5).

3.3 Suggestion of a Photochromic Flavin

Photoreversibility (see Fig. 3B) might be expected if the active form - the semiquinone or fully reduced form of flavin - is
stable for some time. This may be in the order of min in contrast
to the hours of P_{fr}-stability. On the basis of the study of analytical action spectroscopy theory, it has already been concluded
(17) that the blue light receptor action spectra in general could

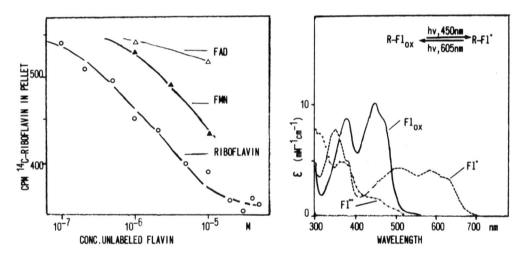

FIGURE 3A (left). Saturation curve of in vitro binding of (reduced)
0.1 µM (^{14}C) riboflavin to microsomes from maize coleoptiles.
Sedimentation tests were carried out in presence of dithionite at
pH 7.8. Increasing amounts of unlabeled riboflavin, FMN or FAD were
added before centrifugation. Data from (15).

B (right). Possible photochromic system for flavins. For comparison, spectra of flavodoxin from Megasphaera elsdenii according to
(16): oxidized (Fl_{ox}), semiquinone (Fl·), and fully reduced form
(Fl··). Bound flavin: R-Fl

be represented in principle as the action spectra of a"photochrome".
This original proposal, however, was made in terms of phytochrome,
but formally the same qualitative predictions can be made for a
flavin photochromic system from the absorption curve of oxidized
versus half or fully reduced flavin (see Fig. 3B). The peak at 360
nm should disappear under certain kinetic conditions. This is found
in Mesotaenium (18, compare Mougeotia). The different relative ef-
fectivities of 410 vs. 480 nm light of the two blue light reactions
inducing polarity in Fucus (19) are also predictable from this
hypothesis. Furthermore, detailed analysis of the fluence-response
curves for polarotropism of Dryopteris protonemata (8) indicates
that the ineffective form of this hypothetical photochrome should
have an absorption maximum at 450 nm and a shoulder at 480 nm. The
active form should absorb predominantly at 370 nm and 520 nm (16).

Finally, it was shown directly that blue light induced phototropic curvature of the Phycomyces sporangiophore could be partially reverted by simultaneous irradiation with 605 nm light (Löser,
G & Schäfer, E., personal communication).

3.4 Light Gradient in Phycomyces

Whatever the chemical nature of the photoreceptor, it is
essential for tropism that a relevant light gradient is obtained.
This problem - although discussed in passing with the above examples - shall be presented again. In general three properties can
produce an effective gradient: focusing advantage of the distal
side by a lens effect, shading or screening by absorbing pigments
or scattering, favouring the proximal side, and, finally, preferential absorption due to dichroism and specific localization of
the photoreceptors.

In Phycomyces a relevant gradient is obtained by a lens effect; but how this lens effect can be translated into asymmetric
growth is still not quite clear (discussion in 1). The lens effect has been proven by immersion in oil or by passing light
through a cylindrical glass lens located so that a divergent beam
strikes the sporangiophore. Under both conditions phototropism is
reversed (see 5).

Under certain conditions, also on Phycomyces, screening by
pigments (plus scattering) may become so large that the focusing
advantage can be overcome and negative phototropism arises. With
UV-light < 300 nm, due to the high concentration of gallic acid

in vacuole (5), absorbance is greatly increased and phototropism
is reversed. Furthermore, in a specific mutant strain C 158, ten
times as much β-carotene occur compared to wild type. Thus, in this
mutant, absorbance of blue light is also increased to overcome the
focusing advantage leading to negative tropism (20).

It should be pointed out, however, that any focusing advantage,
if present, is possibly enhanced by the preferentially radial orientation of peripherally localized photoreceptor molecules (in
fungi, e.g. Botrytis spores, 21).

3.5 Orientation of Photoreceptor in Phycomyces

Analysing physiological responses to polarized light, the orientation of the vector of effective absorption in the blue was
shown to be preferentially horizontal, while at 280 nm the axis of
maximal absorption was 45° in agreement with a flavin molecule
(20; see also Haupt, this A.S.I.).

In our laboratory (U. Wulff, Staatsexamensarbeit, unpublished),
evidence for a radial orientation of the photoreceptor was obtained.
Sporangiophores were preilluminated with blue light symmetrically
from two opposite sides for 30 min. One light was then switched off
and the other one dimmed down (this procedure corresponds to the
one described in Fig. 15-3 of ref.5; Russo, this A.S.I.). At this
point one set of the sporangiophores was turned 90° (some clockwise, some counterclockwise), the other control set was not turned
or turned 180° (no turning gave equal results to 180°). We then
followed the time course of curvature (Fig. 4B) and determined by
extrapolation the onset of bending for various levels of unilateral
light intensities (expressed as % of symmetric preillumination;
Fig. 4A).

As already known and interpreted in terms of adaptation (5,6),
curvature starts later the more the intensity of the unilateral
light is decreased. Surprisingly, however, those sporangiophores
that were turned 90° started curvature 3-4 min later. This can only
be understood if light adaptation and therefore light absorption
during the preillumination was higher at the flanks.

First of all the data indicate that adaptation remains localized (confirming 22) and does not rotate (see below). Furthermore,
the orientation of the photoreceptors must be preferentially radial
rather than tangential to account for this result. The difference
between the 90° and 180° samples should be seen with horizontally

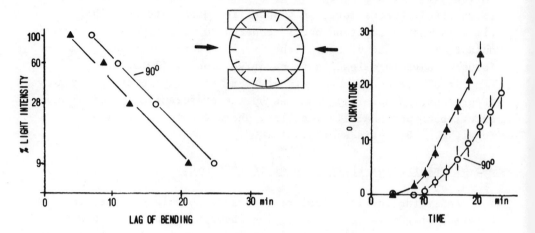

FIGURE 4A (left). Lag periods of curvature (abscissa) after start
of unilaterally illuminating Phycomyces. Intensity was lowered
from the symmetric preillumination to levels given on ordinate.
Each value is the mean from 4-8 specimen; the standard error of
the mean was ca ± 0.5 min. Circles: turned 90°; triangles: turned
180°. Unpolarized light.

 B (right). Time course of curvature. Horizontally polarized light.
Intensity was lowered to 28 %. Otherwise like A.
Inset: cross section with radial photoreceptor orientation. The
flanks with highest presumed absorption during preillumination
(arrows) are encased.

polarized (Fig. 4B) and unpolarized light but not with vertically
polarized light. This prediction was found to be true. In an experiment like that in Fig. 4B, using however vertically polarized
light the 90° set was not delayed at all compared to the 180° set.
      These conclusions agree well with the fact that in another
fungal, translucent cell orientation is preferentially radial (21).
It should be pointed out that the enhancement of any focusing
advantage could explain the quantitative discrepancy found for the
effects of internal screening pigments (see in 5).
      It is impossible that photoreceptors are oriented completely

horizontal (0°) since vertically polarized light has ca. 80 % efficiency on growth and tropism compared to horizontally polarized
one (20). We could not find any difference at the flanks between
e.g. -45° and +45°. Orientation therefore is either not strict -
which is unlikely in view of the UV vector - or else, the photoreceptor consists of a dimer with components both at e.g. -20°
and +20°.

4. THE TROPISM PARADOX IN PHYCOMYCES

4.1 Adaptation Versus Contrast

The tropistic reactions are designed to overcome the following
dilemma which is a general problem of sensory analysis: On one
hand the system must operate and adapt to very different levels of
light intensity. On the other hand the spatial stimulus gradient -
as well as in other cases the contrast in time - must be realized
as ecologically relevant information at different background
levels.

The occurrence of the tropism paradox is therefore not too
surprising, and solving this still unsolved problem might be of
general interest.

Qualitatively it seems quite easy to explain phototropism by
a localized light growth reaction, but in view of the time course
the paradox of phototropism becomes obvious (1,2,5,6).

The light-induced increase in growth rate is transient,
lasting only a few minutes after step-up in light intensity. The
stationary growth rate is independent of light intensity as long
as the intensity is lower than 10^{-4} W/cm^2. Bending, however, does
not stop when the growth reaction is over, but it continues for a
long time. One experiment can illustrate that the light growth
reaction itself is not essential for phototropism. If a symmetrical illumination becomes asymmetrical by switching one light source
180°, the light intensity is not changed and thus no growth reaction results. Continuous bending, however, starts after ca. 6 min.

One way to resolve this paradox was proposed (see in 5):
Adaptation was assumed to spread across the cross-section, while
the stimulus region should remain localized. Then the stimulus
intensity would locally be different from the adaptational state
and a continuous local light growth response would result. This
idea was discarded when it was shown that adaptation is strictly

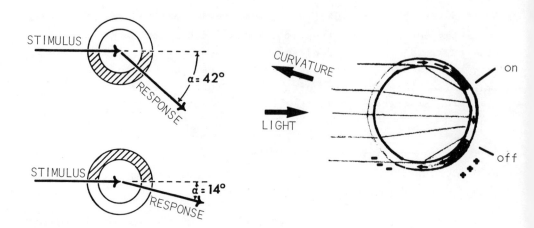

FIGURE 5. Localized adaptation, rotation and tropism in <u>Phycomyces</u>
sporangiophores (cross sections, viewed from above).
  A (left), data of (22): UV-phototropism after UV-preexposure. The
preexposed area is cross-hatched. Above: sporangiophore rotated
90° counterclockwise after preexoposure. Below: rotated clockwise.

  B (right). Blue-light-tropism according to rotation hypothesis
(23,24, graph according to 1). The sensitive outer layer rotates
clockwise; at the edges of the focused light band, the step-up
("on") and the step-down ("off") of light intensity cause a growth
promotion $^{(+)}$ and inhibition$^{(-)}$, respectively. These responses are
expressed with some delay, i.e. after some clockwise rotation.

localized to the region of stimulation, both longitudinally and
azimutally (22; see also above).

4.2 Rotation Hypothesis

    A recent theory (23,24) tries to explain phototropism by the
assumption that photoreceptors are in constant motion around the
sporangiophore as illustrated in Fig. 5. Those photoreceptors that
during motion enter the intense patches of light at the distal side
of the sporangiophore, have travelled through the less intense
light at the flanks and are thus adapted to relatively low intensity. Consequently they react with a light-growth response. Everywhere else on the cross-section the photoreceptors are approximately

adapted to the local intensity. This means that a persistent higher
growth rate at the distal side is predicted, while everywhere else
normal growth continues. This theory is also able to explain the
observed "error angle" of phototropism (Fig. 5) by taking into account the time-lag between stimulation and the beginning of the
growth reaction.

Several serious objections, however, can be raised against the
rotation theory (see in 5,6): Firstly, during phototropic bending,
growth at the proximal side is diminished. The decrease is growth
rate is of the same amount as the increase at the opposite side,
and thus cannot be neglected. It can hardly be explained quantitatively by the shallow negative growth reaction to a step-down
in intensity. The decrease at the "shaded" side seems to indicate
communication across the cell.

Secondly, rotation by spiral growth which could be the morphological basis of a rotation of the photoreceptors, is largest at
the top of the growing zone. Phototropism, however, is confined to
the base of the growing zone.

Thirdly, sporangiophores in stage I rotate at only 0.3° per
min, compared to 10° in stage IV, but show normal phototropism
(see 6).

Fourthly, our data on localized adaptation and photoreceptor
orientation described in 3.5 suggest that the photoreceptors and
the adaptation do not rotate at all. This agrees with data on longitudinal movements (see in 5), where the cell wall is moving longitudinally in relation to stimulation as well as localized adaptation. Therefore the wall structures could as well be twisting over
the plasmalemma structures. Finally, studies reported in support
of the rotation theory (22) analyzed the influence of external
rotation on phototropism. One case reported fits with the theory,
other details, however, do not. Reduction in bending speed by external rotation was to about 30 % relative to the case of no rotation in the transparent cell (stimulation by visible light). The
reduction was not seen in case of the opaque cell (stimulation by
UV-light), contrary to expectation. A large shift in bending directions is expected, when the external rotation just cancels the rotation of the cell. This shift was in fact observed, however, not
at 10°/min, but at 0-1°/min rotation rate.

4.3 Search for an Alternative Model

The sum of the arguments suggests that one has to look for a more satisfying theory of phototropism, a mechanism that can account for localized adaptation as well as for the observed strong continuous tropistic reaction. We propose a mechanim (Siep and Hertel, ms. in prep.) of non-linear amplification of a small spatial gradient, a mechanism which can also be applied to other tropistic reactions to small stimulus gradients, e.g. chemotropism, geotropism, developmental gradients.

The model postulates a macromolecule composed of several subunits and integrated into the plasmalemma. Furthermore, a freely diffusing ligand that binds tightly and in a cooperative way to the former component. The small molecule appears in limited concentration.

Other sensory pathways, e.g. geotropism and autochemotropism merge into this common differential amplifier (see also Fig. 1A and ref. 4,6).

5. PHOTOTROPISM IN MULTICELLULAR ORGANS

In higher plants like the grass coleoptiles phototropism has been well investigated and it may share some of the early transducing steps — photoreceptor, adaptation — with the lower plants. However, the fact that the original relevant light gradient is across the tissue, not across single cells (26), the problem of cell to cell communication, the possible involvement of a lateral auxin transport and the control of growth in higher cells — possibly with auxin involvement — presents additional problems on the output side of the stimulus response chain (see 1,2).

Even experimental approaches e.g. the use of immersion into mineral oil to test the possibility of a lens effect may not be generalized from Phycomyces to Avena.

REFERENCES

1. W. Haupt,"Bewegungsphysiologie der Pflanzen", Thieme Verlag, Stuttgart, (1977).
2. D. S. Dennison, in "Physiology of Movements, Encycl. Plant. Physiol.", W. Haupt & M. E. Feinleib, eds., Springer Verlag, Berlin, New Ser. (1979).

3. R. M. Page, in "Photophysiology", A. C. Giese, ed., Acad. Press, New York (1968).

4. T. Ootaki, E. P. Fischer & P. Lockhard, Molec. Gen. Genet. 131:233 (1974).

5. K. Bergmann, P. V. Burke, E. Cerdà-Olmedo, C. N. David, M. Delbrück, K. W. Foster, E. W. Goodell, M. Heinsenberg, G. Meissner, M. Zalokar, D. S. Dennison & W. Shropshire, Jr., Bact. Rev. 33:99 (1969).

6. V. E. A. Russo & P. Galland, in "Sensory Physiology and Structure of Molecules", P. Hemmerich, ed., Springer Verlag, Berlin, in press.

7. A. M. Steiner, Naturwissenschaften 54:497 (1967).

8. A. M. Steiner, Photochem. Photobiol. 9:493 (1969).

9. B. J. Nebel, Planta 87:170 (1969)

10. H. Kataoka, Plant Cell Physiol. 16:439 (1975).

11. M. H. Weisenseel, in "Physiology of Movements, Encycl. Plant Physiol.", W. Haupt & M. E. Feinleib, eds., Springer Verlag, Berlin, New Ser. (1975).

12. A. Sievers & D. Volkmann, in "Physiology of Movements, Encycl. Plant Physiol.", W. Haupt & M. E. Feinleib, eds., Springer Verlag, Berlin, New Ser. (1979).

13. D. Presti & M. Delbrück, Plant, Cell and Environment 1:81 (1978).

14. D. Presti, W. -J. Hsu & M. Delbrück, Photochem. Photobiol. 26: 403 (1977).

15. R. Hertel, A. J. Jesaitis, U. Dohrmann & W. R. Briggs, Planta in press.

16. S. G. Mayhew, H. L. Ludwig, in "The Enzymes", P. D. Boyer, ed., Academic Press, New York (1975).

17. K. M. Hartmann, in "Biophysik", E. Hoppe et al., eds., Springer Verlag, Berlin (1977).

18. R. Gärtner, Z. Pflanzenphysiol. 63:428 (1970).

19. F. W. Bentrup, Planta 59:472 (1963).

20. A. J. Jesaitis, J. Gen. Physiol. 63:1 (1974).

21. L. F. Jaffe & H. Etzold, J. Cell Biol. 13:13 (1962).

22. D. S. Dennison & R. P. Bozof, J. Gen. Physiol. 62:157 (1973).

23. D. S. Dennison & K. W. Foster, Biophys. J. 18:103 (1977).

24. J. R. Medina & E. Cerdà-Olmedo, J. Theor. Biol. 69:709 (1977).

25. J. Buder, Ber. deut. Bot. Ges. 38:10 (1920).

EXPERIMENTAL DETERMINATION AND MEASUREMENT OF PHOTORESPONSES

Bodo Diehn

Department of Zoology, Michigan State University

East Lansing, MI 48824 (U.S.A.)

INTRODUCTION

In treatises on the measurement of photobehavior, the emphasis in the past has almost exclusively been on how to conduct and record the actual observations of the phenomena under study. The problem of how to apply the photic stimulus is usually discussed in much less detail, probably because this is a fairly straightforward matter. However, a most important area of nearly total neglect has been the question of the preparation and conditioning of the cells prior to an experiment. I firmly believe that this is the reason for much of the often observed incompatibility of results obtained in different laboratories for the same organism. As I have pointed out in a recent review (1), a total of more than 20 published action spectra for photomovement in Euglena have very few details in common - in fact, an untrained reader of this body of literature would probably be able to draw only one firm conclusion, viz. that this cell is particularly sensitive to blue light.

In this contribution, I will attempt to emphasize what I believe has been neglected in the past by me and other reviewers. The topics and subtopics that I will discuss are the following:

I. The biochemical state of the organism being studied:
1. Age
2. Environmental Conditions
a. Temperature
b. Illumination condition and history

c. Composition of extracellular medium
II. The Photic Stimulus:
   1. Geometry
   2. Wavelength and Intensity
III. The measurement of Photomovement
   1. Response Parameters
   2. Experimental Methods
      a. Direct observation
      b. Automatic tracking
      c. Mass-movement recording systems
IV. The Interpretation of Experimental Results.

The discussion of the above topics will be held as general as possible; however, our own work has, of course, been with Euglena, and specific examples will be drawn from our experience with this cell.

I. THE BIOCHEMICAL STATE OF THE ORGANISM

As mentioned above, not very much attention has been given in the past to working under well-defined conditions. This probably stems from the fact that light is a clean and easily manipulated stimulus that usually does not interact with the cell via the medium. Scientists investigating chemobehavior would hardly resuspend their cells in culture medium for their studies, as seems to be customary in our field. Yet, reports on the effect of chemicals on photobehavior of Euglena, for instance, date from 1889 (2). I urge all research groups that work with the same organism to agree on standardized cell preparations, where possible, in photomovement studies. Specifically, the following parameters need to be considered:

1. Age

No cell exhibits exactly the same behavior throughout its life cycle. Euglena, for example, sheds its flagellum preparatory to cell division and thus is unable to respond to any stimuli during this phase. If one were to use a non-synchronized culture (as I have done in some of my earlier work), approximately 10-20% of all cells would be unresponsive at any given time, thus adversely affecting response threshold determinations or action spectral resolution, to give but two examples. This argument mitigates against the use of continuous culturing methods. The argument that samples of continously cultured cells exhibit an "average" behavior is like saying

that white light is preferable to a monochromatic stimulus because
the former represents the averaged sum of all wavelengths.

Most photoresponsive cells are photosynthetic, and synchronized by light/dark cycles. As a consequence, they usually exhibit
pronounced diurnal rhythms in their photoresponses. While these
variations can be compensated for by proper controls (3) it would
appear useful to subject the cultures to light/dark regimes that
will ensure maximal and fairly constant sensitivity of the organisms
during the time that experiments would ordinarily be conducted.

It must further be considered that senescent cultures will
also contain a significant number of aged cells with abnormal responses. For this reason, we harvest our cells while they are still
in the log phase of growth.

2. Environmental Conditions

This is the group of parameters in which one finds the most
variations among research groups investigating the same organisms.
In this context, one has to consider the following factors:

a. Temperature. Traditionally there is, of course, good operational reason to conduct experiments at the given temperature
of the laboratory, provided that it can be kept fairly constant.
Note, however, Fig. 1.

It can be seen that the dose-response curve for the step-down
photophobic response of Euglena at 23°C differs qualitatively and
quantitatively from that at 19°C. The plot has become distinctly
sigmoid at the higher temperature. It is clear that any deviation
from a standard temperature, preferably 25°C, must be well justified. The incubator for the culture should be kept at the same
temperature, to preclude thermokineses or thermophobic responses
of the cells upon a temperature change, such as have been observed
in Paramecium.

b. Illumination condition and history. Since most photoresponsive cells are photosynthetic, they are usually cultured in the
light. On the other hand, photobehavior studies must of necessity
be conducted in the absence of additional photic stimuli; i.e., in
physiological darkness. As a consequence, various investigators
have dark-adapted Euglena for varying times ($\frac{1}{2}$ to 24 hours) before
their experiments (for a summary, see Doughty & Diehn, 1980) (1).
Since Wolken and Shin (4) reported a distinct difference in the
spectral sensitivity of light-adapted vs. short-term dark-adapted
Euglena, and Diehn & Tollin (5) noted a reduction of the rate of

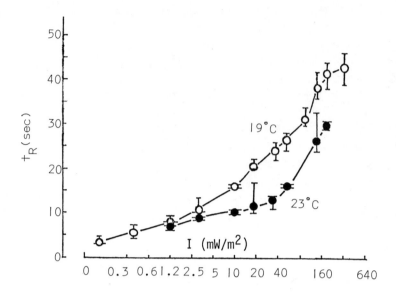

FIGURE 1. Dose response plots for the step-down photophobic response of <u>Euglena</u> at 19°C (open circles) and 23°C (closed circles).
Response duration vs. light intensity decrease ($\Delta I = I$). After Doughty
and Diehn (to be published).

photoaccumulation in cells that had been dark-adapted for 4 hours
or more, it would appear prudent to minimize dark-adaptation. In
agreement with our suggestion (5), that it is the cessation of
photosynthesis in the dark which negatively affects photoaccumulation, we found that fairly dim red illumination of 600-850 nm
("invisible" to the cells, yet sufficiently intense to maintain
photosynthesis: ~ 70 mW/m^2) will stabilize the photoresponses for
long periods of time. We now routinely maintain this red illumination which is interrupted only for a few minutes at a time when the
cells are tested in the "phototaxigraph" (see Section III.2.c). A
constant background illumination provides the additional advantage
of maintaining in a steady state any photochromic pigment that
might be involved in the behavioral responses.
     c. Composition of extracellular medium. That the photosensory
responses of unicells can be strongly affected by the presence of
ions in the experimental solution has been known for nearly a century (2). Ions and other substances can affect the photoresponses
in several different ways: through interference with signal transduction by the photoreceptor pigment (e.g. by iodide ion in <u>Euglena</u>

(6), by modifying the sensory transduction pathway at an interior site, e.g. calcium ion in Euglena (7) or through its effect on motility of the cells, which in Euglena is controlled by magnesium ion (7). The latter effect, i.e., on the motility of the cells, has nothing to do with sensory transduction and should be eliminated from the outset. This requires defining a minimum (buffered) solution for carrying out photobehavior studies, and conducting a matrix of experiments to determine the optimum concentration of each component. Table 1 is a typical example of such a matrix. It can be seen that for Euglena, a buffer containing 1mM Tris, 0.125 mM Mg^{2+}, 0.5mM Ca^{2+} and 1mM K^+ sustains motility in excess of 85% for more than 24 hours. Note that this series of experiments only establishes the most favourable conditions for motility in physiological darkness; this is, however, the most practical medium for studying photobehavior. In a similar manner, buffers can be devised in which certain photoresponses are strongly enhanced (7). Dissolved gases can also interfere with photoresponses, e.g. by activating a chemosensory pathway (8). For a discussion of the interactions of dissolved oxygen with the photosensory transduction system of Euglena see Diehn (9).

II. THE PHOTIC STIMULUS

Since light sources have been covered in a preceding chapter, I will here only discuss how best to deliver the stimulus to the organism, and what stimulus parameters must be considered in general. I will assume the light source to be equipped with a shutter, reproducible means of adjusting the intensity and spectral quality, and the capability of projecting a uniform field on the sample.

1. Geometry

In the mass-movement methods for recording photobehavior, the geometric relationship of the stimulus and the cell sample, is fixed (see IV.2.c), and thus need not be discussed here. Most other methods require delivery of the photic stimulus to a microscope stage. In the simplest approach, this can be done by using the microscope illumination also for stimulation.

The experimental approach to this problem will be determined by whether the stimulus is to be directional or non-directional. Truly non-directional stimuli are very difficult to devise; however, only for the study of kineses and phobic responses can a require-

TABLE 1

EFFECT OF VARIOUS CATIONS ON MOTILITY-SURVIVAL OF EUGLENA
(After Doughty & Diehn, (7))

CATIONS ADDED TO TRIS BUFFER, pH 7.1 (millimoles/litre)									% MOTILE CELLS (time in hours)							
Tris	Mg	Ca	K	Na	Co	Ba	Zn	Ni	0.5	2	3	6	9	15	18	24
1									61	54	41	50	52	57	62	72
1	.031								81	56	54	75	75	79	73	73
1	.062								72	51	54	59	54	59	62	66
1	.125								81	76	68	67	56	69	68	75
1	.25								78	65	51	42	43	49	54	65
1	.5								79	62	58	62	53	59	66	72
1		.5							90	53	45	44	47	52	61	69
1			1						83	37	46	43	44	49	51	57
1				1					86	51	46	47	37	48	57	60
1	.125	.5							81	73	60	71	66	64	59	54
1	.125	.5							85	86	83	87	85	89	95	93
1					.5				88	46	25	19	14	9	8	5
1	.125				.5				84	63	48	32	29	24	22	17
1						.5			83	59	38	33	33	34	40	46
1	.125					.5			63	51	41	49	51	59	62	67
1							.125		81	73	56	50	56	62	62	64
1	.125						.125		79	68	54	60	62	62	63	66
1	.125							.4	0	0	0	0	0	0	0	0

ment for such stimuli be imagined. It would appear that an internally reflective (except for an observation path) sample cuvette would
be necessary, or at least an arrangement with two equal stimulating
beams coming from opposite directions (which most directionally
sensitive organisms integrate to represent a zero-vector stimulus).

However, it turns out that even for studying phobic responses
and kineses, non-directional stimuli are unnecessary. The only
drawback of a directional stimulus would be its ability to cause
the organisms to move out of the field of view. In the case of
phobic responses, this does not happen because only very short
stimuli, typically well below the orientation time of the cells,
are required. Moreover, phobic responses are non-directional in
nature to begin with. In the case of kineses, one can apply the
stimulus parallel to the direction of observation, and restrict
the movement of the cells in this direction by using a shallow
sample cell. In this case one needs to establish whether the inevitable mechanical stimuli, experienced by the organisms upon
touching the interface, do not affect the response being studied.

Given the above simplification that a directional stimulus is
acceptable for nearly all photomovement studies, a single stimulus
is best delivered laterally, such as to not interfere with the
microscope illumination. Small fluorescence cuvettes with five
polished sides are very suitable containers for the organisms to
be studied, and allow the introduction of three independent stimuli.

Vertical stimulation is preferable for "light-trap" experiments and with shallow sample preparations. Semi-transparent mirrors can be used to introduce the stimulating beam coincident with
the microscope illumination, and more elaborate illumination systems
allow the independent control of wavelength and intensity both
inside and outside the trap (10). The simplest, and quite satisfactory, type of vertical stimulation utilizes the microscope illuminator.
Physiological darkness is simulated by placing in the illuminating
beam an interference filter which isolates a wavelength that can
be perceived by the viewer or recording device but not by the
organism (570 nm in Euglena), and for stimulation abruptly displacing it with a filter of the desired "visible" wavelength, or a
neutral density filter.

Slant stimulation may also be used, but presents the problem
of partial reflection at the sample chamber interface. Very careful
determination of the amount of light reaching the organisms is
required.

2. Wavelength and Intensity

One of the first studies conducted with a given organism should be the determination of an action spectrum, such that subsequent work can be carried out at the wavelength of maximum sensitivity. Action spectra, and all the studies at individual wavelengths, should be done via full dose/response plots. Measurements at constant intensity could result in the obliteration of some information because the stimulus did not reach threshold level, and of other detail because of saturation effects. I prefer to plot my data as the intensity (or the inverse thereof) needed to cause a half-maximum response, relative to the intensity required to do the same at a standard wavelength, e.g. the major action spectrum peak.

III. THE MEASUREMENT OF PHOTOMOVEMENT

When all of the considerations discussed up to this point have been taken into account, the time has arrived for generating experimental data - provided that one has decided what to look for.

1. Response Parameters

In this section, I do not intend to discuss the possible types of responses. Not only is this to be covered elsewhere, but an elementary discussion can also be found in the report of a committee on terminology (11). I will, therefore, assume that the reader is familiar with what is meant by kineses, phobic responses, taxes, and accumulation/dispersal. What I do wish to discuss is how, once a response type has been chosen for investigation, it can be quantitated.

The situation is straightforward in the case of kineses. One determines the rate of activity (i.e. swimming speed, rate of directional change, relation, or shape change) in the absence and presence of the stimulus, and expresses the result, as I suggested for action spectra, as a dimensionless number representing the ratio of experimental rate over control rate. A negative kinesis will result in a number smaller than one, and a positive one will yield a number larger than one. It is essential that the time course of the kinesis be determined, since the kinetics of response appearance and decay often suggest possible molecular mechanisms. The same holds for the intensity dependence of the kinesis.

Phobic responses present a more difficult situation. Their characteristics vary from organism to organism, and moreover, these responses can consist of several distinct phases. In Euglena, for instance, we have only recently, upon a careful re-analysis of its phobic behavior, noticed that the "classical" strophophobic behavior, consisting of rotation around a lateral axis and first described in detail ten years ago (3), which we now term "continuous flagellar reorientation" (CRF), is followed by a klinophobic phase marked by a transiently increased rate of directional change (periodic flagellar reorientation, PFR). Lest this be considered trivial, please note Fig. 2 which demonstrates that in Euglena, La^{3+} affects the two response phases differently.

As in all responses, three parameters will allow complete quantitation of all recognized phases of a phobic response: time course (including transduction time), intensity dependence, on I and ΔI independently (12), and action spectrum.

Since phobic responses by their very nature decay even while the stimulus remains applied, its duration (in fact, ideally its kinetic order and rate constant(s), resp. half-life if it were to follow first-order kinetics) would appear to be the major parameter to be reported. We are doing this now, though in the past we have, followed by other research groups, reported the percentage of motile cells exhibiting a response to a given stimulus.

That approach essentially reveals the fraction of cells whose photosensory system is activated to the degree that the stimulus elicits a response, i.e. threshold distribution rather than response rate.

I must confess to some disinterest in phototaxis, since I do not believe this phenomenon to be a primary response. In Euglena, (3) and the other organisms for which there is some evidence for the mechanism of tactic orientation, it is based on phobic responses. If one does wish to quantitate a taxis, the distribution of average cell orientations, i.e. track vectors, and the directness of this path, i.e. track straightness, will characterize the phenomenon. In cases where polar coordinate plots of cell orientations are deemed too demanding of space, mathematical methods are available for data reduction (13).

2. Experimental Methods

The question now arises "how does one procure the data called for in the preceding section?". This is a subject that has been

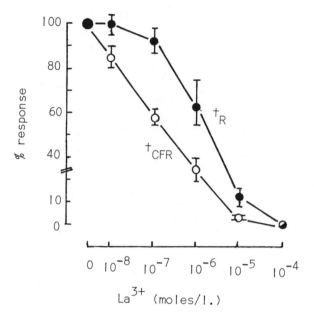

FIGURE 2. Inhibition by lanthanum ion of the strophophobic
(tumbling) phase (t_{CFR}) and the total step-down photophobic
response (stropho-and klinophobic, $t_R = t_{CFR} + t_{PFR}$) of Euglena.
Note that t_{PFR} is actually enhanced at lanthanum concentrations
that strongly inhibit tumbling.

addressed by many reviewers, and I will take that as an excuse
for covering this subject with only the essential detail. Three
major methods can be distinguished, listed here in order of effort
needed on the part the experimenter needed to obtain the data.
Actually, there is little overall superiority of any one method –
the easier it is to procure data, the more effort and caution are
required for their proper interpretation.

a. Direct observation. Observing the organism's behavior under
the microscope is a necessary first step in characterizing its
photoresponses. Real-time evaluation of experiments for data col-
lection is often unproductive because the behavior of only a very
few cells can be observed simultaneously, and inaccurate because
the observer has to make instantaneous judgements that cannot be
re-evaluated. Moreover, several important parameters, such as
transduction time, fall outside the time range that can be evalua-
ted manually.

For analyzing the "elementary motor response" (EMR) of an

organism, i.e. the response of the motor apparatus, be it flagella
or cilia, high-speed cinematography is often necessary (for such
studies with Euglena, see Diehn et al. (14)). Once it has been
established what "gross motor response" (GMR), i.e. alteration in
movement of the organism as a whole, is the result of a given EMR,
the high resolution of cinemicrography is usually no longer required.
In our laboratory, we rely mostly on videomicrography. A videotape
can be replayed as often as necessary for analyzing every cell's
behavior. The frames are 33 msec (in Europe 40 msec) apart and can
serve to establish a time reference in slow-motion replay. By
tracing the cell's position on the monitor screen every $\frac{1}{2}$ sec or
so, kineses, phobic responses, and taxes can be characterized and
quantitated. On the other hand, for very slow processes such as
growth response, time-lapse photography is the most suitable technique.

     b. Automatic tracking. The just-described tracing of the cell
path can be accomplished by automated devices. This is a high-
technology and high-cost approach that has not been embraced
enthusiastically by workers in the field. The "Bugwatcher" of
Davenport et al. (15), accomplishes the computerized evaluation
of the movement in the horizontal plane of several cells simultaneously. This instrument, a copy of which is gathering dust in our
laboratory for lack of software funds, can, in principle, accomplish
the complete characterization of all responses.It has, unfortunately, not yet been taught successfully to distinguish strophophobic tumbling responses from immobilization of the cell.

     Another device in this category was developed by Berg (16).
It tracks a single organism through three dimensions by keeping
the microscope focused on one end of the cell. The latter feature
should facilitate recognition of a tumbling response (the motion
of the microscope stage is recorded and analyzed by computer). The
major drawback of the method is that it is restricted by design to
operation in real-time, and can, therefore, evaluate only one
response event per stimulation sequence. The tracking microscope
is, however, very well suited for studying growth responses.

     c. Mass-Movement methods. The discovery of photomovement in
microorganisms did not require the use of a microscope (17). When
a suspension of photomotile organisms is illuminated, the resulting
movement of many individual cells often results in an accumulation
that is easily detected by the unaided eye. More than 100 years
after the first observation of photoaccumulation, it occurred to
students of photomovement that this phenomenon could be quantita-

ted with the aid of densitometric or turbidimetric techniques. In
fact, this realization so enthralled researchers newly entering
the field that they usually started their careers by divising an
apparatus for measuring photoaccumulation, or by modifying an
existing design. There are two classes of mass-movement methods:
light-trap devices, including some that were not initially recognized as such, i.e. methods in which accumulation is mediated by
step-down phobic responses, and systems in which accumulation is
believed to be based on phototaxis, though even in these, there
may be a phobic component in the accumulation mechanism (9).

In the case of slow-moving cells, evaluation of the photoaccumulation process need not   be carried out in real-time.
Nultsch and his co-workers (10), pioneered the use of densitometric
evaluation of cell densities in projected light traps. Such light
traps can also be used for real-time recording of accumulation
kinetics; in fact, one can for purposes where only part of the
potential information in an accumulation record (see below) is
needed, simultaneously stimulate and observe utilizing the same
beam of light (12).

A more elaborate device was designed by us for studying Euglena
(18). As the name "phototaxigraph" implies, we did not realize at
the time that we had devised a sophisticated light trap apparatus,
only later analysis of the experimental results (see below) convinced us that we should have named it "photoaccumuligraph". Separation
of stimulating and observing beams allows,in this and the other
types of mass-movement recording systems discussed in the following,
the recording of population behavior before and after stimulation
and the quantitation of wavelength and intensity dependence of
accumulation kinetics.

Two other recording systems that are presumed to measure only
phototaxis because they do not incorporate a recognizable light
trap, are the apparatus designed by Throm (19) and the "population
method" of Feinleib & Curry (20).

In both of these devices the sample cuvette is illuminated
unilaterally, at 45° and parallel to its axis respectively, compared
with 90° in the phototaxigraph. Most importantly, light impinges
uniformly upon the entire sample cell, rather than only upon a
defined area. However, slight intensity gradients along the stimulating light path are possible because of screening and light
scattering within the cell suspension. Thus, the possibility of
stimulus gradients leaves open the possibility of a phobic contribution to the accumulation mechanism. Judging from the literature,

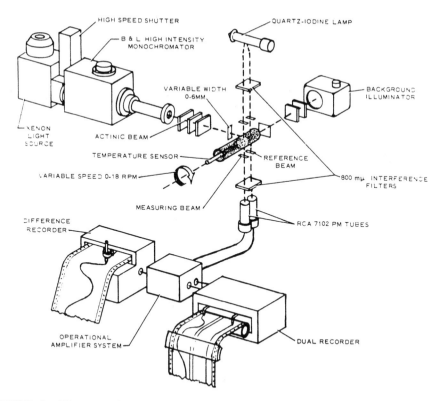

FIGURE 3. Phototaxigraph (after Diehn, 1969 (3)).

neither of these systems has been used recently for photomovement
studies.

A method for measuring motility parameters of cells oriented
in an electric field has recently been described by Ascoli et al.
(21). It is based on a laser doppler "radar" technique. With a
measuring beam at a wavelength that cannot be perceived by the
organism, a stimulation beam could be applied along the same axis,
and any resulting net movement of the population recorded. Since
the technique will yield velocity distributions, it might be possible to determine the degree of phototactic orientation (or
conversely, the contribution of phobic mechanisms) during accumulation.

IV. THE INTERPRETATION OF EXPERIMENTAL RESULTS

"What do these data mean?" is a question that usually has no
simple or perfectly satisfactory answer. This is even more the case

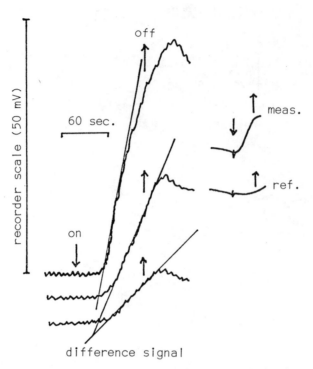

FIGURE 4. Accumulation kinetics of <u>Euglena gracilis</u> in the instrument shown in Fig. 3.

in a general discussion of hypothetical results such as this presentation. It is not my intention, and would indeed be impossible, to discuss here what conclusions could be drawn from certain numerical values of the duration of a phobic response, or of swimming velocities. In that sense, Section III.1 already contains all that I can say in general terms.

What I do wish to discuss is how one could obtain primary response parameters indirectly from photoaccumulation data. I am familiar only with the "phototaxigraph", and will confine my remarks to the information potentially contained in recordings obtained with this instrument. Fig. 4 shows actual "phototaxigrams" at three different light intensities.

Note that the measuring and reference beam traces, shown individually for the middle accumulation curve, indicate that the initial portion of the difference signal is due entirely to accumulation in the stimulated area.

Fig. 5 shows an idealized phototaxigram, divided into characteristic phases for further discussion.

I will begin by reviewing what I believe happens in terms of
individual cells in the phototaxigraph cuvette during a stimulation
sequence. The behavior I shall postulate has not actually been
observed, this would be very difficult to do because of the rotation of the cuvette, but is in full accord with what we know about
the behavior of Euglena and with what we have observed with light
traps in shallow non-rotating preparations (12).

Before stimulation, the cells are uniformly distributed in the
cuvette and move about randomly, changing direction approximately
every 15 seconds. Their behavior under our conditions of cuvette
size and initial population density is similar to the thermal motion of gas molecules, or of solute molecules in a solvent, and can
be treated mathematically according to the theories developed for
the latter systems, including the laws of diffusion (see any textbook of Physical Chemistry) and the kinetic theory of gases (22).

Upon illumination of the actinic volume, step-down responses
at the light/dark border will reflect those cells back into the
illuminated zone whose random movement has been toward the dark
region. The border constitutes an essentially one-way gate from the
dark to the light zone, with only those cells moving in the opposite
direction whose phobic responses are not sufficiently effective
to turn them away from the dark. In terms of a diffusion model, for
the cells in the dark region the apparent concentration of cells
in the lighted area has suddenly become nearly zero since none are
coming in from that direction. Thus, the cells in the dark are
moving down their own apparent concentration gradient when moving
into the light. Behaviorally, cell movement has changed from an
unbiased to a concentration-biased random walk.

As the actual cell concentration in the trap increases, so will
the apparent concentration as perceived from the dark region, since
at higher population pressure inside there will be more cells able
to penetrate the trap barrier. Eventually, a dynamic equilibrium
will be reached again in which equal numbers of cells move across
the border in opposite directions. In this new steady state, the
apparent concentrations, in terms of diffusion theory, are equal
on both sides of the trap border, though the actual concentration
is of course higher inside.

When the light is turned off, the barrier suddenly disappears
(and simultaneously all cells in the actinic region experience a

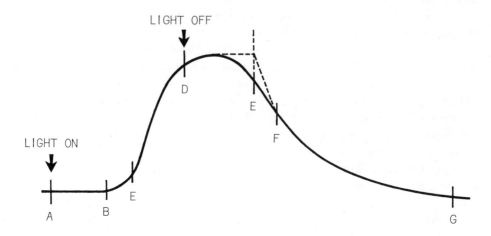

FIGURE 5. Shows an idealized phototaxigram, divided into characteristic phases for further discussion.

step-down stimulus). Net diffusional movement is now controlled
by actual concentrations, and will proceed to re-establish the
initial uniform population density in the cuvette.

Let me now utilize the above description in the quantitative
interpretation of a phototaxigram:

First of all, the phototaxigram can of course be calibrated
with respect to cell concentration, all   it takes is a series
of measurements with a removable partition separating cell suspen-
sions of different concentrations in the measuring and reference
areas of the cuvette. The slope of the accumulation/dispersal curve
then yields the net rate of in- or efflux, expressed for instance
as cells $mm^{-2}sec^{-1}$.

I will start my discussion of the phototaxigram with the tail
end of the curve, since here the situation is not complicated by
the presence of a stimulus. In the region F-G, net movement is
diffusion-controlled. At any concentration, the flux per unit area
as determined from the slope at that point allows calculation of
the diffusion coefficient, from which constant the actual swimming
velocity of the cells can be delivered by mathematical manipulation.

Assuming that no photokinesis occurs after short-term illumination (4), the maximal slope in the initial part of the accumulation curve will represent the flux rate of cells with the just-
calculated velocity from a region with the given initial concentration into a region of apparent lower concentration. The highest

flux rate, i.e. slope, that could ever be observed, would occur if
that apparent concentration were zero, i.e. in behavioral terms,
if the step-down responses were 100% efficient. The actual slope
at point C of Fig. 5 can be used, together with the diffusion coefficient taken from F-G, to calculate the apparent concentration
gradient down which the cells move into the trap. Since the initial
concentration outside the trap is known, the apparent lower concentration inside can be calculated. The initial value of the ratio
$c_{actual}/c_{apparent}$ might be a good measure of the strength of the
step-down responses.

A phototaxigram contains a significant amount of additional
information. Consider, for instance, region D-E, which represents
the time from removal of the light to the point where efflux appears
to have started if one extrapolates from G-F. The reason for the
delay in efflux is that the cells in the actinic region are undergoing step-down responses upon removal of the light, and these
responses preclude translational motion. D-E thus corresponds to the
actual duration of these responses. Note that the phototaxigram
yields two parameters of the response, effectiveness (strength ?)
from the initial slope, and duration from the delay of efflux.

I would now like to propose a new interpretation of the initial
lag period, region A-B. It has been shown by Checcucci et al. (23)
that at the trap border, the light intensity declines not abruptly
but with a measurable gradient. The reason for this is scattering
of light by cells in the stimulating beam. I would like to suggest
that to experience the intensity gradient which is necessary to
trigger the phobic response (12), cells have to move out further
from the lighted region, i.e. the true border of the light trap
lies some distance outward from the stimulating beam. Since the
measuring beam has the exact dimensions of the stimulating beam,
the inward-moving cells will reach it only after they have travelled this additional distance, first the cells moving on the straight-
est path (arriving at time B), then the cells that have travelled
obliquely or with a change in direction (arriving between B and C).
If this interpretation is correct, then the phototaxigram also
contains information about the light distribution around the trap
(A-B) and about the directness of the cells' path (B-C).

For the sake of balance, I must finish with a "caveat" regarding
the usefulness of the phototaxigraph. Thus far, I have discussed
only the effects of creating and removing a light trap on the
distribution of a cell population. These descriptions do not apply
to comparisons of the effects of other agents. We have, for instance,
established by direct observation that raising the calcium ion

concentration in the test solution from 0.5 mM to 10 mM increases
the duration of the step-down response, while simultaneously
increasing the rate of directional change of Euglena (7). In the
phototaxigraph, however, these two effects are antagonistic: a
stronger photophobic response enhances the rate of accumulation,
while more frequent directional changes reduce the influx of cells
into the trap. As a consequence, we found that calcium has very
litlle effect on photoaccumulation. Clearly, one should not rely
solely on the phototaxigraph when studying the effects of external
agents. Phototaxigraph studies can only supplement, but not replace
direct observation methods.

REFERENCES

1. B. Diehn & M. J. Doughty, in "Structure and Bonding"
   (in press) (1980).
2. J. Massart, Archiv. Biologie 9:515 (1889).
3. B. Diehn, Biochim. Biophys. Acta 177:136 (1969).
4. J. J. Wolken, & E. Shin, J. Protozool. 5:39 (1958).
5. B. Diehn, & G. Tollin, Arch. Biochem. Biophys. 121:169 (1967).
6. E. Mikolajczyk, & B. Diehn, Photochem. Photobiol. 22:269
   (1975).
7. M. J. Doughty, & B. Diehn, Biochim. Biophys. Acta, (in press),
   (1979).
8. G. Colombetti, & B. Diehn, J. Protozool. 25:211 (1978).
9. B. Diehn, in "Handbook of Sensory Physiology",H. Autrum, ed.,
   Springer Verlag, Berlin, (1979).
10. D. P. Häder, Dissertation Marburg (1973).
11. B. Diehn, M. Feinleib, W. Haupt, E. Hildebrand, F. Lenci, &
    W. Nultsch, Photochem. Photobiol. 26:559 (1977).
12. C. Creutz, G. Colombetti, & B. Diehn, Photochem. Photobiol.
    27:611 (1978).
13. E. Batschelet, Am. Inst. Biol. Sci. 57   (1965).
14. B. Diehn, J. R. Fonseca, & T. L. Jahn, J. Protozool. 24:492
    (1975).
15. D. Davenport, G. J. Culler, J. O. B. Greaves, R. B. Forward,
    & W. G. Hand, IEEE Trans. Biomed. Eng. BME-17:230 (1970).
16. H. C. Berg, Rev. Sci. Instrum. 47:868 (1971).
17. L. C. Treviranus, in "Vermischte Schriften Anatomischen und
    Physiologischen Inhalts von G.C. und L.C. Treviranus",
    (1817).

18. D. Lindes, B. Diehn & G. Tollin, Rev. Sci. Instr. 36:1721 (1965).
19. G. Throm, Arch. Protistenk, 110:313 (1968).
20. M. E. Feinleib & G. M. Curry, Physiol. Plantarum 20:1083 (1967).
21. C. Ascoli, M. Barbi, C. Frediani & A. Mure, Biophys. J. 24: 585 (1978).
22. K.Brinkmann, Z. Pflanzenphysiol. 59:12 (1968).
23. A. Checcucci, L. Favati, S. Grassi & T. Piaggesi, Monitore Zool. Ital. 9:83 (1975).

THE PHOTORECEPTIVE APPARATUS OF FLAGELLATED ALGAL CELLS: COMPARATIVE MORPHOLOGY AND SOME HYPOTHESES ON FUNCTIONING

Pietro Omodeo

Istituto di Biologia Animale dell'Università di Padova
35100 Padova, Italy

INTRODUCTION

1. Purpose of this paper

The phototactic behavior of unicellular organisms has for long attracted the attention of researchers who hoped that the simplicity of the material would make this phenomenon easy to investigate.

In fact the undertaking turned out to be more difficult than expected, so that, even if considerable progress has been made in recent years, a clearer understanding of the phenomenon has still to be attained.

In this paper I begin with a survey of the photosensitive and motor apparatus over a wide yet homogeneous group of organisms, for I am convinced that the comparative approach to sets of systems which carry out the same function makes possible an understanding of which parts are essential to the basilar functioning, and of which are accessory; the essential parts, always present, are much the same; the accessory parts vary and may even disappear according to the habitat and the circumstances. For instance, the retina is a constant component of the eye, while the eyelids and lacrimal glands are present in all terrestrial vertebrates but not in the aquatic. The crystalline lens is never lacking in vertebrates, but its curvature changes according to environment and specific habits.

The comparative method can give invaluable help to solve the

127

most entangled problems provided that it is supported by sound
inductive logic. Hence we resort to system analysis which, when
dealing strictly with the anatomy (microanatomy in our case) of
the organisms concerned, allows a great economy of effort, so that
the number of hypotheses concerning the functioning of such and
such a structure is drastically reduced and so more easily checked.

Thus I shall conclude attributing, tentatively, the most appropriate function to the cellular structures here considered, so
that it becomes possible to pose fruitful questions to the behavioral study and to plan crucial biochemical experiments. Let it be
clear that I shall be suggesting working models, not definitive
solutions.

2. Organisms considered

The eukaryotic algae here considered are contested by botanists
and zoologists who disagree on classification and assign to the
same taxa different ranks. In Table I there is a summary of the
classifications most commonly adopted in recent treatises on botany

TABLE I
Classification of eukaryotic algae/flagellate protozoa, according
to botanists (links) and zoologists (right)

	Classes	motile unicellular stage: monad (M), gamete(G) swarmer (S)			non motile pluricellular stage	Orders
	Euglenophyceae	M			Yes	Euglenoidini
	Dinophyceae	M	G	S	No	Dinoflagellates
Chromophyta	Chrysophyceae	M	G	S	Yes	Chrysomonadini
	Xantophyceae		G	S	Yes	Xantomonadini
	Phaeophyceae		G	S	Yes	--
	Bacillariophyceae		G		No	--
	Eustigmatophyceae			S	Yes	--
	Cryptophyceae	M			No	Cryptomonadini
	Haptophyceae	M			No	Haptomonadini
Chlorophyta	Prasinophyceae	M			No	Chloromonadini
	Chlorophyceae	M	G		Yes	
	Oedogoniophyceae		G		Yes	--
	Bryopsidophyceae		G		Yes	--
	Rhodophyceae	-	-	-	Yes	--

and zoology. In this paper I am using the botanical terminology not
because I judge it the best, but because most work in phototactic
behavior has been done by botanists.

In the classes Cryptophyceae, Euglenophyceae, Dinophyceae and
Haptophyceae the flagellated phototactic cells are the vegetative
ones. In the Chlorophyceae, Crysophyceae and Prasinophyceae, either
the vegetative cells or the gametes and the swarmers can be phototactic. In the Xantophyceae, Phaeophyceae and Eustigmatophyceae
both the gametes and the swarmers can be phototactic, but not the
multicellular vegetative stages.

As a rule the heterotrophic species do not possess phototaxis
but many predacious Dinophyceae are exceptions.

The photosensitive apparatus has been studied with the electron
microscope (EM) in about 150 different species, a sample big enough
from which to generalize.

3. Classification of photosensitive apparatuses

The essential components of a cellular photosensitive apparatus
are the stigma (called also eyespot) and the photoreceptor. At times
the photoreceptor cannot be identified even with the most powerful
instruments, though the stigma can be seen easily because of its
size and its red colour. In fact, in 1838 Ehrenberg (1) described
it in many different species and employed its size and position
as a useful diagnostic character of the "Infusionsthierchen".

The stigma is a roundish shield, inwardly or outwardly concave
and made up of one or more layers of lipidic globules packed closely
together. These globules, containing carotenoids (2), have a strong
optical density, and through them the stigma is believed to act as
a shade for the photoreceptor (Jennings (3); Mast (4); Halldal (5)).
The globules of the stigma are always synthetized into a plastid
(6) and are found near its surface, and sometimes in a sac connected
to the plastid by a thin isthmus (Chromulina placentula (7)); and
other times the stigma is completely separated from the plastid
(Euglenophyceae, Eustigmaphyceae and a few species of other classes).

The photoreceptor, when visible, appears as a swelling of the
basal portion of one flagellum, and hence it is commonly called
the paraflagellar body (PFB). The PFB was seen for the first time
in Euglena by Wager (8) who correctly interpreted it as a photoreceptor.

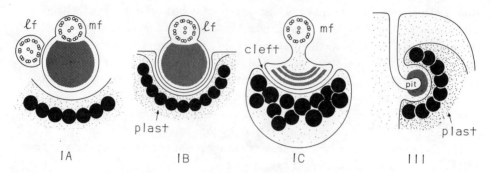

FIGURE 1. Scheme of the photosensitive apparatus of type I (A,B,C) and of type III

Dodge (9, 10) put the different kinds of stigma into four categories. Considering this classification and the structure of the photoreceptor (when known), I propose to divide the photosensitive apparatuses of the algae into four types:

Types I: the photoreceptor appears as a small sac on one side of a flagellum; its surface is closely apposed to the surface of some other part of the cell, I shall call this junction the synapsis. The stigma is outwardly concave and near a flagellum (Fig.1).

Type II: the photoreceptor is, perhaps, the patch of plasma membrane which covers the stigma; the stigma is independent of the flagella, generally multilayered and inwardly concave.

Type III: the photoreceptor is a mere modification of the cytoplasm that is comprised between the bottom of a cell invagination and the stigma (Fig.1).

Type IV: the photoreceptive apparatus is so highly specialized as to be considered as an ocelloid (or ocellus). The stigma is a pigment cup wrapping a "retinoid"; it is surmounted by a lens of complex design; between retinoid and lens is interposed a chamber lined by the plasma membrane (Fig. 2).

Photoreceptive apparatus of type I can be conveniently subdivided into three categories.

I A: the PFB is carried by the main flagellum and the synapsis connects its plasma membrane with that of the stigma.

I B: the PFB is carried by the shorter flagellum and the synapsis connects its plasma membrane with that of the stigma.

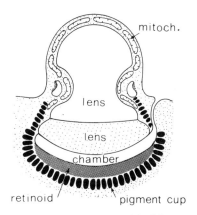

FIGURE 2. Scheme of the ocelloid of a Warnowiacea

I C: the PFB is carried by the main (and often single) flagellum and the synapsis is like that of category I B.

COMPARATIVE MORPHOLOGY

4. Cells with type I A photoreceptive apparatus

This apparatus, characteristic of autotrophic Euglenophyceae,
has been studied by EM in Euglena granulata (11), in E. spirogyra
(12) and chiefly in E. gracilis (Gibbs (13); Leedale (14); Kivic &
Vesk (6,15,16); Wolken (17); Piccinni & Mammi (18)). Its ultrastructure is quite uniform.

The stigma, independent of the chloroplasts and composed by
one layer of lipidic (osmiophilic) globules, is placed near the tip
of the cell, its concave side facing the reservoir from which the
two flagella emerge.

The composition of the lipid globules, of 300 nm in average
diameter, has been studied by Walne & Arnott (11).

The photoreceptor is a small swelling near the basis of the
main flagellum. It is a crystal measuring about 1x0.7x0.8 μm, whose
monoclinic unit cell has the following dimensions: a = 8.9 nm,
b = 7.7 nm, c = 8.3 nm; $\beta = 110°$; the a^+ axis is directed towards
the axoneme, the b^+ axis toward the lesser flagellum and the c^+
axis is nearly parallel to the axoneme (18). The crystalline mass
appears separated from the plasma membrane that envelops it.

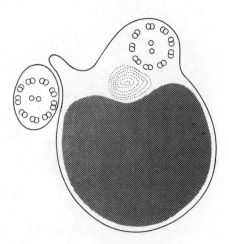

FIGURE 3. Diagrammatic section through the PFB and the lesser
flagellum of Euglena

Benedetti & Checcucci (19) have identified in it a flavinic pigment
whose fluorescent light is polarized.

The distal part of the lesser flagellum adheres to the main
flagellum on the side of the PFB (Fig. 3); cf. 16,20,18. In those
preparations whose anatomical relations appear best preserved, the
two plasma membranes are separated by a cleft 30-35 nm wide and in
the midst of which a thin lamina of fuzzy material can be seen.

During the swimming the main flagellum of Euglena pulls the
cell which advances spinning and screwing itself in the water aided
by its helicoidal ridges (Fig. 4). When the cell swims towards the
light the stigma describes a helicoid that encircles the path of
the photoreceptor which is thus continuously illuminated (4,20,21).
When the cell swims away from the light the stigma shades the PFB
continuously, otherwise the stigma shades the PFB intermittently.

In artificially bleached Euglena the photoreceptive apparatus
remains intact (16,22) and so it appears in some wild species devoid
of chlorophyll, but as a rule heterotrophic Euglenophyceae have
neither stigma nor photoreceptor and they have no synapsis between
the two flagella (23,14). In some species of Kinetoplastida (which
are related to Euglenoids) a junction between the single flagellum
and the cell body is present (cf. 24) having apparently a mechanical role.

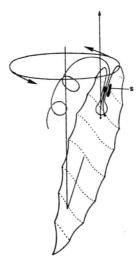

FIGURE 4. Morphology and movement mechanics of Euglena. In front
of the stigma (s) on the longer flagellum there is the photoreceptor
(from 20)

5. Cells with type I B photoreceptive apparatus

This type of apparatus is common to many species of the Chromophyta classes (Chryso-, Xantho-, and Phaeophyceae); it has been
described also in Diacronema vlkianum of the class Haptophyceae,
order Pavlovales, but not in related species (Fig. 5).

In these organisms the stigma is unilayered, outwardly concave,
and is formed by lipid globules of 100-150 nm in diameter (Fig. 6).
It is enclosed in a plastid or in a plastid diverticle and is
situated under the frontal platform of the cell (Fig. 7) or on one
side of the cell.

The presumptive photosensitive material, slightly osmiophilic,
is contained in a swelling (PFB) at the base of the lesser smooth
flagellum. It appears as a mass of granules, fine or rough, that
occupies the whole swelling or the part which adheres to the stigma;
sometimes this material has a lamellar structure (Fig. 6D). The
PFB measures about 500x500x750 nm.

Appearances of the PFB vary, at times they are artifacts due
to faulty fixation during preparation for EM at times they are
true.

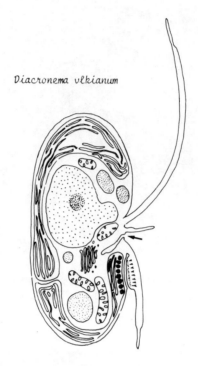

Diacronema vlkianum

FIGURE 5. Cell morphology of <u>Diacronema vlkianum</u>. The arrow points to the "pit", an invagination which is characteristic of the Pavlovales. In front of the stigma there is the paraflagellar body (PFB)

In some Chrysophycean genera (<u>Ochromonas</u>, <u>Spumella</u>, <u>Anthophysa</u>, <u>Chrysococcus</u>) on one side of the PFB, the so-called left side, there is a dense osmiophilic lump (Fig. 6 A,B), which does not appear to be connected with photoreceptive structures.

Chrysococcus cordiformis and <u>Mallomonas papillosa</u> (25) have a PFB but no stigma. <u>Ch. cordiformis</u>, however, bears, scattered in the chloroplast, many osmiophilic globules (26) which appear to be identical with those of the stigma of related species.

The PFB of the Chromophyta always rests in a concavity of the cell surface corresponding to that of the stigma, where the two facing membranes form a synapsis. The junction surface is rather wide, far wider than in <u>Euglena</u>. The synapsis surface may be widened by lateral expansions of the PFB, eventually strengthened by microtubules, so that the swelling in transverse sections appears fan-shaped (Fig. 6). In some species a wider contact results from

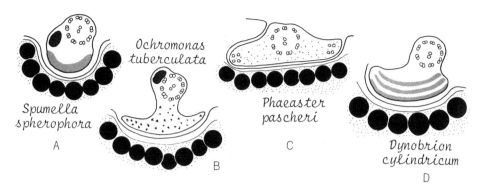

FIGURE 6. Paraflagellar body and stigma in some Chromophyta.
Spumella spherophora and *Ochromonas tuberculata* bear on a side
of the flagellum an osmiophilic mass. Note the wide surface of
contact between PFB and cell body

the fact that the PFB, brought by a rudimentary flagellum, is embedded in a pocket of the cytoplasm (Fig. 7 B). The synaptic cleft is
similar to that described for Euglena.

In those flagellated cells of Chromophyta that are devoid of
photoreceptive apparatus there is no synapsis.

A detail of possible significance, concerns the axoneme of the
lesser flagellum which, when clearly visualized shows a precise
orientation in relation to the stigma. In fact the plane that passes through the two central tubules crosses the central part of
the stigma (Fig. 1).

It is noteworthy that the type I B photoreceptive apparatus
continues to exist in some species which are mainly (Ochromonas
tuberculata, cf. 27) or wholly (Anthophysa vegetans (28)) heterotrophic.

Flagellated cells of the Chrysophyceae swim rotating along
their longitudinal axis,pulled by the main "Flimmer" flagellum,
which is inserted anteriorly or slightly laterally. When the cell
moves towards the light source the frontal PFB is steadily illuminated; when the cell trajectory makes an acute angle with the light
beam the stigma shades the PFB intermittently, but if it makes a
wider angle the stigma shades the PFB continuously.

Flagellated cells of Xanto- and Phaeophyceae have a pulling
anterior flagellum, but a lateral stigma (Fig. 8). Their behavior
has not been thoroughly investigated, but it seems to be similar
to that of Euglena.

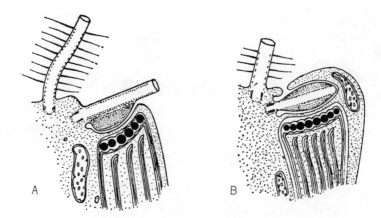

FIGURE 7. Photoreceptor and stigma of Dinobryon (A) and of Chromulina (B). Note the close contact of the plasma membranes of the two organelles (Redrawn from Dodge (10), and Rouiller and Fauré-Fremiet, (29)).

6. Cells with type I C photoreceptive apparatus

These cells are algal zoospores of the small Eustigmatophyceae class (30). Their PFB is situated on the proximal segment of the pulling "Flimmer" flagellum; it contains several osmiophilic layers which resemble those of some Chrysophycean species (Dynobriom cylindricum, Anthophysa).

The stigma is formed by a cluster of lipidic globules, independent from the chloroplast, enclosed in a protuberance which is in close contact with the photoreceptor.

7. Cells with type II photoreceptive apparatus

The photoreceptive apparatus of type II has been described for some species of the Prasinophyceae and for many species of all the classes forming the great division of the Chlorophyta, except those of the Zygnemaphyceae, which never have flagellated motile cells. This type of photoreceptive apparatus (Fig. 9) characterizes also Pavlova lutheri (31) and P. calceolata (32) of the Pavlovales order (Haptophyceae); but closely related species of this order have a quite different photoreceptive apparatus.

In these cells the stigma, inwardly concave and generally multilayered, is formed by lipidic globules with diameters of about

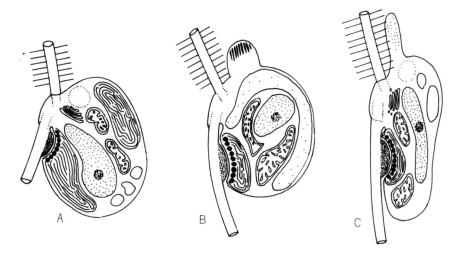

FIGURE 8. Cell morphology and phototactic apparatus in a Xanto-
phycean zoospore (A) and in the gametes of the Phaeophyceae Focus
(B) and Ascophyllum. (A, redrawn from 10, B and C drawn after
micrographs of 10, and 44)

50-120 nm. It is enclosed in the superficial layers of a chloroplast and is situated on the cell equator or nearby in one of the
quadrants between the plane passing through the two flagella and
the orthogonal plane (7,33) (Fig. 10). In cells endowed with two
pairs of flagella the stigma is situated likewise (34).

In these cells the photoreceptor has never been recognized
with certainty. Since this structure in other algal cells appears
invariably connected with the stigma, and since in all Metazoa the
photosensitive elements are always surrounded by a cup of pigment
which functions as a shade, it has been presumed that the photoreceptor of these cells is likewise situated. In fact, in some
species the plasmalemma which covers the stigma is evidently somewhat thickened (35,11,36); in other species however there is no
such modification (37,38,39).

The two (or four) pulling flagella of the chlorophyceae make
symmetrical movements comparable with those of a breast-stroke
swimmer; as the flagellar beating is not perfectly planar the cell
moves forward gyrating slowly on its own axis (40,41).

When the cell moves toward the light source or away from it,

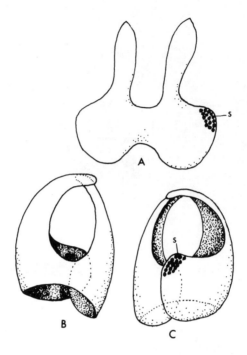

FIGURE 9. Chloroplast and stigma (s) of Pavlova calceolata.
A, the chloroplast expanded in a plane; B and C, two views of the
chloroplast as present inside the cell. (Redrawn from 32).

the stigma offers to the beam only its anterior or posterior border,
and so produces the least shade. When the cell moves obliquely in
respect to the light beam, there comes a moment when the stigma
projects its shade against the overhanging plasma membrane.

8. Cells with type III photoreceptive apparatus

This type of photosensitive apparatus is well known only in
some species of Pavlova, and especially in P. gyrans (42) and in
P. granifera (43).

In these algae near the flagellar base there is an invagination,
the so-called "pit", that, curving gradually, comes near the concavity of the stigma. The stigma lies in the chloroplast, near the
internal surface, in the central anterior part of the cell. Between
the bottom of the pit and the stigma (but out of the chloroplast)

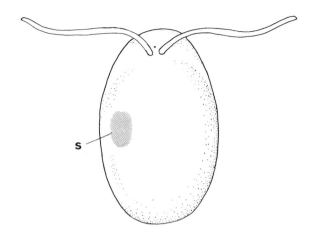

FIGURE 10. Position of the stigma in a biflagellar Volvocale.

there is a layer of electron-opaque material (Fig. 1), which I
interpret, tentatively, as photosensitive.

Pavlova has two flagella inserted laterally under the anterior
tip; the main flagellum points forward and pulls the cell.

Dodge & Crawford (45) have described the stigma of Glenodinium
foliaceum, a Dinophycea. It is bilayered, rectangular and has a deep
notch in correspondence with the root of the longitudinal flagellum (Fig. 11). The photoreceptor, according to these authors,
should be a body located under the stigma body whose lamellar
structure actually resembles that of the modified flagellum of
chordate photoreceptors. Yet the lamellar body of G. foliaceum is
included in the cytoplasm and such an arrangement does not suit a
cell receptor which must be connected with the plasma membrane. In
any case lamellar intracytoplasmic organules not involved in photoreception have been described in other microorganisms.

Greuet (46) remarks that in the concavity of the stigma of
G. foliaceum there is a large vacuole that may be homologous to
the ocellar chamber of the Warnowiaceae (see below). I agree with
Greuet and I add that it may be also homologous with the "pit" of
the Pavlovales.

The motile cells of the Dinophyceae have two flagella emerging
near the equator: one is straight and directed backward; the other
is helicoidal and encircles the cell, running in an equatorial
groove (annulus) (Fig. 11). The straight flagellum pushes the cell,
while the other imparts to it a rotatory motion. The stigma, when

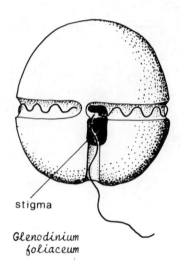

stigma

*Glenodinium
foliaceum*

FIGURE 11. Morphology of Glenodium foliaceum (from 10).

the cell has one, is situated in the angle delimited by the annulus
and the ventral groove. The photobehavior of Gymnodinium dorsum,
a stigmaless Diniphycea, has been dexcribed by Hand (47).

Chroomonas mesostigmaticum is a Cryptophycea which, as its
name suggests, has the stigma in the middle of the cell. I think
that this species may have a type III photosensitive apparatus
like Pavlova; indeed in micrograph 6-2, published by Dodge (10),
a narrow flattened canal may be seen near the stigma.

It is interesting, from a phylogenetic point of view, that
in Pavlova and in the related genus Diacronema we find like species
endowed with photoreceptive apparatus of different types. The transition from type II to type III may be understood by considering
the organization of Pavlova calceolata, whose stigma is outwardly
facing and independent of the pit (32): a different twisting of
the chloroplast would suffice to bring the stigma near the bottom
of the pit (Fig. 9).

More interesting is the phylogenetic relation between type
III photosensitive apparatus and the ocelloid of the Warnowiaceae
shown by Greuet, on which I shall comment further on.

9. Cells with type IV photoreceptive apparatus

When in 1885 Vogt (48) saw the ocelloid of a Warnowiacea,

deceived by its complexity, suspected it was a medusa eye engulfed
by the cell. But in 1884-1885 Hertwig (49,50) had already made
clear the true nature of this cellular organelle which characterizes
all the species of this heterotrophic predacious family of Dinophyceae (Fig. 2).

The ocelloid of Nematodinium has been studied by EM by Mornin
& Francis (51), but its ultrastructure is known especially through
a series of excellent papers by Greuet (52,53,54,55,56,46).

This cell organelle (globular, about 30x30 μm) can be briefly
described: there is a thick "retinoid" whose geometry varies according to the species; it has a very singular ultrastructure and
is surrounded by a cup of dark lipid globules (melanosoma) tightly
packed. The melanosoma according to Greuet is of plastidial deri-
vation and is homologous with the stigma of other algal cells. The
retinoid is surmounted by a lens made of two principal pieces whose
form is specific and always strange. A diverticulum of the plasma
membrane penetrates amid retinoid and lens, there spreading to form
a flattened chamber as wide as the retinoid. Both the pigment cup
and the lens are enveloped by a huge fenestrated mitochondrion
(Fig. 16).

Nematodinium and Warnowia swim as other dinoflagellates do by
means of the two flagella, the helicoidal one being more developed.
Erythropsidinium (= Erythropsis) and Leucopsis (53,54) move from
the coming and going of a piston emerging from the posterior part
of the cell (Fig. 15).

Warnowiaceae bring their ocelloid where other Dinophyceae have
the stigma.

Warnowiaceae have special structures and organelles for the
capture and ingestion of their prey.

THE FUNCTIONING

10. Methodological approach

It seems unlikely that more research will disclose important
novelties in the field of the photosensitivity of flagellated algal
cells. Thus it is opportune to make an effort to obtain from the
data so far acquired, which are mostly microanatomical, the maximum
of information on the functioning of the photosensitive apparatus.

To this end, technological considerations on trajectory control will help, especially those derived from system analysis (57,

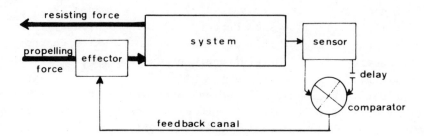

FIGURE 12. Block diagram of a self-directing system, endowed with
a single receptor. By means of a delay device the system reacts to
variations of light flux in time.

58): schemes of cybernetics can provide reliable guide to observation and experimentation (59) and to a rational reconstruction of
the causal chain by which stimulus and response are coupled.

11. Trajectory control

A trajectory can have either ballistic of feedback control.
The former instance, which includes jumps and throws, is quite rare
among simpler living beings because it requires complex sensory and
motor structures and because it is less reliable: only one case
will be considered here (sect. 16). The latter is, however, very
common and can be conveniently dealt with.

Models of feedback control on trajectories which can be reasonably expected to occur in living organisms are few and in the case
of the cells here considered can be reduced to only two by taking
into account the special architecture of their motor and photoreceptive apparatuses.

Any feedback control device implies peremptorily the comparison
between two signals (60,61), and its schemata change according to
the signal origin.

In the case of a trajectory control the comparison may concern:
1) simultaneous signals transduced by two symmetric equal sensors,
2) signals transduced in two different instant by a single sensor,
3) a signal transduced by a single sensor and an endogeneous steady
signal. Among Metazoa control by symmetrical sense organs is
quite common, but among the cells here considered no one is obviously endowed with paired photoreceptive apparatuses, and thus,
even if someone has accepted as true that Euglena is so equipped

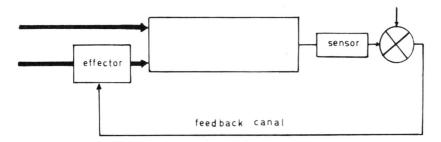

FIGURE 13. Block diagram of a self-directing system, endowed with two symmetrical propellers: note that one imput of the comparator is a constant value (down pointing arrow).

(62), common sense suggests to consider here only the points 2) and 3) schematized in Figs. 12, 13.

Other control characteristics, and thus behavioral peculiarities, are connected with the functioning of the propelling structure of the cell.

If the cell is asymmetric and advances spinning along its axis it can correct its trajectory only by sudden steering (by jerks); such behavior has been clearly observed in Euglena (4,21) and can be confidently attributed to all heterokont or monokont cells of the Chromophyta. The cell can accomplish a gradual smooth correction of its trajectory only if it goes forward without spinning, or rotation with a very long period: it is for steering in this way that torpedoes are fitted with two contrarotating propellers and a gyro stabilizer. Gradual smooth correction of the trajectory toward the light has been described in Chlamydomonas (63) and a similar behavior can be confidently attributed to all isokont cells of the Chlorophyceae.

Now I want to add that a monokont cell can steer only by changing the insertion angle of the flagellum, i.e. in a way quite analogous to that of an automobile. An isokont cell instead can more opportunely steer by displacing the barycentre of the motor couple, i.e. in a way analogous to that of a tank or of a caterpillar tractor. A heterokont cell can steer in both ways provided that the lesser flagellum be motile, otherwise it will behave as a monokont cell.

Now let us consider each element of the cybernetic scheme and establish its correspondence to the anatomical parts.

12. The sensor

In principle, any sensory process depends upon the interaction between some kind of energy and a sensor, i.e. a structure that tranforms the impinging energy into a signal energy, suitable to operate in the cell.

The signal energy, as far as we know, has always the same nature: any cell pumps to the outside some kind of ions and to the inside some others, so a transmembrane electrical potential (resting potential) is generated, whose value is of some scores of mV. Any force which acts as a stimulus directly or indirectly opens or closes the passage through the membrane to Na^+, K^+ and Cl^-: the flow variation of these ions engenders a variation of potential which tends to propagate along the cell and can trigger complex reactions.

Restricting our considerations to the transduction of radiant energy in the cell, we can suggest two different possibilities:

1) Photon flux, acting on an appropriate pigment mass, engenders a flux of electrical charged particles that in turn engenders a signal in the form of an action potential.

2) Photon flux, acting on a pigment embedded in the plasma membrane modifies its conformation and so a variation in the membrane permeability follows which gives rise to an action potential.

The second possibility presents a favourable circumstance because the stimulus energy is promptly magnified at the expense of the resting potential of the cell, and, at last, at the expense of the energy delivered by the membrane ATPase.

It is perhaps for this advantage that in Metazoa the second outcome is quite common. In eukaryotic Protista, however, where even a moderate signal energy can warrant an adequate response, the first outcome seems to be frequent.

When the photopigment is not embedded in the plasma membrane a large amount of it is necessary for obtaining a sufficiently intense signal, since in this case amplification may be absent.

In the photoreceptive apparatus type I the sensor is generally and opportunely identified with the PFB. Wolken (62) suggested that the light transduction may occurr in the stigma, but the idea has little likelihood. In any case the signal origin is as mentioned in 1).

In the photoreceptive apparatus type II, as above mentioned, the sensor may be identified with the plasma membrane surmounting the stigma. This opinion is open to criticism. I have little doubt,

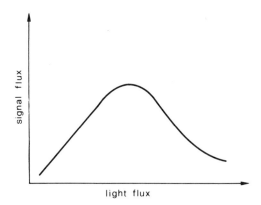

FIGURE 14. The bleaching of the photopigment and the pigment regeneration are competitive processes: beyond a crucial value of
the illumination the bleaching prevails and the signal flux begins
to decrease. (From 20).

however, that the signal origin in as mentioned in 2).

In the photoreceptive apparatus type III the light transduction occurs as in type I: in the case of the Warnowiaceae ocelloid that is almost sure.

13. Photopigment turnover

The sensor functioning is not linear because of a saturation
process, and that obscures some aspects of photobehavior.

When a photopigment has been modified (bleached) by the light
it must be regenerated to function again. The regeneration requires
some source of energy and some enzyme(s). That means that the photopigment undergoes a turnover and that the process has its own
kinetics.

In order to warrant a normal functioning of the photoreceptor
the rate of bleaching must not be higher than the rate of regeneration: beyond a crucial illumination value the inner signal flux
does not increase but decreases with increasing photon flux (Fig.
14). The critical value is proportionate to the receptor size and
has some interest because it gives information both on the bleaching and on the regenerating kinetics.

Piccinni & Omodeo believe (20) that negative phototaxis appears
in Euglena when the photopigment is bleached more rapidly than it
is regenerated.

14. The comparator

The signals starting from the sensor are proportional to illumination intensity and not to illumination variations, thus they must be adequately elaborated before getting operative values. Such elaboration corresponds to that accomplished by a driver who, wanting to know if his vehicle accelerates, compares two consecutive readings of the speedometer.

Piccinni & Omodeo (20) pointed out that in Euglena the delay required for a two instant analysis may originate at the level of the synapsis between the two flagella. The data piled up on the anatomy of the Chromophyta, confirming the constant mutual relation between PFB and synapsis, have given support to this hypothesis. Furthermore, the results of the study on the structure and orientation of the flavoprotein crystal in the PFB of Euglena allow us to be more precise on this topic and to complete the model as follows.

The illuminated photosensitive material would pump from the surrounding cytoplasm electrically charged particles, forcing them through the synapse. As a result, the PFB membrane would be depolarized (or hyperpolarized according to the sign of the charged particles) and the postsynaptic membrane would undergo an identical change but of opposite sign. When the photon flux is steady, depolarization and hyperpolarization cancel each other; when the photon flux varies either signal would prevail and trigger a reaction.

The model may be generalized for all the cells having photoreceptive apparatus type I. The absence of vesicles in the flagellar membrane leads to suppose that the junction is an electrosynapsis: the supposition is strengthened by the common occurrence of some device which allows for the widest contact between the two membranes (Figs. 6,7).

It is noteworthy that in Vertebrates synapses between flagella and axons have been described (64).

Cells with photoreceptive apparatus of type III have no microanatomical structures which can effect a delay in the signal and thus an analysis of the light flux variations. Forward & Davenport (65) and Forward (66,67), however, described a Dinophycea having two kinds of photopigments.

To the Chlorophyta cells endowed with the simple photoreceptive apparatus type II we can apply the cybernatic model in Fig. 14.

Considering this scheme, there is no doubt about what the sensor
communicates to the comparator, but it is less evident what the
second input tells to the comparator. In order for the control to
function this input must be a constant with which the signal coming
from the photoreceptor shall be compared, and therefore it must
have the same nature. The most plausible solution is that this input consists of a hyperpolarization of the plasma membrane at the
site of the stigma.

When the illumination is optimal, the hyperpolarization is
exactly balanced by the depolarization caused by the illuminated photopigment. When the light dims the hyperpolarization
prevails and the opposite occurs when the light increases. The net
difference changes gradually, and so does the intensity of the
feedback signal which acts on the motor apparatus, that is probably
the flagellum closer to the photoreceptor.

15. The feedback canal and the effector

In all cells investigated till now the signals travel as a wave
of depolarization or hyperpolarization which propagates along the
plasma membrane. When two signals collide their sum (or their difference if they have opposite signs) is effective. In Protozoa the
process is identical (68,69). The opinion that in the algal cells
considered here the process is different would be hardly sustainable,
especially as Ascoli and coworkers (verbal communication) have reg-
istered action potentials evoked by light in the Chlorophycea
Haematococcus pluvialis.

In all the cells motor responses are triggered and regulated
by a flow of Ca^{++} produced by a depolarization wave at the end of
its travel (cf. 70 and 71 for the Prokaryota, 69 for the Ciliates).

In eukaryotic cells motor responses are effected either by a
tubulin-dynein system or by an acto - myosin system. Tubulin and
dynein are organized, except in some special cases, in the axomene
of cilia and flagella. Actin and myosin, in the unspecialized cells,
are situated under the plasma membrane, but in the specialized
muscle cells they form fibrils which are always in contact with
the plasma membrane or with its invaginations.

No other contractile protein systems are known in eukaryotic
cells but these.

Motile cells of phototactic algae (T neglect Diatoms) swim by
means of flagella and crawl by means of acto-myosin systems;

acto-myosin also controls cell division and, through the contractile
vacuoles, the osmotic pressure of cytoplasm.

The phototactic behavior of a cell requires as an obvious
condition that it can steer in an appropriate manner.

As I have mentioned in sect. 11, the best hypothesis we can
make on the steering of isokont cells is that the new trajectory
is obtained by the temporary slowing down, or reinforcement, of the
beat of one of the two flagella, but it is hard to understand how
the regulating signal acts only on one flagellum of the couple as
their bases are in contact.

The steering in heterokont cells of Chrysophyceae and Phaeophyceae and in monokont Eustigmaphyceae can be attributed (20) to
a strange "proboscis" that bulges from the anterior pole of the
cell (Fig. 8 B,C): changing in shape it can modify the angulation
of the pulling flagellum.

As an alternative we can suppose that the steering is effected
by some movement of the lesser recurrent flagellum, but such hypothesis does not fit when, as often occurs, this flagellum is a mere
stump.

In the Euglenophyceae the main flagellum emerges from an invagination of the anterior pole of the cell. It has been suggested
by Piccinni & Omodeo (20) that the fibrils surrounding the walls
of this invagination can straighten the pulling recurrent flagellum, thus modifying the cell trajectory. This hypothesis has been
checked by Coppellotti et al. (72), but the results were ambiguous.
Checcucci et al. (73) point to a possible involvement, in the flagellar erection, of the paraflagellar rod; this suggestion is supported by the fact that in this accessory structure an ATPase activity has been demostrated (Piccinni et al. 74).

Dinophyceae are endowed with two flagella specialized in quite
different ways. One of them is straight and forward directed, the
other is helicoidal, surrounds the cell equator and imparts   a
rotatory movement. Such equipment allows many different motor ap-
titudes, but at present it is scarcely possible even to hazard a
guess on the way the cell controls its path toward the light.

16. The ocelloid of the Warnowiaceae and its use

My survey would be incomplete if I did not consider the role
of the ocelloid of Wanrowiaceae, whose microanatomy has been illustrated in section 9. This extraordinary structure represents,

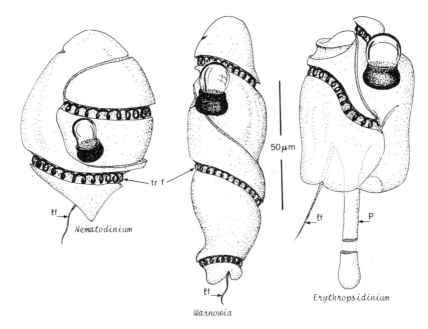

Fig. 15 : Three representatives of the Warnowiaceae. Note the large fraction of the cytoplasm occupied by the ocelloid. 1f = longitudinal flagellum; p = piston; tr f = trans verse flagellum. (From drawings and photographs of Greuet).

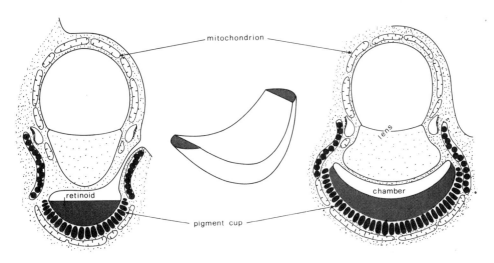

Fig. 16 : Two orthogonal sections through the Nematodinium ocelloid. In the center the form of the retinoid is indica ted, its shorter axis is directed toward the cell center. (From drawings and micrographs of Greuet).

without a doubt, the culminating point reached by the evolution of the photosensitive apparatus of algal cells (Fig. 15).

I will restrict, however, my considerations to Nematodinium which is so far the best known among the Warnowiaceae (Fig. 16).

Francis (75) in a careful analysis of the optical properties of the lens of N. armatum, showed that only objects about 50 μm far are focused in the retinoid, and that because of the pyriform lens there is no single point of focus but rather a zone of focus. He suggests that this ocelloid functions as an image forming eye, although with low resolution power.

This hypothesis, however, is weak as the cell lacks devices suitable for analysing forms. Moreover the retinoid has not a plane surface, as Francis believed, but a cylindric one (Fig. 16), as was clearly shown afterward by Greuet (56,46), the shorter axis of this surface being perpendicular to the ventral side of the cell. Instead it is likely that the characteristics of the output, owing to the retinoid form and to the optic properties of the lens,will vary according to the distance, the dimension, and the direction of the object which projects its shade.

One can assume, as a working hypothesis, that when the shade is focused in the retinoid and moves in a particular direction along its longer axis, then and only then a signal modulated in a special way shall be emitted. Then, the nematocysts, with which the cell is fitted (Mornin & Francis (51); Greuet (76)) are released against the prey. In other words, Nematodinium fires its batteries only when the target is at the right direction and distance.

It may be that in Erythropsidinium and Warnowia too the photoreceptive apparatus signals variations of light flux in space and not in time as in the case of other algae.

It is interesting to note that Erythropsidinium does not catch the prey with the nematocysts as Nematodinium does but by means of a wounding appendix, the "stomopod", which injects lysing substances into the prey. This harpoon-like appendix originates from a region, specialized for absorbing food, which is called "stomatopharyngean complex" by Greuet (53).

REFERENCES

1. C. G. Ehrenberg, Die Infusorientierchen als Wollkommen Organismen Leipzig (1938).

2. P. P. Batra & G. Tollin, Biochim. Biophys. Acta 79:371 (1964).
3. H. S. Jennings, "Behavior of the Lower Organisms", Columbia U. P., New York (1906).
4. S. O. Mast, "Light and Behavior of Organisms", Wiley, New York (1911).
5. P. Halldal, in "Biochemistry and Physiology of Protozoa", S. H. Hunter, ed., Academic Press, New York (1964).
6. P. A. Kivic & M. Vesk, Planta 105:1 (1972).
7. J. H. Belcher & E. M. F. Swale, Br. Phycol. Bull. 3:257 (1967).
8. H. Wager, J. Linn. Soc. Zool. London 27:463 (1900).
9. J. D. Dodge, Br. Phycol. J. 4:199 (1969).
10. J. D. Dodge, "The Fine Structure of Algal Cells", Academic Press, New York (1973).
11. P. L. Walne & H. J. Arnott, Planta 77:325 (1967).
12. G. F. Leedale, B. J. D. Meeuse & E. G. Pringsheim, Arch. Mikrobiol. 50:68 (1965).
13. S. P. Gibbs, J. Ultrastructure Res. 4:127 (1960).
14. G. F. Leedale, "Euglenoid Flagellates", Prentice-Hall Englewood Cliffs, New York (1967).
15. P. A. Kivic & M. Vesk, J. Exp. Bot. 23:1070 (1972).
16. P. A. Kivic & M. Vesk, Cytobiologie 10:88 (1974).
17. J. J. Wolken, J. Protozool. 24:518 (1977).
18. E. Piccinni & M. Mammi, Boll. Zool. 45:405 (1978).
19. P. A. Benedetti & A. Checcucci, Plant Sci. Letters 4:47 (1975).
20. E. Piccinni & P. Omodeo, Boll. Zool. 42:57 (1975).
21. A. Checcucci, Naturwissensch. 63:412 (1976).
22. E. Kronstedt & B. Walles, Protoplasma 84:75 (1975).
23. J. P. Mignot, Protistologica 2:51 (1966).
24. K. Wickerman, J. Protozool. 16:54 (1969).
25. J. H. Belcher, Br. Phycol. J. 4:105 (1969).
26. J. H. Belcher & E. M. F. Swale, Br. Phycol. J. 7:53 (1972).
27. D. J. Hibberd, Br. Phycol. J. 5:119 (1970).
28. J. H. Belcher & E. M. F. Swale, Br. Phycol. J. 7:335 (1972).
29. C. Rouiller & E. Fauré-Fremiet, Exptl. Cell Res. 14:47 (1958).
30. D. I. Hibberd & G. F. Leedale, Nature 225:758 (1970).
31. J. C. Green, J. Mar. Biol. Ass. U. K. 55:785 (1975).
32. J. Van der Veer, J. Mar. Ass. U. K. 56:21 (1976).
33. H. Ettl & J. C. Green, J. Mar. Biol. Ass. U. K. 53:975 (1973).

34. J. D. Pickett-Heaps, "Green Algae: Structure, Reproduction in Selected Genera", Sunderland, Mass. (1975).

35. H. J. Arnott & R. M. Brown Jr., J. Protozool. 14:529 (1967).

36. M. J. Hobbs, Br. Phycol. J. 7:347 (1972).

37. H. E. Gruber & B. Rosario, J. Cell Sc. 15:481 (1974).

38. M. Melkonian, Protoplasma 86:391 (1975).

39. L. R. Hoffman, Protoplasma 87:191 (1976).

40. V. Ulehla, Biol. Zentralbl. 31:645 (1911).

41. T. L. Jahn & E. C. Bovee, in "Research in Protozoology", T. T. Chen, ed., Pergamon Press, Oxford (1967).

42. J. C. Green & I. Manton, J. Mar. Biol. Ass. U. K. 50:1113 (1970).

43. J. C. Green, Brit. Phycol. J. 8:1 (1973).

44. M. Cheignon, C. R. Acad. Sci. Paris (gr. II) 258:676 (1964).

45. J. D. Dodge & R. M. Crawford, J. Cell Sci. 5:479 (1969).

46. C. Greuet, Cytobiologie 17:114 (1978).

47. W. G. Hand, J. Exp. Zool. 174:33 (1970).

48. C. Vogt, Zool. Anz. 8:153 (1885).

49. R. Hertwig, Morph. Jahrb. 10:204 (1884).

50. R. Hertwig, Zool. Anz. 8:108 (1885).

51. L. Mornin & D. Francis, J. Microscopie 6:759 (1967).

52. C. Greuet, C. R. Acad. Sci. Paris 261:1904 (1965).

53. C. Greuet, Protistologica 3:335 (1967).

54. C. Greuet, Protistologica 4:209 (1968).

55. C. Greuet, Protistologica 4:481 (1969).

56. C. Greuet, "Ultrastructure de l'Ocelle du Dinoflagellé Nematodinium comparée à Celles d'Autres Représentants de la Famille des Warnowiidae", 7^Congrés Intern. de Micr. Electro nique, Grenoble 3:385 (1970).

57. J. Milsum, "Biological Control and System Analysis", McGraw-Hill, London (1965)

58. D. J. McFarland, "Feedback Mechanisms in Animal Behavior", Academic Press, New York (1971).

59. M. D. Mesarovic, in "System Theory and Biology", M. D. Mesarovic, ed., Springer Verlag, Berlin (1968).

60. R. N. Clark, "Introduction to automatic Control Systems", Wiley, New York (1964).

61. P. Omodeo, Enciclopedia del 900, 6:902 (1980).

62. J. J. Wolken, "Invertebrate Photoreceptors", Academic Press, New York (1971).

63. M. E. H. Feinleib & G. M. Curry, Physiol. Plant. 20:1083
    (1967).
64. M. Kemali, D. Gioffrè, Cell Tiss. Res. 195:527 (1978).
65. R. B. Forward & D. Davenport, Science 161:1028 (1968).
66. R. B. Forward, Jr., Planta 92:248 (1970).
67. R. B. Forward, Jr., Planta 111:167 (1973).
68. Y. Naitoh & R. Eckert, in "Cilia and Flagella", M. A. Sleigh,
    ed., Academic Press, New York (1974).
69. D. L. Nelson & C. Kung, in "Taxis and Behavior", G. L.
    Hazelbauer, ed., Chapman & Hall, London (1978).
70. E. Hildebrand, in "Taxis and Behavior", G. L. Hazelbauer, ed.,
    Chapman & Hall, London (1978).
71. J. Adler, M. F. Goy, M. S. Springer, S. Szmelcman, in "Membrane
    Transduction Mechanisms", R. A. Cone & J. E. Dowling, eds.,
    Raven Press, New York (1979).
72. O. Coppellotti, E. Piccinni, G. Colombetti, F. Lenci, Boll.
    Zool. 46:71 (1979).
73. A. Checcucci, G. Colombetti, R. Ferrara & F. Lenci, in "Proc.
    3rd Natl. Congr. Cybernetics & Biophys.", Lito Felici,
    Pisa (1974).
74. E. Piccinni, V. Albergoni & O. Coppellotti, J. Protozool.
    22:331 (1975).
75. D. W. Francis, J. Exp. Biol. 47:495 (1967).
76. C. Greuet, Protistologica 7:345 (1971).

LOCALIZATION AND ORIENTATION OF PHOTORECEPTOR PIGMENTS

Wolfgang Haupt
Institute of Botany
University of Erlangen-Nürnberg
8520 Erlangen
Federal Republic of Germany

Light is a basic factor in life. It acts in either of two
fundamental ways:
i) In photosynthesis, light is the source of energy for the anabo-
lism of photoautotrophic plants and hence most important for the
energetics of the whole biosphere.
ii) Light acts as a signal on autothrophic and heterothrophic organisms, giving information about the environment. For this function,
its energy is of minor importance. The best known phenomena are
vision in animals and man, photomorphogenesis in plants, and orientation behavior of plants and animals.

In spite of this fundamental difference - energy versus information - there is at least one common step in all biological light
effects: before becoming effective, light has to be absorbed by a
pigment and its energy has to be transformed into chemical energy.
Photobiologists in all fields, therefore, are interested in identifying the photoreceptor pigment for the light response in question,
and in fact, this very often is the first step in elucidating the
reaction chain, i.e. the causal relationship between the impinging
light and the observed response. Chlorophyll is well established
as the photoreceptor in photosynthesis, and rhodopsin is known to
receive the light signal in vision. More recently, phytochrome has
been discovered as the main pigment in photomorphogenesis of green
plants, whereas an additional hypothetical photoreceptor pigment,
both in green and heterotrophic plants, has not yet been identified
definitely. This "cryptochrome" very probably belongs to the flavins.

155

Finally, bacteriorhodopsin is responsible for a dual effect of
light in <u>Halobacterium</u>; in this organism light is used as a source
of energy and as a signal for behavior.

In photosynthesis as well as in vision, the photoreceptor
pigment is strongly associated with a biomembrane or even with a
well-defined part of it (Amesz, Hildebrand, this vol.). This association is an absolute requirement for the photobiological functions of these pigments. Since this is true for both an energy effect and a signal effect of light, it is tempting to generalize
to all photoresponses. The question, therefore, is posed whether
in photomorphogenesis and orientation behavior to light in aneural
organisms biomembranes are required as well. As a first step along
this line, the intracellular localization of the different photoreceptor pigments has to be investigated. In the present lecture,
four systems will be referred to as examples.

Beforehand, a methodical remark has to be added. There are at
least three kinds of experimental approaches for investigating the
intracellular localization of photoreceptor pigments:
i) In cells of a proper size and shape, partial irradiations by
microbeams can be promising.
ii) Use of linearly polarized light sometimes results in specific
effects which are called "action dichroism", i.e. dependence of the
response on the orientation of the electrical vector. Most cases
of action dichroism have to be interpreted in terms of oriented
pigment molecules. This in turn, points to a close association of
the pigment with stable subcellular structures.
iii) If it can be demonstrated that the earliest effect of the
light is concerned with membrane properties or membrane functions,
this is strong evidence in favor of a pigment-membrane association.
Examples are the light effect on ion pumps in animal vision and
the chemiosmotic effect of photosynthetic light in the thylakoids
of green plants. In our system to be referred to we will make use
of these approaches whenever possible.

1. HALOBACTERIUM HALOBIUM

From a theoretical point of view, <u>Halobacterium halobium</u> is
a particularly interesting organism in photobiology: this bacterium
can use the light as an energy source for its metabolism and as a
signal for orienting itself in the environment as well. Both these
fundamental light effects are mediated by the same photoreceptor
pigment, viz. by bacteriorhodopsin. This pigment is closely related

to the visual pigment rhodopsin in animals and man (1-3).

1.1. Localization of Bacteriorhodopsin

Halobacterium is unique insofar as the localization of the
pigment can be recognized directly in the microscope (Fig. 1).
It is concentrated in the cell membrane at particular regions, and
in these regions the membrane consists exclusively of bacteriorhodopsin. Hence, bacteriorhodopsin is not bound to the membrane,
but it is the membrane at the particular regions, and this is called
the purple membrane.

By application of the methods of protein chemistry, information
has been obtained about the molecular array in the purple membrane.
We can consider it as a crystal-like lattice (1-3).

1.2. Energy Harvesting

Upon light absorption, bacteriorhodopsin is deprotonized to
the so-called 412 complex. This complex returns without the need
of additional energy to the ground state of bacteriorhodopsin, of
course by uptake of a proton. Importantly, this exchange of protons
is a vectorial process: during deprotonization the proton is released to the surrounding medium, but for reprotonization a proton
is taken up from the cell's inside (Fig. 2a). Thus, in effect, the
pigment cycle acts as a proton pump (1-5).

The resulting electrochemical proton gradient can be used by
the cell in different ways, from which only two will be mentioned:

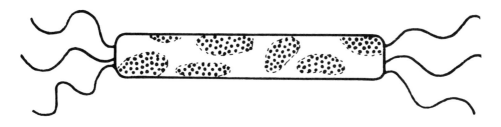

FIGURE 1. Halobacterium halobium with purple membrane regions.
After (1).

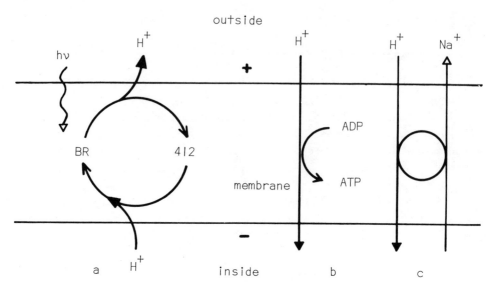

FIGURE 2. Purple membrane: a. Bacteriorhodopsin (BR) cycle upon irradiation, proton release and uptake. Intermediates between BR and
412 complex not shown. b. Proton-gradient driven ATPase. c. Proton-
gradient driven Na$^+$ extrusion. After (1, 2).

i) By means of a proton-driven ATPase the re-entering proton can
synthesize ATP and thus transfer its energy to the cell metabolism
(Fig. 2b) (4). This can be considered as an early phylogenetic
"invention" of photophosphorylation.
ii) By means of a proton-driven sodium pump, excess of Na$^+$ ions
can be transported out of the cell (Fig. 2c) (5). Both these effects derive their energy totally from the proton gradient. Thus,
light acts as the energy source for the observed response.

1.3. Perception of Light Signals

    Halobacterium is a free moving organism, due to the activity
of its flagella (cf. Nultsch, this vol.). Under constant environmental conditions, locomotion can be observed more less along a
straight path. If, however, the light intensity is suddenly increased or decreased ("step up","step down"), the cell performs
a typical photophobic response, i.e. reversal of the swimming direction. By action spectroscopy and some additional experiments,

the photoreceptor pigment for the step-down response has been shown
to be bacteriorhodopsin with a peak of sensitivity at about 565 nm.
The step-up response, on the other hand, is mediated by a precursor
of bacteriorhodopsin and has its peak near 370 nm (1,2).

Perception of step-down signal by bacteriorhodopsin results
in a transient depolarization of the cell membrane, i.e. in a decrease of the membrane potential, and this probably triggers the
reversal of the flagellar activity, maybe via a change in calcium
fluxes (1). Although we have not yet complete knowledge about the
photophobic reaction chain, it can be considered as certain that
the sensory transduction in Halobacterium is inevitably linked to
membrane effects of the photoreceptor pigment and hence to its
being an integral part of the purple membrane.

Thus, bacteriorhodopsin is a membrane effector both as an
energy-harvesting pigment comparable to chlorophyll, and as a
photoreceptor for sensory transduction comparable to the visual
pigment in animals and man. This is concluded from its localization,
from its highly-ordered molecular pattern, and from the analysis
of the reaction chains.

2. MOUGEOTIA SPEC.

In contrast to Halobacterium, in the green alga Mougeotia
the response to light is not concerned with motility of the whole
organism, but with reorientation of the chloroplast within the cell
(Fig. 3). In the cylindrical cell, the big flat chloroplast turns
its face to low-intensity light and its profile to high-intensity
light. The following report will be restricted to the low-intensity
movement, which is less complicated and better understood than the
high-intensity movement (6,7).

By action spectroscopy and by the classical red far-red anta-
gonism, phytochrome has been proven as the photoreceptor pigment.
This pigment is characterized by its two forms Pr and Pfr which
are interconvertible by the proper irradiation with red (r = 660
nm) and far-red (fr = 730 nm) light. Several sets of experiments
can provide information about its localization and orientation in
Mougeotia.

2.1. Localization and Orientation of Phytochrome

Because of the size of the cell (ca. 150 x 20 μm), partial
irradiations with microbeams are feasible. Such experiments show

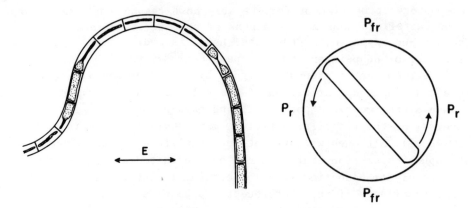

FIGURE 3. <u>Mougeotia</u> filament with chloroplasts in face and profile position, respectively. Result of irradiation with polarized red light (E=electrical vector). After (6).

FIGURE 4. Cross section through a <u>Mougeotia</u> cell with the chloroplast orienting in a cytoplasmic gradient of Pfr. After (6).

that the chloroplast can orient to the light even if only the cytoplasm is irradiated. Thus, the chloroplast orients in a cytoplasmic gradient of the active phytochrome Pfr such that its edges always avoid the regions of highest Pfr concentration (Fig. 4).

More detailed information is obtained by using polarized light. <u>Mougeotia</u> exhibits a typical action dichroism: response is found only in those cells which are oriented perpendicularly or at least obliquely to the E-vector, but no response is induced if cell and E-vector are parallel to each other (Fig. 3). This points to a certain orientation of dichroic phytochrome molecules.

By combining the two approaches, viz. microbeam irradiation and polarized light, and considering the orientation in the Pfr gradient, we arrived at the conclusion that the red absorbing form of phytochrome Pr is oriented parallel to the surface of the cell and probably preferentially along helical lines around the cell (Fig. 5a). The far-red absorbing form of phytochrome, however, is oriented normal to the cell surface (Fig. 5b). This means that during photoconversion Pr → Pfr or Pfr → Pr the pigment molecule

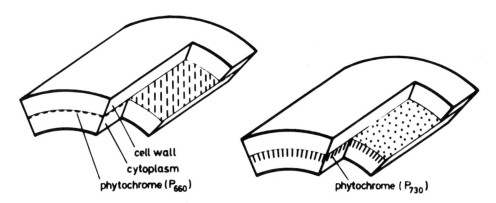

cell wall
cytoplasm
phytochrome (P_{660})

phytochrome (P_{730})

FIGURE 5. Dichroic orientation of phytochrome in the Mougeotia cell. After (6).

has to change its spatial arrangement (flip-flop mechanism).

From the well-defined pigment orientation and especially from the precise flip-flop mechanism, it has to be concluded that phytochrome is closely associated with a stable structure of the cell which does not change much during the chloroplast movement. This structure most probably is the plasma membrane (plasmalemma) (8).

2.2. Perception of Light Direction

In order to orient its chloroplast to the light, the cell has to perceive the light direction, i.e., unidirectional light has to result in an absorption gradient in the photoreceptor layer near the cell surface. This can be understood on the basis of the surface-parallel dichroic orientation of the photoreceptor molecules Pr (6, 7): polarized light with the E-vector normal to the cell axis is absorbed mainly at the cell surface, but little absorption is found at the flanks; hence a strong gradient of phototransformation (Pr → Pfr) is formed between front/rear and the flanks. The same is true for unpolarized light, because it always contains a component which vibrates normal to the cell axis. In saturation, different photostationary states Pfr:Pr are established at front/ rear and at the flanks; this is due to the flip-flop mechanism. Accordingly, a permanent Pfr gradient ensures the response even in continuous saturating light (6).

2.3. Phytochrome as a Membrane Effector

After we have found that in Mougeotia phytochrome is localized and oriented close to the membrane, we may ask whether this might be accidental or might have some direct bearing on the reaction chain induced by the light. More specifically: acts phytochrome as a membrane effector when triggering the chloroplast orientation?

As a first approach, it was found that plasmolysis of Mougeotia cells in mannitol is influenced by phytochrome: after red light (i.e. phytochrome in the Pfr form), plasmolysis and deplasmolysis start earlier and proceed faster than after red followed by far-red (i.e. phytochrome in the Pr form). This points to some properties of the membrane being under phytochrome control, e.g. water permeability (8).

More specific results have been obtained recently (9): the uptake of $^{45}Ca^{2+}$ into the cells is facilitated or stimulated by phytochrome in the Pfr form (Fig. 6). Moreover, a strong correlation has been found between Ca^{2+} content of the cells and chloroplast response to phytochrome (10). Ca-starved cells lost their ability for chloroplast orientation parallel to the intracellular concentration of that fraction of Ca which cannot be extracted by water. This Ca fraction is localized mainly in the small tannin vacuoles in the cytoplasm near the edge of the chloroplast (11). Since very probably actomyosin-like proteins provide the motive force for the movement (7); since actomyosin generally is controlled by Ca^{2+} and since the Ca-containing tannin vacuoles are localized close to structures which are presumed to be actomyosin microfilaments, the role of Ca in the transduction chain has become very probable in Mougeotia. Phytochrome acting on the Ca^{2+} uptake, then, can be considered as a membrane effector in the light-controlled chloroplast movement in this alga (12). Consequently, the above mentioned membrane association of phytochrome would be an obligatory requirement for this response.

Since phytochrome is the most important photoreceptor pigment for signal effects of light in green plants, the question may be asked whether these conclusions in Mougeotia may be generalized to all phytochrome effects in all green plants. However, we have to point to the limits of our observations and conclusions. It is true, phytochrome in Mougeotia is membrane-associated (and probably a membrane effector) as far as chloroplast orientation is concerned.

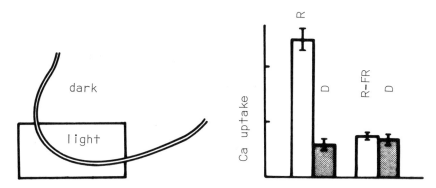

FIGURE 6. Uptake of $^{45}Ca^{2+}$ into Mougeotia cells. a. Draft of partial irradiation of a filament. b. Radioactivity uptaken during 1 min after the irradiations as indicated, compared in the irradiated and dark area. R = 659 nm, FR > 700 nm, 30 seconds each. After (9).

However, nobody has excluded the possibility that in Mougeotia other cell compartments may contain ample phytochrome which is not involved in our particular response - maybe only because it cannot establish a gradient in Pft. Thus we may clarify our statement, saying that in Mougeotia phytochrome is likely to be a membrane effector as far as it is involved as a photoreceptor pigment in chloroplast orientation. No generalization to other systems is feasible yet.

3. THE PHYCOMYCES SPORANGIOPHORE

Unlike the locomotion of Halobacterium and the intracellular movement in Mougeotia, Phycomyces exhibits an oriented bending of its sporangiophore towards the light (Fig. 7a) or away from it, i.e. a positive or negative phototropism, due to differential growth of opposite regions (13). The photoreceptor pigment with its two maxima in the blue and near uv region is a classical example of "cryptochrome" and there are strong arguments in favor of a flavin compound (Hertel, this vol.).

A second light response in Phycomyces is the light-growth

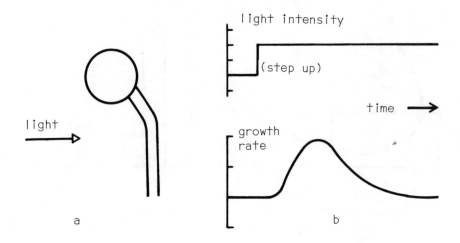

FIGURE 7. Photoresponses of the _Phycomyces_ sporangiophore. a. Positive phototropism resulting from differential growth. b. Positive
light-growth response (below) upon a sudden increase in light intensity (above).

response, i.e. a transient increase or decrease of the growth rate
upon a step-up or step-down of the stationary light intensity (Fig.
7b). This response makes use of the same photoreceptor pigment as
the phototropism, and there are casual relationships between these
two responses (13).

The longitudinal distribution of the photoreceptor pigment can
be tested by partial irradiation of the sporangiophore with microbeams. The sensitivity to light is restricted to the few millime-
ters below the sporangium where growth occurs (13). This finding,
however, does not contain much information, and we cannot discern
between the two following possibilities: i) The pigment really is
restricted to the growing zone; ii) the pigment is distributed
evenly along the whole sporangiophore, but lack of signal transfer
to the growing zone makes it impossible to recognize a physiological
effect of light absorption outside the growing zone.

More conclusive data are available about the distribution of
cryptochrome over the cross section of the growing zone. Nevertheless, no general agreement has been obtained yet insofar as not all
observations and arguments point to the same direction. The following

hoop axis

longitudinal axis

FIGURE 8. Dichroic orientation of cryptochrome in the sporangiophore of _Phycomyces_, concluded from physiological responses for
the transition moments at 456 nm (left) and 280 nm (right). After
(13).

sections will deal with these problems.

3.1. Action Dichroism

     If a light-growth response or a photophobic bending is evoked
by polarized light, the light is slightly more effective if its
E-vector is oriented normal rather than parallel to the length of
the sporangiophore, both in phototropism and in light-growth
response. The advantage amounts up to 24% (14). Theoretically and
experimentally it has been excluded that this action dichroism is
due to differential reflection at the surface, and hence a certain
dichroic orientation of the cryptochrome molecules has to be concluded (15). Careful analytical work (16) has shown that the main
direction of the dipole moments in the blue range is oriented about
5° from the hoop axis, whereas in the uv it is oriented about 35°
from the hoop axis (Fig. 8). By the way, this agrees very well with
the biophysical properties of flavin molecules.An additional strong
radial component(cf.Hertel,this Vol.)is not shown in the Figure.
     Considering this action dichroism and the underlying orienta-
tion of the cryptochrome molecules, we may conclude an association
of the pigment with a stable cell structure. Taking into account
the well-known cytoplasmic streaming in the sporangiophore, which
excludes a stable orientation in the cytoplasm, the cell membrane
is the most likely site of the pigment.

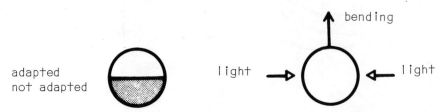

FIGURE 9. Schematic cross section through a <u>Phycomyces</u> sporangiophore, one half of which is light-adapted by a proper preirradiation
(left). Upon a uniform test irradiation, the light signal is preferentially perceived in the non-adapted half, which results in
bending by unequal growth.

It may be added that there are many other blue-light responses
in the plant kingdom, which are attributed to cryptochrome and which
exhibit an obvious action dichroism. In cryptochrome-controlled
chloroplast orientations, e.g., the pattern of assembly/disassembly
is strongly related to the E-vector of polarized light, and this
can be true even in a plasmolyzed cell (7,17). Moreover, control
by light of cytoplasmic streaming in <u>Elodea</u> and <u>Vallisneria</u> and
of fastening or loosing of cytoplasm at the cell wall depends on
the polarization of the cryptochrome-absorbed light (18). Since in
these cases there is nearly no doubt about a close association of
the pigment with the cell membrane, we may take this as additional
support for our conclusion in <u>Phycomyces</u>.

3.2. <u>Local Azimutal Autonomy</u>

Phototropic bending is due to azimutal differences in cell
wall growth. These differences are controlled by differences in
light absorption. Consequently, azimutal light absorption and
azimutal growth are closely correlated. Moreover, the well-known
adaptation processes remain restricted to part of the cross section
if only part of the sporangiophore is illuminated with adaptation
light (Fig. 9); i.e. adaptation occurs in local azimutal autonomy

(13). These results were difficult to interpret if the photoreceptor pigments were spread over the whole cross section; rather they strongly suggest a close spatial neighborhood between photoreceptor pigment and site of response, i.e. a localization of the pigment close to the growing cell wall.

3.3. Microbeam Irradiation

In addition to unilateral total irradiation an absorption gradient across the sporangiophore can be established by partial irradiation at the flank; accordingly, lateral bending will result. If very small laser beams are put to different parts across the cell diameter, maximal response is not restricted to the marginally located beam, but it is found even if the beam is located halfway the distance between the periphery and the cell axis (19). This result seems to be inconsistent with the photoreceptor's localization near the cell membrane as well as near the tonoplast of the central vacuole; it suggests,therefore, a more less uniform distribution throughout the cytoplasm.

Since this is in strong contrast to the conclusions from action dichroism and local azimutal autonomy, sources of error in the experiments or interpretations may be considered. Thus, light scattering in the microbeam experiments might obscure the differences expected from a pigment localization close to the surface (13).

3.4. Extraction of Active Fractions

In the cellular slime mold Dictyostelium a light-induced absorbance change has been detected at 430 nm, which is due to reduction of a b-type cytochrome (20,21). The same absorption change has been found upon irradiation in Phycomyces. The action spectrum of this light effect is very similar to the phototropic action spectrum of Phycomyces and thus points to cryptochrome. It therefore seems justified to take the photoreceptor pigment for this absorbance changes at least as a reasonable model for the phototropic photoreceptor, whenever this system is observed in an organism. As an important step along this line, the absorbance--change system has been found in membrane-rich fractions from Neurospora and from Zea (22,23), and this again points seriously to the plasma membrane as the site of cryptochrome in general and hence also in Phycomyces.

Summarizing all facts known to-date, we can favor the asso-

ciation of cryptochrome with membranes, probably with the cell
membrane (plasmalemma). Although most evidences are indirect,
they strongly support each other. There seems to be only one con-
tradictory observation, viz. the result of partial irradiations of
the Phycomyces sporangiophore; but the conclusions from this result
should be reevaluated, considering carefully all possible light-
-scattering effects.

3.5. Generalization

It is true, we have used similarities between as different
response systems as phototropism and intracellular movement, to
support the conclusion about cryptochrome localization and orientation. However, there are many other cryptochrome responses for
which no such information are available yet. Induction of carotenoid synthesis may serve as an example (24). It seems reasonable
to generalize our conclusion to those systems as long as no contra-
dictory results are known.

4. EUGLENA GRACILIS

As it has been shown earlier (Feinleib, this vol.), the phyto-
flagellate Euglena performs a typical photophobic response upon a
proper step-down or step-up stimulus of the light intensity. This
response also underlies the phototactic orientation: due to the
rotation of the cell, the strongly pigmented "eyespot" (stigma)
periodically shadows the basal part of the flagellum; as a result,
the motor response of the flagellum slightly corrects the course
of the organism toward the light source. This is periodically
repeated until the proper course is reached (25).
It has been assumed, therefore, that the photoreceptor pigment
is localized in the apparent swelling called paraflagellar body
(pfb; Fig. 10; see also Omodeo, this vol.), in contrast to last
century's suggestions which ascribed this role to the stigma
(hence the outdated term eyespot).
Again, additional evidence for the pfb as the site of the
photoreceptor pigment came from experiments with polarized light.
Under certain conditions, Euglena exhibits a remarkable action
dichroism in its phototactic response (25,26). No dichroism has
been found in the stigma, but a strong birefringence in the pfb
points to a highly ordered structure in this very small organelle
(27).

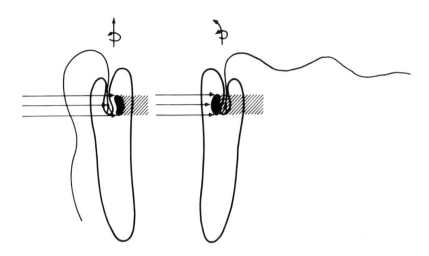

FIGURE 10. Euglena cell with lateral irradiation.In the second posi-
tion(right),the photoreceptor (pfb) is shaded by the stigma;as a result the flagellum shows the motor response.

Since the action spectra of photophobic response and phototaxis are
very similar to the absorption spectrum of a flavin compound(28),and
since the flavin quencher potassium iodide specifically inhibits photoresponses of Euglena (29,30),it appeared promising to search for a
flavin or flavoprotein in the pfb.Indeed, fluorescence studies revealed remarkable flavin-like signals in the pfb,and additionally the
dichroism found in these experiments clearly points to the pigment
being highly ordered in the pfb (31,32).Thus, prediction from earlier
experiments and recent results coincide remarkably. It seems reasonable to attribute the high order of the pigment molecules to membrane structures in the pfb. But final proof is still lacking.
        Since there are several taxonomic groups of flagellates with
many species reacting phototactically or photophobically,the question
can be asked to what extent generalization is possible. Unfortunately,
there are several reasons which exclude such an extrapolation.
i) Different groups of flagellates very probably use different photoreceptor pigments for their phototaxis or,more general,for their photomovement responses (25,32,33).Besides the flavoprotein in Euglena,
carotenoids are suggested (in Volvocales and Dinoflagellates)as well
as phycoerythrin (in Cryptomonadales).It would be very strange to find
such a diversity of pigments in homologous structures.
ii) Although many experimental approaches have been tried to localize the photoreceptor structure or the photoreceptor pigment

(32,34; cf. also Omodeo, this vol.), in no case besides <u>Euglena</u>
were these efforts really successful. It is true, there exist
some reasonable suggestions: in <u>Tetracystis</u> as well as in <u>Volvox</u>
a particular site of the cell membrane is assumed to contain the
photoreceptor pigment (35,36), and in dinoflagellates the photoreceptive structure is tentatively localized near the base of the
flagellum (37). But in no case can this be taken for certain.
iii) The pfb may be peculiar to <u>Euglena</u> and its relatives. There
is no structure in other groups of flagellates which is homologous
to the pfb beyond doubt.

As a result, our knowledge about localization of the photoreceptor pigment for photomovement in flagellates is strictly limited to <u>Euglena</u>. Nevertheless, taking together all results referred
to in the preceding sections, it seems reasonable to anticipate
also in all other flagellates a close association of the photoreceptor pigment to membranous structures, whatsoever their localization.

5. CONCLUSIONS

It is amazing to realize how fundamentally different photoreactions in as well fundamentally different groups of organisms
share common features. For a photoreceptor pigment to become
physiologically functioning, an absolute requirement seems to be
its structural association with a biomembrane. This is true for
the energy-harvesting light process in photosynthesis as well as
in very different signal effects of light, e.g. vision in animals
and man, orientation movements in motile and in bending plants,
control of intracellular movements, and probably also in the wide
field of photomorphogenesis in plants.

Moreover, this requirement is shared by as different types of
photoreceptor pigments as rhodopsin and bacteriorhodopsin, chlorophyll, phytochrome, and flavoprotein. And if we consider what is
known about blue-green algae, which have not yet been mentioned
in this lecture, we easily may add phycocyanin and phycoerythrin
to that list (Nultsch, this vol.).

If membranes are inevitably connected to the physiological
function of a photoreceptor pigment, one might assume membrane
effects as an early step in all transduction chains started by
light. Indeed, transport of ions across a membrane is among the
earliest steps in vision as well as in photosynthesis and in light
responses of <u>Halobacterium</u>. In addition, there is increasing evi-

dence for membrane effects of phytochrome being a common step in
at least some reaction chains, and photoresponses in blue-green
algae seem to have electrophysiological membrane effects as a
transduction step as well. Cryptochrome-regulated chloroplast accumulation in the alga Vaucheria seems to be correlated with membrane currents (Blatt and Weisenseel, personal communication), and
there are also reasons to assume an electrophysiological sensory
transduction in photomovements of flagellates.

This lecture tried to center to the uniformities underlying
all light reactions in organisms. These uniformities become more
amazing, the more we realize the enormous diversity of light reactions and their transduction chains in different organisms. On the
other hand, the diversities will become really fascinating more
than ever on the background of the fundamental uniformity of membranes as early transducers of all kinds of photoreception processes.

REFERENCES

1. E. Hildebrand, in "Taxis and Behavior", G. Hazelbauer, ed.,
   Chapman & Hall, London (1978).
2. G. Wagner, Biologie in Unserer Zeit (in press) (1979).
3. W. Stoeckenius, R. H. Lozier, & R. A. Bogomolni, Biochim.
   Biophys. Acta 505:215 (1979).
4. D. Oesterhelt, Angew. Chem. (Internat.Ed.) 15:17 (1976).
5. J. K. Lanyi, Microbiol. Revs. 42:682 (1978).
6. W. Haupt, in "Phytochrome", K. Mitrakos, & W. Shropshire Jr.,
   eds., Acad. Press, London-New York (1972).
7. S. J. Britz, in "Encyclopedia of Plant Physiology, Vol. 7:
   Physiology of Movements", W. Haupt, & M. E. Feinleib, eds.,
   Springer, Berlin, Heidelberg-New York (1979).
8. M. H. Weisenseel, & E. Smeibidl, Z. Pflanzenphysiol. 70:420
   (1973).
9. E. M. Dreyer, & M. H. Weisenseel, Planta 146:31 (1979).
10. G. Wagner, & K. Klein, Photochem. Photobiol. 27:137 (1978).
11. G. Wagner, K. Klein, & R. Rossbacher, Cytobiologie 18:198
    (1978).
12. W. Haupt, & M. H. Weisenseel, in "Light and Plant Development",
    H. Smith, ed., Butterworth, London-Boston (1976).
13. D. Dennison, in "Encyclopedia of Plant Physiology, Vol. 7:
    Physiology of Movements", W. Haupt, & M. E. Feinleib, eds.,
    Springer, Berlin,Heidelberg-New York (1979).

14. W. Shropshire Jr., _Science_ 130:336 (1959).
15. L. F. Jaffe, _J. Gen. Physiol._ 43:897 (1960).
16. A. J. Jesaitis, _J. Gen. Physiol._ 63:1 (1974).
17. J. Zurzycki, _Acta Soc. Botan. Pol._ 37:11 (1968).
18. K. Seitz, _in_ "Encyclopedia of Plant Physiology, Vol. 7: Physiology of Movements", W. Haupt, & M. E. Feinleib, eds., Springer, Berlin,Heidelberg-New York (1979).
19. M. L. Meistrich, R. L. Fork, & J. Matricon, _Science_ 169:370 (1970).
20. K. L. Poff, & W. L. Butler, _Nature_ 248:799 (1974).
21. K. L. Poff, & W. L. Butler, _Plant Physiol._ 55:427 (1975).
22. R. D. Brain, J. A. Freeberg, C. V. Weiss, & W. R. Briggs, _Plant Physiol._ 59:948 (1977).
23. A. J. Jesaitis, P. R. Henert, R. Hertel, & W. R. Briggs, _Plant Physiol._ 59:941 (1977).
24. W. Rau, _Planta_ 72:14 (1967).
25. D. P. Häder, _in_ "Encyclopedia of Plant Physiology, Vol. 7: Physiology of Movements", W. Haupt, & M. E. Feinleib, eds., Springer, Berlin,Heidelberg-New York (1979).
26. K. E. Bound, & G. Tollin, _Nature_ 216:1042 (1967).
27. C. Creutz, & B. Diehn, _J. Protozool._ 23:552 (1976).
28. A. Checcucci, G. Colombetti, R. Ferrara, & F. Lenci, _Photochem. Photobiol._ 23:51 (1976).
29. B. Diehn, & B. Kint, _Physiol. Chem. and Physics_ 2:483 (1970).
30. E. Mikolajczyk, & B. Diehn, _Photochem. Photobiol._ 22:269 (1975).
31. P. A. Benedetti, A. Checcucci, _Plant. Sci. Lett._ 4:47 (1975).
32. F. Lenci, & G. Colombetti, _Ann. Rev. Biophys. Bioeng._ 7:341 (1978).
33. W. Haupt, _in_ "Biophysics of Photoreceptors and Photobehaviour of Microorganisms", G. Colombetti, ed., Lito Felici, Pisa (1975).
34. E. Piccinni, & P. Omodeo, _Boll. Zool._ 42:57 (1975).
35. H. J. Arnott, & R. M. Brown Jr., _J. Protozool._ 14:529 (1967).
36. K. Schletz, _Z. Pflanzenphysiol._ 77:189 (1976).
37. W. G. Hand, & J. A. Schmidt, _J. Protozool._ 22:494 (1975).

IDENTIFICATION AND SPECTROSCOPIC CHARACTERIZATION OF PHOTORECEPTOR

PIGMENTS

Giuliano Colombetti and Francesco Lenci

C.N.R. Laboratorio Studio Proprietà Fisiche Biomolecole
e Cellule
PISA (Italy)

INTRODUCTION

From the previous chapters it should be quite clear by now
that living organisms interact with the external world via specialized molecules that are able to undergo certain physico-chemical
modifications in response to environmental stimuli, thus initiating
the process of signalling to the organism that something outside is
now different. How this signal is thereafter processed and eventually transformed into a biological response (such as e.g. a modification of body shape or of motile properties) is a very important
question to answer and, in fact, many of the lectures in this book
deal with this aspect of the problem. Of the same importance, however, is the question of which molecules are devoted to the detection
of external stimuli and what physico-chemical changes take place
thereon in their molecular structure. In fact, sometimes it may be
very difficult to build up a reasonable model of a sensory transduction chain if its very first steps, that is stimulus perception
and subsequent molecular modifications are not known.

It is therefore very easily understood that one of the first
goals of research on sensory trandsuction is the identification and
characterization of these receptor molecules. In particular, when
light is the external stimulus, it is necessary to identify those
molecules that are able to absorb luminous energy and to study
their subsequent reactions in order to be able to suggest possible
molecular mechanisms of signal transduction. This chapter will deal
with the problem of the identification of the pigments responsible
of photon absorption in the photomotile responses of some Eucaryo-

tes. Halobacterium halobium, blue-green algae, desmids and diatoms
have already been dealt with by Prof. Nultsch. Only those responses
will be taken into account, which are commonly and shortly described by the term photomotile responses, excluding, however, responses like photokinesis and phototropism. The former is a photocoupling process where the light energy is directly utilized in the
photophosphorylation process and the pigments involved are photosynthetic pigments, as discussed by Nultsch. Phototropism was
already considered by Dr. Hertel.

1. IDENTIFICATION OF RECEPTOR MOLECULES

Now the question arises as to the determination of the nature
of the pigments involved in a certain response. Which are the tools
at our disposal to solve this problem?

1.1 Isolation

It is clear that when the photoreceptor molecules are known
to be localized in determined cellular substructures the most
natural approach to identify the photoreceptor pigments is to isolate those structures and study "in vitro" their spectroscopic,
chemical and biochemical properties. There are not very many
examples of this in the literature, one can be found in Dr. Song's
chapter, describing the results obtained on the pigment involved
in the photoresponse of Stentor. Unfortunately this is quite an
exception and usually it is much easier to follow other routes than
to isolate the photoreceptor pigments, given also that in only a
few cases have the photoreceptor structures been identified with
reasonable certainty; this makes,of course, the solution of the
problem even more complex, if possible at all. However, when photoreceptor isolation is not feasible, two conditions have to be ful-
filled in order to identify unambiguously the photoreceptor pigments
first, the pigment is shown to be present in the photoreceptor,
and second, the action spectrum of the photoprocess is proportional to the absorption spectrum of the candidate pigment. We will
return later to discussing action spectra determination.

1.2 Microspectrophotometry

When the position of the presumed photoreceptor in the cell
is known (once again this is not very common), an alternative route

to the isolation of that structure can be that of absorption microspectroscopy "in vivo". However, also in this case, difficulties
exist. Let's take for instance a case where we have direct experience, that of Euglena gracilis. For many years it was thought that
a special structure lying near the base of the emerging flagellum,
the so called paraflagellar body, was the true photoreceptor structure. With the help of a microspectrophotometer built in our Lab,
we tried to investigate the optical absorption properties of the
PFB by scanning the region around the flagellar base (the scanning
was necessary because it is very difficult to localize the PFB under
the optical microscope). The only detectable signal was clearly
from the stigma, which was thought once to be possibly the photoreceptor structure. Nowadays, we think that all the workers in the
field agree that this is not true. The main evidence against this
hypothesis is that cells without the stigma show a clear light
response.

The spectra obtained from the stigma indicate the presence of
carotenoids and this confirmed results previously described in the
literature; moreover, from the measurements it was also possible to
conclude that the stigma was not dichroic, an information very
interesting in itself, as we will see later on, but all of this unfortunately did not give any information about the photoreceptor pigments,or,to say it better,this gave an indirect evidence that the
absorbance of the receptor was quite small.

A rough estimate gives a maximum absorbance of the order of
10^{-3}. Approximating the photoreceptor organelle with a cube of
1 μm side this gives a concentration of the order of 10^{-3}M; for a
pigment with an extinction coefficient of the order of 10^4, this
is equivalent to $\sim 6.10^5$ molecules in a volume of $\sim$ 1 μm^3. The
following step was suggested by two facts: first flavins were a
possible candidate as photoreceptor pigments in Euglena, as it had
been put forward by Diehn and Kint (1), and second, flavins are
highly fluorescent pigments in comparison to their forever compe-
titors, carotenoids. So, putting together these two facts, it was
natural to try and look for a fluorescent organelle in the region
of the PFB. When the first measurements were performed, the microspectrophotometer in our Lab was still at its very beginning as
microspectrofluorometer; therefore, the method chosen was necessarily visual observation of our specimens under fluorescence microscope. Selecting appropriate wavelengths for excitation and emission

it was possible to observe a weak fluorescent light coming exactly
from the region of the PFB (2). This emission had indeed a maximum
(as judged by eye), around 520 nm, was maximally excited by light
at around 450 nm, and moreover, it was polarized. This last fact
was really exciting since it could be related to the ordered
structure of the PFB (2). This was good evidence that
flavins were indeed contained in the photoreceptor organelle of Euglena. However, it was only qualitative evidence
after all it is difficult to judge by eye the emission spectrum of
a pigment. Fortunately, the microspectrophotometer was modified
and turned into a microspectrofluorometer coupled to a computer.
We will not go into technical details of this apparatus, that is
described elsewhere (3),we only want to mention that, in order to
be able to record the emission spectrum of the PFB, high spatial
resolution and photometric efficiency are required. Moreover, it
is necessary to clean up the emitted light from cytoplasmic stray
emission, as well as residual exciting radiation. A dual beam
technique is therefore utilized and the emission spectra are recorded by computer averaging the difference between the emission
of the PFB and that of the neighbouring cellular regions. Furthermore, the photomultiplier had to be cooled down to $-40°C$, in order
to reduce the dark current to below three photoelectrons per sec.

Multispectral fluorescence/absorption maps of the cell regions
absorbing at 480 nm and emitting at 530 nm allow the determination
of the relative positions of the stigma and of the PFB, thus
confirming the presence of the fluorophores in the latter organelle.
By comparing the emission from the PFB with that of a typical flavin, like FMN for example, it can be attributed to a flavin chromophore. These fluorometric measurements make also possible to give
a rough estimate of the number of fluorescing centers in the PFB.
The light emitted at 530 nm has an intensity of approximately
$10^{-13}W$, corresponding to about 2.5×10^5 photons. If we assume that
the fluorescing pigments have a quantum yield of the order of .25,
like riboflavin for instance, then the number of fluorophores can
be figured at around 10^6. This value should be compared with the
estimate given above from the absorption measurements.

1.3 Quenchers of Excited States

Another possible method of identifying photoreceptor pigments
could be the use of substances which are known to act as specific
quenchers of the excited states of the presumed photoreceptor

molecules. This was the philosophy of the experiment of Mikolajkzyk
and Diehn (4), in which they used potassium iodide, a quencher of
flavin triplet, and looked for a possible effect on the phobic
reaction of Euglena. Their results seem to suggest that flavins
may be involved at least in the case of the step-down response.
The idea is present also in the work of Stavis (5) on Chlamydomonas,
where a possible interpretation of the results obtained is that
azide might quench the excited states of the photoreceptor molecules.

It should be noted that the results obtained using drugs must
always be considered with great care, since the specificity of those
substances "in vivo" is much less well defined than "in vitro".

1.4 Action Spectra - General Considerations

Let us now come back to the second point, the determination
of the action spectra. This kind of approach does not allow in general a positive identification of the pigments involved. This is
due to several reasons: for instance, action spectra are usually
poorly resolved, many pigments have similar absorption spectra in
the visible range and moreover the absorption properties of a pigment "in vivo" can be deeply modified by the environmental conditions, so that shoulders or even absorption bands can be present
or absent. Action spectra determination is, however, in most cases,
the only technique available that can give an idea, maybe a very
rough one, of the pigments involved. In the following we will
discuss briefly how an action spectrum should be determined and
thereafter we will give a few examples. There is one point to stress
before going on: it is the importance of a quantitative and reproducible measurement of the response under a given stimulus. This
part of the problem has already been dealt with by Dr. Diehn, and
we do not insist any longer on it. Just a small note: cell population methods should only be used to measure the strength of a photomotile response, when the relation between the response(s) of single
cells and the macroscopic behavior is well understood. So, now that
we know how to measure a response, we are left with the problem of
the measurement of the stimulus, light in our case. The problem has
already been considered by Dr. Montagnoli. Just a few remarks; the
light source should have sufficient energy in the near UV, given
the importance of this spectral region in discriminating between
flavins and carotenoids. It should, however, be kept in mind that
the ratio $A(370)/A(450)$, which in principle allows this discrimina-

tion, can be altered for instance in carotenoids under certain
experimental conditions (6). Different wavelengths can be selected
either by means of monochromators or by using interference filters.

A very important improvement in the type of light sources
available to photobiologists was represented by the introduction
of dye lasers, which can be tuned at every wavelength in the near
UV, visible red and far-red regions, and whose output intensity is
only limited by the power of the pumping laser. The intensity of
the exciting light ought to be expressed in IS units, i.e. in W/m^2.
Units such as lux or lumen should be absolutely avoided, first
because such units are matched on the spectral sensitivity of the
human eye, which a priori has not much to do with that of microorganisms, and second because they do not give any idea of the
number of quanta falling on the system under examination, unless one
wants to go into quite complicated calculations. The determination
of the number of quanta is quite important, in fact action spectra
should always be expressed in terms of response per equal quantum
flux since light is absorbed in quanta.

Some authors usually express action spectra in terms of response
per equal energy. As we shall see below, this leads to a deformation
in the action spectrum, that can be as high as a factor of 2 going
from 300 to 600 nm. In fact, to convert from light energy measured
in W/m^2 to quanta/m^2 the following formula is generally utilized
in calculations: $n = I/h\nu$, where I is in W/m^2, h is the Planck
constant and ν is the frequency of light that is related to the
wavelength λ through the relation $\nu\lambda = c$, where c is the speed of
light in vacuum. A formula that can be utilized for practical purposes is $n = 5.034 \ 10^{15} I \lambda$ where I is in watt and the wavelength in
nanometers. In principle all the previous considerations hold true
only when the bandwidth of the light is very narrow, theoretically
for monochromatic light. For large bandwidths, like those normally
available with interference filters or monochromators it can anyhow
be utilized with a sufficient degree of approximation, since the
error made is much smaller than that made in measuring light intensities. At this point, we will discuss a few considerations on action
spectra determination. A more extensive treatment can be found in (7).
In what follows we will assume that: i) only one photopigment is
present; ii) the system under consideration is weakly absorbing
(i.e. the number of incident photons $N(\lambda)$ does not vary significantly with depth of penetration). We can now define the response as:

(1) $R\left[N(\lambda)\right] = f\left[\alpha(\lambda) \cdot N(\lambda)\right]$

where f is an arbitrary function, which must be the same at every wavelength. When this is not verified, one should consider the possible presence of more than one photoreceptor pigment or of shading pigments. $\alpha(\lambda)$ is the absorption coefficient for the actively absorbed quanta. $\alpha(\lambda) \cdot N(\lambda)$ is therefore the number of active quanta absorbed. $\alpha(\lambda)$ is proportional to the absorption coefficient $\beta(\lambda)$ of the active pigment $\alpha(\lambda) = \Phi \, \beta(\lambda)$, Φ is the quantum efficiency of the phenomenon.

A plot of $\alpha(\lambda)$ vs λ is defined as the action spectrum of the phenomenon under investigation. It should be noted that Φ may depend on the electronic transition involved. In this case, no proportionality is expected between action and absorption spectrum. If the photoreceptor pigment is already known, information on Φ can be obtained from the action spectrum.

A variable Φ has been reported for some photochemical reaction, but it has never been documented in the literature on photomovements.

We will now discuss briefly some cases of practical importance. It should be kept in mind that the two experimentally determined quantities are R and N.

Case 1. f is a linear function of $N(\lambda)$; this means that:

$$(2) \quad R\left[N(\lambda)\right] = k \, \alpha(\lambda) \cdot N(\lambda)$$

then:

$$(3) \quad \alpha(\lambda) \sim \frac{R\left[N(\lambda)\right]}{N(\lambda)}$$

The action spectrum is therefore determined but for a constant factor, simply dividing the measured response by the number of photons that cause the response.

Case 2. No linearity. We can adjust the light intensity at two different wavelengths: λ_i, λ_j, so that:

$$(4) \quad R\left[N(\lambda_i)\right] = R\left[N(\lambda_j)\right]$$

from (1) it follows that:

$$(5) \quad f\left[\alpha(\lambda_i) \cdot N(\lambda_i)\right] = f\left[\alpha(\lambda_j) \cdot N(\lambda_j)\right]$$

Now provided that f is strictly monotonous we have:

$$(6) \quad \alpha(\lambda_i) \cdot N(\lambda_i) = \alpha(\lambda_j) \cdot N(\lambda_j) = K$$

This means that:

$$(7) \quad \alpha(\lambda_i) = \frac{K}{N(\lambda_i)}$$

The action spectrum is obtained by plotting $I/N(\lambda_i)$ vs λ_i for a given response.

The condition of strict monotony for f means in practical terms that in order to avoid misleading results those parts of the intensity-effect curves where the function is almost constant (lag or saturation portions e.g.) should not be chosen to determine the action spectrum. It is therefore advisable to determine the intensity-effect curves at every wavelength tested.

A practical way of checking whether f is the same at every λ is to plot the response vs log $N(\lambda)$. If the curves thus obtained at different wavelengths are parallel, then f is the same at every λ.

An action spectrum measured according to these precautions ought to be proportional to the absorption spectrum of the pigment responsible for the response, provided that only one pigment is involved in the response and that Φ is constant. We will see later on a case where more than one pigment is involved in the measured response.

1.5 Action Spectra in Photomovements

It is not our intention to bore the reader with a long list of all the action spectra reported in the literature. People who are interested can find many good reviews in the literature. We will only discuss briefly the action spectra determined for some eucaryotes, choosing as examples systems with probably different receptor pigments, with particular emphasis on our own work.

The phototactic action spectrum for Chlamydomonas was determined some years ago by Prof. Nultsch (8) and co-workers using population methods. The spectral range investigated was between 350 and 550 nm, longer wavelengths being completely ineffective. Intensity-effect curves were determined and from these curves the authors calculated the energy necessary to give a certain reaction value and took the reciprocal of these values to indicate the relative quantum efficiency.

The action spectrum shows maxima at around 440 and 503 nm. The resolution is unfortunately not very good and according to the authors it is not possible to decide whether carotenoids or flavins

are involved in this photoresponse. However, the peak at 503 nm
could suggest a possible role for carotenoids.

As a general comment on Chlamy, it should be noted that the
sensory transduction chain in this alga has been widely investigated as will be discussed in our next chapter. However, the main
attention of the researchers has been toward the steps following
light reception rather than on light reception itself and in comparison only little work has been done for the identification of
both photoreceptor structure and photoreceptor pigments. It would
be desirable that the attention of the workers pointed to this
direction, since a better knowledge of the photoreceptor system in
Chlamy could be of much help in the elucidation of the primary steps
of its photosensory transduction chain and very likely also of that
of other microorganisms.

A second example of action spectra determination by means of
a population technique is offered by Watanabe and Furuya (9), with
the flagellated alga Cryptomonas. Their apparatus differs from that
used by other authors, the principle is, however, still to determine
variation in cell density under the action of light. Intensity-effect
curves were determined, the responses obtained were usually linear
with the logarithm of the stimulus intensity at least up to 1 W/m^2.
The decrease in response at 570 nm could indicate a possible negative reaction, but unfortunately the authors do not even mention
this fact. The action spectrum was then calculated using equal incident quanta (5 x $10^{13}/cm^2s$). In order to exclude a possible influence of photokinetic effects, the speed of single cells in function
of the wavelength of the exciting light was measured. The authors
do not give any information on the light intensity used in these
experiments, but it is presumable that it was comparable to that
used in determining accumulation rates. No photokinetic effect
could be detected, and therefore the reaction rate differences were
considered to be only due to a true phototactic response. The action
spectrum calculated shows a sharp peak at about 560 nm, with a
shoulder at 490 nm. No effect could be found for wavelengths longer
than 640 nm. This complete ineffectiveness of red light seems to
exclude any possible role for chlorophyll. On the other hand, a
phosphate buffer pigment extraction from the cells showed a single
peak at 565 nm. This suggested that the pigment present was mainly
a phycoerythrin probably with a small quantity of phycocyanin. The
apparent coincidence of the maximum in the action spectrum with the
absorption maximum of phycoerythrin seems to indicate a possible
role of phycoerythrin in the photomotile response of Cryptomonas,

although some other pigment might be involved in the response in
the blue region. Also in the case of Cryptomonas the information
contained in the action spectrum is unfortunately not enough to
characterize unambiguously the photoreceptor pigment(s) of the alga.
Anyhow, assuming that phycoerythrin is the photoreceptor pigment we
are facing once more the case of a photosynthetic pigment which acts
as photoreceptor pigment independently of its photosynthetic role.
That phycoerythrin is a photosynthetic pigment is confirmed by the
high efficiency of O_2 evolution under light at 565 nm. On the other
hand, photosynthesis is not directly involved in the photomotile
response of this alga as shown by experiments with DCMU (10), that
blocks completely light induced O_2 evolution but has no effect on
the photomotile response.

A very interesting case is offered by Volvox, a colonial flagellate. Its phototactic orientation is supposed (11) to be the result
of a periodic shading of the photoreceptor caused by the rotation
of the colony during movement. If this holds true the phototactic
action spectrum should represent a superposition of the absorption
spectrum of the photoreceptor pigments and those of the shading
system. Now the question arises as to whether it is possible to
separate the contribution of the different pigments in the photoresponse of the colonies, in order to identify those pigments and
to confirm or disprove the shading hypothesis. Now, whenever facing
a situation like this, one must find some trick in order to eliminate the contribution either of the shading system or of the
photoreceptor itself, being thus left with a "pure" spectrum,
hopefully identifiable. There is always this problem when more
than one pigment can be present in a photoreceptor structure, and
we will find it again for instance in Euglena. Well, coming back
to Volvox, there is a very interesting characteristic of this
microorganism: it does not swim in a straight line, but in a circular path bending toward the left for an external observer look-
ing from above. When given a directional stimulus, Volvox swims
in a straight path toward the source but its swimming direction
does not coincide with that of the stimulus and there is a certain
deviation to the left of the vector joining organism and source
when viewed from above (deviation angle). This is because of the
combination of both the swimming characteristics of the cells and
the trend to go toward the light source. In a very nice and elegant
piece of work, Schletz (12) has been able to utilize all these
facts in order to solve the problem. First of all, it could be shown
that the deviation angle defined above did not depend on steady-

state light intensity but depended only on the efficiency of the
shading system, provided that photokinetic effects could be
excluded, as was the case. In other words, higher shading corresponds to smaller angles; so the dependence of the deviation angles
on the wavelength was determined; from it, an action spectrum of
the shading system was calculated. It has a maximum between 420
and 450 nm without a fine structure. The stigma of Volvox was the
most likely candidate for such a shading action, and therefore its
absorption spectrum was determined by means of a microspectrophotometer. The result was not very similar to the action spectrum obtained, and it was thought that also the chloroplast could play the
role of shading device. In fact, when the chloroplast absorption
spectrum was added to that of the stigma a curve was obtained,
whose shape was very similar to the action spectrum of the shading
system. The second step was the determination of the spectral sensitivity of the actual photoreceptor. To do this, the author measured the percentage of colonies undergoing a step-up response when
illuminated (stop reaction). The action spectrum was obtained with
an experimental procedure that is a bit complicated. Without going
into too many details: the points at the different wavelengths were
determined on a relative scale comparing the effect of each wavelength with that at 482 nm. The reason for this quite complicated
procedure is not completely clear and at first sight the usual
technique (intensity-effect curves and so on) would appear more
desirable. The main ground for this choice of the author was probably to avoid completely a possible intervention of the stigma using
a frontal illumination on colonies which were oriented phototactically. The resulting action spectrum shows a clear maximum at 480-490
nm, no fine structure and no effect of red light beyond 620 nm. The
last step was the determination of the action spectrum for phototaxis. The method chosen was mainly to calculate the percentage of
colonies orienting their motion toward the light source. Intensity-
effect curves were calculated, a threshold value for light intensity was defined as the amount of light that induces a response in
50% of colonies and the action spectrum obtained plotting 1/S vs
wavelength. Comparing the three action spectra, i.e. for deviation
angle, for phobic response and phototactic orientation, it is quite
clear that the maximum of the phototactic action spectrum coincides
with the maximum of the phobic response, whereas the shoulder at
about 440 nm corresponds to the maximum in the action spectrum of
the shading system. From these results it turns out that the hypothesis that phototaxis in Volvox is mediated via a periodic shading

of the photoreceptor is reasonable. From the spectral characteristics of the action spectrum of the photophobic reaction, that is
no effect of near UV light and position of the maximum at 480-490
nm, it was concluded that the photoreceptor pigment is most likely
a carotenoid or a caroteno-protein, as suggested for Chlamy.

The photobehavior of Euglena has been widely investigated in
the last twenty years, not to mention the pioneering work of
Jennings and Mast at the beginning of the century. In fact, some of
J&M observations are still today matter of debate. Historically, the
first action spectra published in the literature were obtained by
a German group working with Bünning in Tübingen. We will not review
this work here, however, because we think it has the characteristics
of good observational biology, rather than of a quantitative and
reproducible research. In the following years, in the sixties to
be more precise, the problems concerning the photobehavior of
Euglena were handled almost exclusively by Dr. Diehn, to whom we
owe a lot of interesting experiments and stimulating discussions
on that topic. When our group started the investigation of the
photomotile behavior of Euglena there were still many questions
unanswered. The first problem we faced was the effectiveness of red
light in inducing a behavioral response in our cells. The question
was not a trivial one, because in case of a positive answer it could
indicate a role for a red light absorbing pigment, maybe chlorophyll
itself. We faced the problem in three different ways (13): first,
we determined the action spectrum of photoaccumulation in red light
for green cells. No accumulation could be obtained with dark-bleached cells, which showed, however, a "normal" accumulation under
blue light. The action spectrum coincides pretty well with the
absorption spectrum of chlorophyll. Secondly, we showed that the
time course of accumulation paralleled a regreening in dark-grown
cells. Then, we used specific inhibitors of photosynthesis, in
particular of photosynthetic O_2 evolution like CMU and NH_2OH. No
effect of the inhibitors could be found on the blue light induced
accumulation, whereas an almost complete cancellation of the response
to red light occurred.

Our conclusion was,therefore, that red light induced accumulation was not a true photoresponse, but rather a response mediated
via the photosynthetic system; in particular we thought of a form
of chemophobic reaction toward photosynthetically evolved oxygen.
This was at that time a reasonable hypothesis, later, working on
single cells it was possible to show that Euglena actually presents
a chemoresponse toward oxygen. When the question of red light was

clarified, we devoted our attention to the determination of the
action spectrum of the true photoresponse of the cells. At that
time there was a certain discussion among people working in the
field, and within our research group as well, on the possible role
of the stigma in the photomotile response of Euglena. The question
was clearly related to the hypothesis of J&M and it was necessary
to clarify the point in order to be able to decide whether the
action spectrum had to be considered as a composite spectrum as in
the case of Volvox, or not. Moreover, there was nothing like the
leftist deviation of Volvox to help separating the contribution of
the different pigments. Some attempts, however, had been made to
solve the problem. So, for instance, Diehn (14) thought he had
eliminated the shading action of the stigma by using polarized light;
but his argument that was based on the presumed dichroic nature of
the stigma, was not convincing to us. By the way, as mentioned
before, the measurements of Benedetti et al. (15) show that the
stigma in Euglena is not dichroic. Therefore, we decided to investigate further the question by using another approach. If shading
organelles (of course also the choroplast was a possible candidate)
do play a role in the process of photoaccumulation in Euglena, it
should be possible to get different action spectra using cells with
different potentially shading structures. We had then available
several types of Euglena, in particular green cells possessing stigma
and chloroplasts, dark-bleached cells with only the stigma and
streptomycin bleached cells having neither stigma nor chloroplasts.
All these cells kept the PFB, whose structure was similar in all
cases, as shown by fluorescence microscopy. So we determined action
spectra for photoaccumulation for the three types of cells, following
the standard procedure. These action spectra strongly resemble
each other and very interestingly they are similar to the absorption
spectrum of a flavin pigment, in the particular case a flavoprotein,
D-aminoacidoxidase. One thing we should mention; in the case of
streptomycin bleached cells the action spectrum reported refers to
the effectiveness of different wavelengths in causing a deaccumulation of the cells under light, whereas in the other two cases an
accumulation is involved. The reason for this different behavior is
at present not known. It might be that the streptomycin-bleached
mutant has a different threshold for the step-up photophobic response which is thought to be involved in deaccumulation under light.
In fact, single cells of the streptomycin-bleached mutant do not
show any step-down response, whereas they do show the step-up
reaction at every intensity checked.

From our results we concluded therefore that the presence of
screening organelles does not influence the structure of the action
spectrum in Euglena, and that therefore it only reflects the presence of flavins in the photoreceptor structure. However, we were
not yet completely satisfied with our results. In fact, the measurements with the phototaxigraph are in some ways ambiguous, since
many different phenomena can contribute or at least influence the
accumulation in the lighted region and it is not always possible
to keep all the interfering effects under control. Therefore, we
decided that it was worthwhile to determine the action spectrum of
the photophobic step-down response on single cells. The step-down
response was chosen because it is thought to cause accumulation of
cells under light and to be also involved in the actual phototaxis.
Our idea was also to obtain information on the mechanisms of accumulation in the phototaxigraph. The experimental apparatus used
(16) is essentially as already described by Dr. Feinleib.   We
record on videotape the photobehavior of the cells, then the
swimming paths are traced on transparent plastic sheet placed over
the TV monitor while replaying the video-tape frame-by-frame. The
physical parameters that could be taken as a measure of the strength
of the step-down response, are in principle several: for instance,
one can choose the percentage of cells showing the response (that
is what we have done); one could also measure the tumbling or rotation time after light-off and also the number of rotations per unit
time. The reason why we chose the percentace of responding cells
is that it is a clearly defined parameter, undoubtedly related to
the photoresponsiveness of the cells. Other parameters, such as
for instance tumbling time could possibly be only a measure of
the adaptation rate to the new illumination conditions. Well, once
we had chosen what to measure, we followed the standard procedure,
determining first intensity-effect curves at the different wavelengths tested. We found that the experimental points could be
nicely fitted by exponential functions. Plotting the slopes at the
origin of these curves in function of wavelength yielded an action
spectrum very similar to the action spectrum obtained by means of
the phototaxigraph with two major peaks at 370 and 450 nm, this
latter showing a good degree of fine resolution, very likely
ascribable to flavins embedded in a hydrophobic environment or
strongly immobilized. This evidence, together with the results
obtained with the microspectrofluorometer strongly supports the
idea that flavins are the photoreceptor pigments in Euglena. It
should be noted that this conclusion can be drawn only because the

two different and completely independent experimental techniques
point to the same pigment. From just the spectroscopic measurement
one could only conclude that flavins are located in the photoreceptor structure and not of course that they are the photoreceptor
pigments. The similarity of the action spectra from population and
single cell techniques gives further evidence that the absorbing
properties of the stigma do not interfere with the spectral respon-
siveness of the cells and also indicates that a single photomotile
reaction is the main cause of accumulation in the phototaxigraph.
The next step in this direction will be the determination of the
action spectra for the photophobic responses of single cells in
dark- and streptomycin-bleached cells. These latter especially
offer a very interesting and promising system to study, since as
already mentioned, they seem to be deprived of the capability of
undergoing the step-down response, and may therefore give a unique
opportunity of understanding the molecular pathways involved in
this response.

   To conclude, we have seen just a few examples of action spectra
determination in photomotile behavior of microorganisms, with all
the potentialities and limits of this technique to characterize the
photoreceptor pigments. It is apparent that in different microorganisms different pigments are involved in the photomotile responses.
In our few cases, out of four microorganisms, we have at least
three different photopigments, not to mention all the other cases
investigated,such as for instance blue-green algae or Stentor.
In summary, we can say that there are roughly six different types
of receptor pigments in photomotile responses: flavins, carotenoids,
chlorophylls, phycobilins, bacteriorhodopsin, stentorin. How to
interpret this is not clear. Perhaps, it is reasonable to think
that nature has tried to evolve photoreceptor pigments several
times, and this is possibly reflected in the large variety of
pigments responsible for photoresponses in microorganisms.

REFERENCES

1. B. Diehn, & B. Kint, Physiol. Chem. and Physics 2:483 (1970).
2. P. A. Benedetti, & A. Checcucci, Plant Sci. Letters 4:47
      (1975).
3. P. A. Benedetti, & F. Lenci, Photochem. Photobiol. 26:315
      (1977).
4. E. Mikolajczyk, & B. Diehn, Photochem. Photobiol. 22:269
      (1975).

5. R. L. Stavis, Proc.Natl. Acad. Sci. USA 71:1824 (1974).

6. P. S. Song & T. A. Moore, Photochem. Photobiol. 19:435 (1974).

7. L. N. M. Duysens, in "Photobiology of Microorganisms", P. Halldal
   ed., Wiley, London (1970).

8. W. Nultsch, G. Throm, & I. von Rimscha, Arch. Mikrobiol.
   80:351 (1971).

9. M. Watanabe, & M. Furuya, Plant and Cell Physiol. 15:413
   (1974).

10. M. Watanabe, Y. Miyoshi, & M. Furuya, Plant and Cell Physiol.
    17:683 (1976).

11. K. Huth, Z. Pflanzenphysiol. 62:436 (1970).

12. K. Schletz, Z. Pflanzenphysiol. 77:189 (1976).

13. A. Checcucci, G. Colombetti, G. Del Carratore, R. Ferrara, &
    F. Lenci, Photochem. Photobiol. 19:223 (1974).

14. B. Diehn, Biochim. Biophys. Acta 177:136 (1969).

15. P. A. Benedetti, G. Bianchini, A. Checcucci, R. Ferrara, S.
    Grassi, & D. Percival, Arch. Microbiol. 111:73 (1976).

16. C. Barghigiani, G. Colombetti, B. Franchini, & F. Lenci,
    Photochem. Photobiol. 29:1015 (1979).

PRIMARY PHOTOPHYSICAL AND PHOTOCHEMICAL REACTIONS: THEORETICAL BACKGROUND AND GENERAL INTRODUCTION

Pill-Soon Song

Texas Technical University
Department of Chemistry
Lubbock, TX 79409 (U.S.A.)

The photoreceptor bound to a protein or membrane absorbs a quantum of light. The light-energized photoreceptor then relaxes to the ground state via several modes of relaxation processes. Such relaxation modes can be coupled to dynamic as well as static conformational changes of protein and/or membranes to which the photoreceptor is bound. The coupled relaxation may then lead to trigger changes or generation of membrane potentials, induction or structural specificity of the photoreceptor relative to its effector site, and regulation of growth hormones, etc.

We envision several coupling modes between the photoreceptor's electronic relaxation and the macromolecular conformation. Molecular relaxations of the excited photoreceptors include (a) photophysical processes, (b) acid-base equilibria, (c) changes in the electronic structure of the photoreceptors accompanied by dipole and polarizability reorientation, (d) conformational changes of the chromophores and isomerization, (e) photodissociation of the photoreceptors, and (f) photoredox reactions. It is unlikely that photochemical reactions of irreversible nature can be significantly involved in photosensory responses of organisms, except for those cases involving photodynamic action.

In this lecture, these relaxation processes will be discussed in terms of specific photoreceptor molecules such as flavins and carotenoids.

A. INTRODUCTION

One of the most fundamental processes in nature involves interactions between organisms and light ranging in wavelength from 260 nm to 900 nm. Photosensory transduction in many organisms exemplifies such interactions in terms of direct utilization of light energy into biological responses. In particular, blue light is widely perceived by a number of organisms in their photosensory behaviors (e.g., phototaxis and phototropism). The aim of the present paper is to elucidate the excited states of blue light photoreceptors (mainly flavoproteins and carotenoproteins) in terms of their spectroscopy and photoreactivity. Thus, this paper represents an up-dated version of our previous work (1). We also suggest various molecular mechanisms possible with these photoreceptors on the basis of spectroscopic and photochemical properties of their electronic excited states. To do so, we start by reviewing some of the molecular relaxation processes of photoreceptors which may be coupled to the primary trigger system of photosensory responses of organisms toward blue light.

A photoreceptor bound to a protein and/or membrane absorbs a quantum of blue light. The light-energized photoreceptor may then relax to the ground state via several modes of transitions. Such relaxation modes can be coupled to dynamic as well as static conformation changes of protein and/or membranes to which the photoreceptor is bound. The coupled relaxation may then affect the membrane potential, structural specificity of the photoreceptor relative to its effector site, and regulation of growth regulators or hormones, etc. Several modes of coupling between photoreceptors' relaxation processes and the macromolecular conformation can be envisioned in terms of the following schemes.

(a) Photophysical processes. Internal conversion, intersystem crossing and radiative emission from the lowest electronic excited state (scheme 1) are not by themselves useful for effecting conformation changes of protein or membrane. However, heat produced from the radiationless transitions can be used for "local heating" of the photoreceptor proper, depending on the thermal conductivity of the photoreceptor binding environment. In some instances, it is feasible to use the local heating to induce reorientation of the photoreceptor itself. For example, Albrecht (2) has shown theoretically that local heating may cause local orientation of the solute molecule due to thermally induced rotational diffusion. However, there is no evidence for or against the possibility of local heating

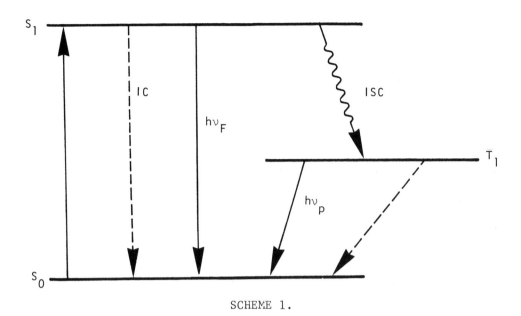

SCHEME 1.

being a factor in the blue light-induced photoresponses (vide infra).

(b) Polarizability. The polarizability, both isotropic and anisotropic, of photoreceptor changes upon excitation by light. Because of its instantaneous evolution and disappearance in concert with the light absorption process, the utility of polarizability change as a driving force for conformational change is probably very limited, although anisotropic polarizability may have a significant effect on the microenvironmental dynamics of the chromophore binding site.

(c) Excited state dipole moment. The permanent dipole moment of a photoreceptor usually changes upon light excitation, entailing reorientation of the photoreceptor and/or surrounding residue dipoles of the protein or membrane.

The change in dipole moment of a photoreceptor upon light excitation entails reorientation of the photoreceptor and/or surrounding residue dipoles of the protein or membrane. From the solvent Stokes' shifts of absorption and fluorescence spectra of flavin, the fluorescent state of riboflavin tetrabutyrate was found to be 1.75 times greater than that of the ground state. Again, this does not seem particularly significant to be used as an efficient mode of the primary photoprocess of the flavin photoreceptor. In fact,

several flavoproteins including D-amino acid oxidase, lipoamide
dehydrogenase and coleoptile plasma membrane preparations showed
no significant degree or rate of relaxation in the nanosecond time-
resolved spectroscopic measurements (3).

(d) Acid-base equilibria. The pK_a of acidic and basic centers
of photoreceptor can change dramatically upon light excitation.
Aside from the obvious implication in the light-induced proton
flux and transmembrane potential associated with such a change,
the pK_a change due to pK_a^* may also induce static and dynamic conformational changes of interacting proteins and/or membranes. In
order to generate a proton gradient, it is assumed that an efficient
proton conducting network is established in association with the
acid-base groups, as found in Halobacterium halobium (4), and as
has been suggested for the photophobic response of Stentor (5).
Phototautomerism involving proton transfer in a photoreceptor
molecule can also be considered in the same context as in the acid-
base equilibria.

Light excitation changes acidity and basicity of photoreceptor chromophores dramatically. Thus, it is conceivable that a transient proton gradient, transmembrane potential and dynamic and
static conformational changes of interacting proteins and/or membrane can be induced by changes in pK_a upon excitation (pK_a^*). The
pK_a^* of riboflavin was estimated to be ca. -4 from changes in the
absorption spectrum for the neutral and protonated species, since
the proton quenching prevented the direct titration using the
fluorescence lifetime method (6). An approximate pK_a^* was calculated
from the lifetime data, however, assuming that the N_1 - protonated
flavin does not fluoresce, yielding pK_a^* of -2. The pK_a^* of N_3 was
determined from lifetime measurements, which yielded a value of
10.52 for the fluorescent state compared to $\sim$ 10 for the ground
state. It seems unlikely that these extreme pK_a^* values would be
particularly effective as a primary photoprocess of the flavin
photoreceptor at physiological pH.

(e) Conformational change and isomerization. Electronic excitation can bring about conformational changes in the photoreceptor
chromophore. Photoisomerization in rhodopsin is well known. These
chromophore changes are often accompanied by conformation changes
of the interacting protein and/or membrane. Photoinduced viscosity
changes of a polymer solution arising from the photoisomerization
of bound azo dyes serve as a good illustration of this effect (7).

(f) Photodissociation. Photodissociation equilibria of molecular complexes are well known. These include not only dissociation

of molecular complexes upon light excitation, but also the formation
of excimers and exciplexes during the excited state lifetime. The
former may temporarily dissociate in the excited state or in the
metastable state, resulting in conformational changes in the protein
or membrane. The reverse is also possible (e.g., binding of phytochrome P_{fr} produced from the excitation of P_r (8), the dissociation
may be brought about after photoreactions, as in the photoisomerization of 11-cis retinal rhodopsin to all-trans retinal and release
of opsin.

(g) Photoredox reactions. It is possible that a photoredox
reaction may have a direct role in producing light-induced membrane
potentials, as found in photosynthetic electron transport.

For the purpose of this chapter, we will use flavin and carotenoids as our illustrative examples to elucidate the role of excited
states in photophysical and photochemical events involved in photobiological responses.

B. BACKGROUND

I. Flavoprotein Photoreceptor: Photophysical Criterion

In contrast to the well characterized photoreceptors such as
chlorophylls, phytochrome, and rhodopsin, the identity of blue-
light photoreceptors is neither certain nor unique in terms of the
two likely chromophore candidates, flavin and carotenoid (9).
However, in many systems such as Neurospora (10), Phycomyces (11,
12), Euglena (13) and others, flavin now seems to be the logical
photoreceptor, based on various types of information including
spectroscopic and photochemical considerations (9). The difficulty
in the isolation of the flavoprotein photoreceptor lies in the
fact that flavin is ubiquitous in living organisms and is the
coenzyme for numerous flavoenzymes, thus making the task of finding
a unique photoreceptor flavin extremely difficult. We report herein
an attempt to find a flavin photoreceptor from corn coleoptiles, on
the basis of photophysical considerations (i.e. fluorescence lifetime).

Coleoptiles have long been known to be the photoactive portion
of plant phototropism, as the coleoptiles with their tips shielded
from light do not phototropically respond to blue light. In Avena
coleoptiles, the phototropic response is maximal for the upper 50
μm and it is ca. 4000 times less sensitive 2 mm further down the
tip (15).

The photoreceptor for phototropic response in <u>Avena</u> and other plants seems to be plasma membrane bound (16). This is suggested by the fact that auxin is laterally translocated across the coleoptile on irradiation with blue light, presumably as a result of light-mediated transport across cell membranes, and that in <u>Phycomyces</u> the photoreceptor is thought to be oriented with respect to the axes of the organism (17).

In spite of the apparent implication for the involvement of a flavoprotein as the blue light photoreceptor in corn and other organisms, no flavoprotein photoreceptors have ever been isolated and characterized. There are two formidable difficulties which cannot be readily overcome; these are, (a) the concentration of flavin photoreceptors is likely to be quite low and (b) once isolated, the assay for the flavin photoreceptor function is not available, unlike other photoreceptor systems such as phytochrome and rhodopsin.

(a) Characterization of a membrane -bound flavoprotein from corn coleoptile. We have studied flavins specifically to ascertain whether or not their excited state dynamics are compatible with the primary photoreactivity required for initiating the phototropic transduction mechanism. We have also examined a preparation from the corn coleoptile tips that is enriched in what is probably plasma membrane (16), and photoresponse with respect to the light-induced reduction of a specific b type cytochrome, which has been proposed to be involved in the blue-light response (16,18).

We have solubilized 21000 and 50000 xg membrane fractions of corn coleoptile with SDS and examined them utilizing SDS gel electrophoresis according to Fairbanks et al. (19). The solubilized membranes (21 and 50 kp) gave essentially the same pattern on 5.6% acrylamide gels with respect to protein band, indicating that they were predominantly of the same plasma membrane composition. Since 50 kp has been shown to contain substantial endoplasmic reticulum (16), the plasma membrane is represented by the solubilized fractions. The SDS gel electrophoresis also showed a yellow band at Rf 0.7 in both 21 and 50 kp fractions. This corresponds to mol wt of 15,000 $\pm$ 2,000 daltons. This flavoprotein is not NADH dehydrogenase. Is it the phototropic receptor flavoprotein or some unknown protein (including proteolytic degradation product)? The only known flavoproteins with mol wt in this region are flavodoxin (ca. 17,000) found in anaerobic bacteria. A proteolytic product can probably be eliminated since the PMSF treatment and all steps in the procedure were carried out at 4°C, but other possibilities must await further work, including elution from the gel for further characterization. However, it has not escaped our attention that "the yellow flavin chromophore on the SDS gel is likely to be covalently bound. This is of considerable interest since other well characterized photo-receptors of sensory transducing organisms possess covalently linked chromophores",presumably to achieve the maximum efficiency and stability in their photochemical and photobiological functions (20). Periodic acid–Schiff reagent staining for carbohydrates of the SDS gel revealed numerous glycoproteins, but whether or not the yellow band is a glycoprotein remains to be established. There was faint staining corresponding to the yellow band, however.

The absorption spectra of 21 and 50 kp showed peaks at 422, 449 and 481 nm, consistent with previously reported absorption spectra (21). These spectra are in reasonable agreement with the light-minus-dark difference spectrum reported by Galston et al. (22) for the analogous preparations from pea epicotyls.

We have recorded the uncorrected fluorescence emission spectrum of 21 kp in a triangular cuvette obtained by front-surface excita-tion and detection on a single photon counting spectrofluorometer (3). Both the fluorescence emission and excitation spectra are attributable to the flavin. The same spectrum was obtained for 50 kp.

Table 1 shows the fluorescence lifetimes of various flavins. Table 3 shows the fluorescence lifetimes for the corn coleoptile preparations. It can be seen that both 21 and 50 kp fractions con-tain fluorescence components of lifetime $\lesssim$ 1 nsec within experi-mental errors, while 9 kp and 50 kp show somewhat longer lifetime $\sim$ 1.4 nsec. All fractions are seen to be heterogeneous, indica-ting presence of more than one fluorescent flavin in these prepa-rations. It should be noted that the lifetimes shown in Table 2 were measured at 30 MHz which is best suited for short lifetime components. Independent measurements on a nanosecond pulse fluoro-meter (box car averager) also suggested that 21 and 50 kp prepara-tions show a fluorescence lifetime < 1 nsec as the major component (Fig. 1). Auxin showed a 30% fluorescence quenching of 50 kp and 21 kp at 10 mM, again indicating that the singlet excited state of the flavin chromophore in the plasma membrane preparations undergoes an efficient reaction with auxin. This quenching is par-ticularly significant, since auxin does not quench the fluorescence of D-amino acid oxidase in a similar concentration range.

(b) Primary photoreactivity of flavins. Table 3 shows quenching data for flavin in the presence of various quenchers. The quenching

T A B L E   1

Fluorescence lifetimes of flavins and flavoenzymes at room temperature measured on the phase-modulation fluorometer (average of 20-30 values).

Flavin	Medium	τ_F, nsec
Riboflavin tetrabutyrate	EtOH	5.65
	CCl_4	6.94
	CCl_4 + 3 mM TCA*	4.27
	CCl_4 + 20 mM TCA	2.55
FMN	H_2O	4.65
	90% Glycerol:H_2O	5.40
	H_2O	2.35
	33% Sucrose:H_2O	3.40
	90% Glycerol:H_2O	5.00
Lipoamide Dehydrogenase	30 mM Pi Buffer, 0.3 mM EDTA, pH 7.2	3.50
D-Aminoacid oxidase	PP Buffer, pH 8.5	2.10[†]
	+ 10 mM Ø-pyruvate	0.44
	+ 2 mM Indole-carboxylic acid	2.65

* Trichloroacetic Acid.
[†] The steady state polarization degree of fluorescence for this enzyme (absorbance at 450 nm = 0.9) was 0.36, indicating that the major fluorescence intensity was due to the bound flavin.

T A B L E   2

Fluorescence lifetimes of various fractions of the corn coleoptile at room temperature and at 30 MHz modulation frequencies. The excitation and emission wavelengths were 441.6 nm and 534.2 nm, respectively (10 nm bandpass). The samples were excited from the front surface of a triangular cell. The protein concentration was ca. 4 mg/ml.

Centrifugal Fraction	Fluorescence lifetime (nsec)[†]			
	By phase shift		By demodulation	
9 kp	1.357	0.425	2.177	0.414
21 kp	1.018	0.349	2.572	0.381
50 kp	0.767	0.262	1.544	0.381
50 ks	1.400	0.164	3.020	0.395

† The phase and modulation modes yield two different sets of lifetimes, indicating the heterogeneity of the flavin fluorescence.

* Dilution of the sample yielded even shorter lifetimes. For example, 10 x dilution of 50 kp gave 0.63 nsec, indicating that the longer component contributed to these lifetimes to some extent.

T A B L E   3

Quenching constants of flavins with various quenchers

Flavin	Quencher	Solvent	K_Q, M^{-1}*	$k_Q, M^{-1} sec^{-1} \times 10^{-9}$
Riboflavin	H$^+$	H$_2$O-Perchlorate	43.4	7.9
			77.4	
Riboflavin	Azide	H$_2$O	24.6	5.1
			35.5	
RFTB	Auxin	Pyridine	24.9	3.6
			46.6*	
Riboflavin	Auxin	H$_2$O	49.9	10.0
			54.7**	

* First value as determined from lifetime data; second values from ϕ_F data.

** Upward deviations due to static quenching.

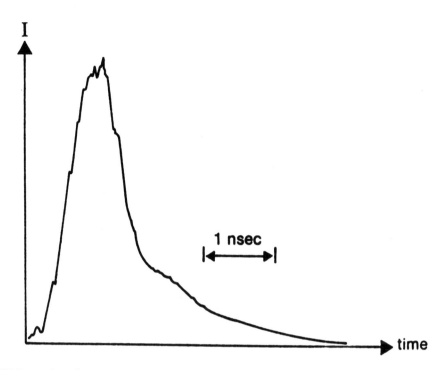

FIGURE 1. The fluorescence decay curve (not deconvoluted) of 21 kp recorded on a Box-car averaging nanosecond time-resolved spectro-fluorometer.

by indole acetic acid is not linear (Fig. 2), indicating that static (complexation) quenching occurs at higher concentrations of auxin. From the lifetime decrease measurements (Table 3 and Fig. 2), it is clear that the dynamic quenching of flavin by auxin is essentially diffusion-controlled. Azide and other quenchers listed in Table 3 also showed dynamic quenching. Tryphtophan showed a slight dynamic quenching, as the static quenching predominated.

In an attempt to learn more about the singlet excited state reactions of flavin in solution and plasma membrane preparations, the photoreaction was followed both aerobically and anaerobically by measuring photobleaching of riboflavin by auxin.

From Stern-Volmer kinetic results, the maximum quantum yield was found to be 0.72, indicating an efficient photoreaction between the excited flavin and auxin. The photoreaction was affected by neither oxygen nor KI which are both excellent quenchers for the

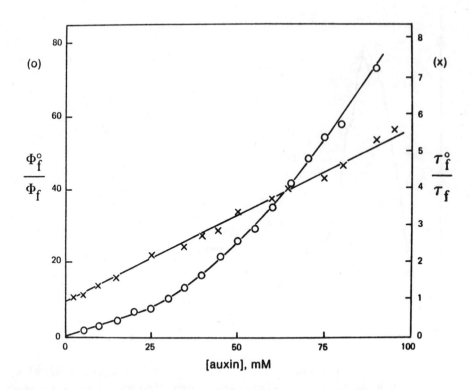

FIGURE 2. The Stern-Volmer plot of the riboflavin fluorescence
(-o-) and lifetime (-x-) in water at room temperature as a function
of the indole acetic acid concentration.

flavin triplet. This suggests that the photoreaction of flavin with
auxin proceeds via the singlet state of the former, consistent with
the quenching data shown in Table 3 and Fig. 2. However, oxygen
clearly participates in secondary reactions, since the tlc autora-
diogram patterns for the aerobic and anaerobic photolysis are different (23).

II. Carotenoprotein Photoreceptor

A number of action spectra for photoresponses of organisms
resemble the absorption spectra of carotenoids. The most recent
identification of the action spectrum in terms of a carotenoid is
for the photoinduced carotenoid biosynthesis in Neurospora crassa
(24). In the case of phototaxis in Euglena (13) and phototropism

in Phycomyces (25,11,12) and Avena coleoptile (26), the photoreceptor
is likely to be a flavoprotein. In our previous analysis (9), it was
concluded that carotenoids were not suited as photoreceptors in
these photoresponses. Among other reasons, the short lifetime due to
efficient internal conversion from the lowest excited singlet state
was regarded as the most critical barrier for a carotenoid or carotenoprotein to overcome as a primary photoreceptor. This viewpoint
still holds, as far as the primary photoreceptor for sensory transduction is concerned. Nonetheless, it is worthwhile to reexamine
carotenoids as a possible photoreceptor in light of recent findings.
First, we summarize results on carotenoproteins which act as the
secondary (or antenna) photoreceptor.

The problem of the short lifetime of the excited state in carotenoids can be partly overcome by resonance (exciton) interactions
among more than one carotenoid molecule and between the carotenoid
and acceptor molecule (primary photoreceptor), as shown recently in
the photoreceptor pigment complex composed of 4 peridinin, 1 chlorophyll a and 1 protein isolated from marine dinoflagellates (27,28).
The molecular topography of the dimeric exciton pairs of peridinin
is shown in Fig. 3. Energy transfer from the exciton state to the
excited state of the primary photoreceptor molecule (i.e., chlorophyll a) can account for the phototactic action spectra of several
dinoflagellates (29). For example, the action spectrum for phototaxis of Gyrodinium dorsum (30) is consistent with the organism's
in situ absorption spectrum which is largely attributable to the
carotenoprotein and chlorophyll a. We have also shown that the
organism photoaccumulates within the lighted area of the optical
cuvette illuminated through the single beam spectrophotometer slit
even after pretreatment of the algae at 77 K for 30 minutes (Song,
unpublished observation). An approximate action spectrum of this
photoaccumulation includes contribution from the chlorophyll absorption band at 660 nm. Thus, accurate action spectra of dinoflagellates containing peridinin-chlorophyll proteins may coincide with
the fluorescence excitation spectra which reveal a high efficiency
(approaching 100%) of energy transfer from the carotenoid to chlorophyll a in vitro (27,28).

There are no definitely established carotenoproteins as the
primary photoreceptors which have been isolated and characterized.
Retinal Schiff's base in rhodopsin and in bacteriorhodopsin acts as
the primary photoreceptor chromophore. However, these are photoreceptors of the short chain polyene. Although it is possible for
the protonated Schiff's base in solution absorbs maximally at 450 nm

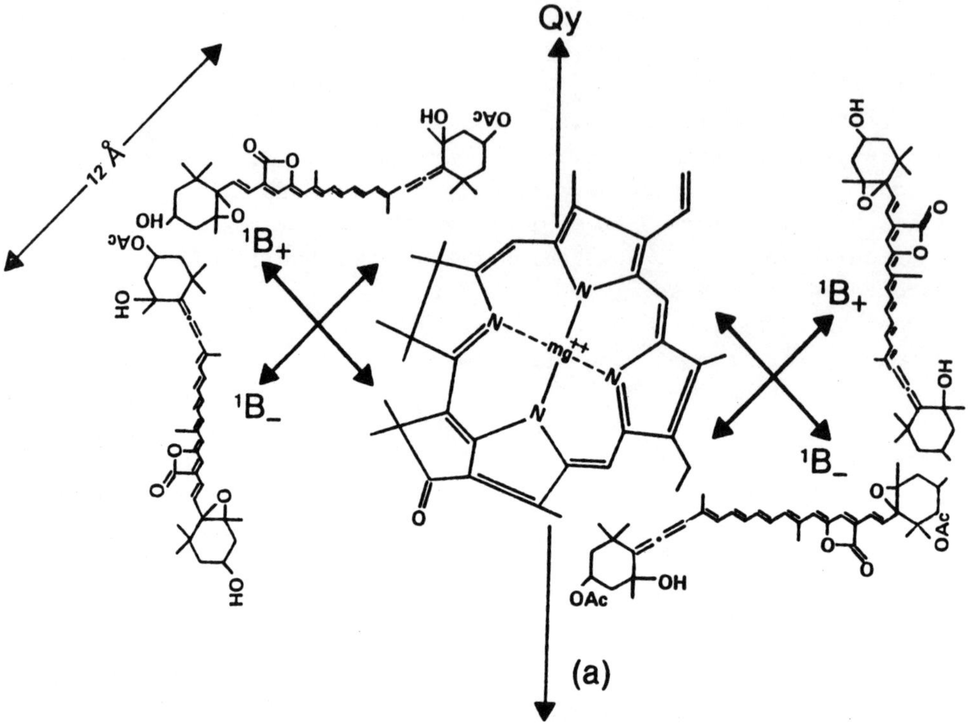

FIGURE 3. The molecular topography of antenna photoreceptor complex, peridinin-chlorophyll a-protein, isolated from marine dinoflagellates. Two peridinin molecules form exciton pairs, resulting in two exciton states, B_+ and B_-.

region, there is no evidence that this type of polyene is actually involved in blue light responses.

The absorption spectra of β-apo-8'-carotenal Schiff's base has the most likely absorption spectrum in fitting several phototropic and phototactic action spectra, with resolution in the blue light region, except for the lack of a strong near UV peak (31). The latter difficulty can be resolved if the carotenoid is isomerized to the cis form. The protonated Schiff's base of apo-carotenal shows its λ_{max} shifted too far to the long wavelength region to be comparible with blue light action spectra. Reticulataxanthin and citranaxanthin Schiff's bases show similar spectral results (31).

Carotenals and their Schiff's bases were found to be nonfluorescent, suggesting that the excited state lifetimes are extremely short as in other carotenoids (31). However, the permanent dipole moment of the Schiff bases of carotenals increases by a factor of 4 upon excitation of the Schiff's bases to their l_B excited state. The pK_a^* of the Schiff's base nitrogen is also predicted to increase upon light excitation.

C. DISCUSSION

1. Photophysical Requirements of Flavins as the Photoreceptor

With the notable exceptions of D-amino acid oxidase and lipoamide dehydrogenase, most flavoproteins are nonfluorescent or only very weakly fluorescent. This is most probably due to strong interactions between FMN or FAD and aromatic residues, particularly trp and tyr. Flavoproteins of this type are not suitable as the photoreceptor, since the excited state is effectively quenched via static and/or dynamic quenching processes. Furthermore, the highly efficient intersystems crossing in free flavins is almost completely suppressed in flavoproteins (32).

From the brief consideration given above, it is suggested that flavins in the blue light receptors not be bound at a site where the excited state is quenched via static charge transfer interactions. Thus, the flavoprotein photoreceptor is likely to be fairly fluorescent, particularly when the photoreceptor is functionally disrupted (e.g., isolated or frozen). It is anticipated that the photoreceptor in a fully operational form in vivo is nonfluorescent or very weakly fluorescent owing to the efficient primary photoprocesses responsible for triggering blue light responses. In this connection, a covalently bound flavin is of special interest (vide supra).

2. What is the Most Likeky Mode of the Primary Photoprocess in the Blue Light Receptor?

From the fact that the plasma membrane bound flavoprotein (of 21 and 50 kp) possess a short fluorescence lifetime compared to other fluorescent flavins and flavoproteins, we come to the conclusion that the photochemical role of flavin in triggering the phototropic event is probably photoreduction (3).

Flavin is readily photoreduced by auxin (Galston, 1974; vide supra). The action spectrum for the photooxidation of auxin in an

etiolated pea homogenate is consistent with flavin as the photoreceptor (33).

Auxin is apparently laterally translocated in coleoptiles from
the lighted to the shaded side without significant photooxidation of
auxin (26,34,35). Previously, we proposed a photochemical scheme
whereby auxin bound to a macromolecule is released for translocation induced by the photoexcitation of the flavin photoreceptor (1).
This model accounts for the lateral auxin transport in coleoptiles
without photodecomposition. We now attempt to further refine this
model by considering specific photoprocesses which may be responsible for the phototropic transduction of blue light energy.

Let us assume that auxin is either bound to a protein (receptor)
or membraneous compartment analogous to the acetylcholine receptor
sites on synaptic membrane. This assumption is experimentally supported (36,37). To trigger release of the bound auxin, the excited
flavin photoreceptor changes the conformation of the auxin-bound
protein and/or membrane, resulting in the release of the bound
auxin in a nonstoichiometric ratio (i.e., the number of auxin
released is much greater than the number of photoreceptor excited,
and the photosignal is thus amplified). This scheme is analogous
to the rhodopsin-induced release of the visual excitation transmitter (Ca^{2+}) from the rod outer segment discs (38).

The auxin release can be triggered by the photoreduction of
the flavin photoreceptor itself by a hydrogen donor. In this respect,
it is interesting that indole acetic acid itself is an excellent
donor. Furthermore, the flavin photoreduction by auxin is nearly
oxygen-independent (23), in contrast to other hydrogen donors
examined. The photoreaction by either singlet or triplet flavin
is, therefore, faster than or comparable in its rate to the quenching of the singlet or triplet flavin by oxygen. For these reasons,
the auxin release from the membrane or membraneous compartment may
be triggered by the photoreduction of flavin. This mechanism does
not result in significant photodecomposition of auxin, since the
local concentration of flavin photoreceptor can be much lower than
that of auxin at the primary photoreaction site where auxin is
accumulated. This speculative model also accounts for the necessary
amplification factor to effect the phototropic curvature, as discussed previously (26,1).

It is noteworthy that phototropic auxin transport in coleoptiles
is oxygen-dependent (26,36,39,40). This could be accounted for by
the reoxidation of the photoreduced flavin photoreceptor by oxygen,
which could also give rise to the post auxin transverse electrical

potential.

Similar photochemical mechanism(s) are possible for other blue light responsive organisms (fungi and algae, etc), as long as a good hydrogen donor is available for the primary photoreaction, i.e., photoreduction of flavin photoreceptor, for phototropism and phototaxis.

3. Can the Plasma Membrane-bound Flavoprotein from Coleoptiles be the Phototropic receptor?

Most flavoproteins are nonfluorescent or only very weakly fluorescent due to static quenching at their binding sites. Thus, their fluorescence lifetimes are not necessarily short compared to those of free flavins. In fact, even weakly fluorescent D-amino acid oxidase, lipoamide dehydrogenase and several other flavoproteins show fluorescence lifetimes considerably longer than those expected from the dynamic quenching and from the relationship between the fluorescence lifetime and quantum yield ($\emptyset_f = \tau_f/\tau_f^o$, where τ_f^o is the radiative lifetime). In view of this observation, it is safe to conclude that the plasma membrane bound flavin with $\tau_f < 1$ nsec represents a unique flavoprotein, which may well be the phototropic receptor for corn. It is clear that the flavin in these preparations is subject to an efficient photoprocess due to its binding environment. Thus, the fluorescence lifetime of plasma membrane-bound flavin is given by $\tau_f = [k_f + k_r + k_{IC} + k_{ISC}]^{-1} < 1$ nsec where k_r represents the primary photoreaction via the singlet excited state, and k_f, k_{IC} and k_{ISC} are rate constants for fluorescence, internal conversion and intersystem crossing, respectively. Such a photoprocess is likely to be a reaction with an endogenously bound donor in order to shorten the fluorescence lifetime substantially. The identity of the reacting donor is not known, but several model reactions (Table 3) indicate that the singlet excited state of flavin can undergo an efficient photoreaction[†]. The short fluorescence lifetime of the plasma membrane bound flavin reflects an efficient dynamic photoreaction of the flavin in its binding environment. The plant growth hormone, auxin, involved in the phototropically coupled lateral transport in vivo serves as an excellent donor model.

[†] Xe did not specifically inhibit phototropic responses of corn, suggesting that the triplet flavin is probably not involved (R. Vierstra, K. Poff, and P.S. Song, unpublished results).

In this connection, it is interesting to note that the plasma
membrane bound flavin is photochemically reactive, as shown previously (3). On the other hand, there was no comparable quenching
of D-amino acid oxidase by auxin. The nature of the fast photoprocess in the singlet excited state of flavins and plasma membrane-
bound flavoprotein is yet to be elucidated mechanicistically.
It is possible that an encounter complex between auxin and the
excited flavin (exciplex) contributes to the quenching, followed
by nucleophilic addition of the former to the N_5 or C_{4a} position
of the excited flavin. The primary photoreaction product may well
be of a covalent type described by Hemmerich et al. (41), thus
reducing the flavin.

Although 21 and 50 kp preparations contained b-type cytochrome,
it is not possible to ascertain whether the photoreduced flavin
produced above is the electron donor to cytochrome. However, it is
noteworthy that some type of efficient photoreduction of the flavin is the most likely mode of the primary photoprocess, because
other modes of electronic relaxation processes were found to be
rather sluggish making it difficult to account for the efficiency
of phototropism in general and the short fluorescence lifetime of
the plasma membrane bound flavin in particular.

4. Light Harvesting by Carotenoproteins as the Blue Light Photore-
ceptor.

To overcome the most critical kinetic difficulty due to the
short lifetime of the excited state of carotenoid, we have shown
that the excited state of peridinin can be stabilized by dimeric
resonance interaction (exciton) with another molecule of peridinin
(vide supra). This is illustrated in Scheme 2.

Further resonance stabilization is possible by exciton interactions among degenerate or near degenerate excited states of peridinin
and chlorophyll a. It was shown by the picosecond spectroscopy that
an efficient energy transfer occurs from the resonance stabilized
dimeric state of peridinin to chl a, while the transfer is halted
when the dimeric arrangement is destroyed (28). The rate constant
for the transfer was found to be $k_t > 10^{12}$ l mol^{-1} sec^{-1}.

Many blue light response action spectra of photosynthetic algae
and dinoflagellates can be explained in terms of energy transfer
from the secondary (peridinin) to the primary (chl a) photoreceptors
(Scheme 2). For those action spectra lacking red sensitivity, it
is necessary to assume that only a fraction of or topographically

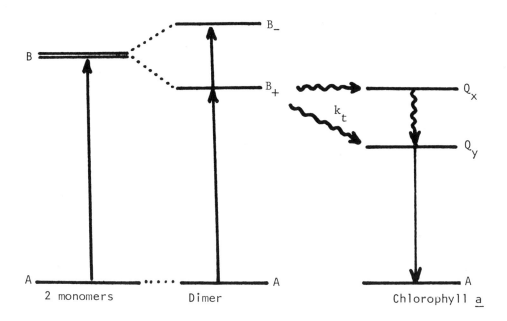

SCHEME 2. Stabilization of the B state of peridinin by exciton interactions between two peridinin molecules. k_t; energy transfer rate constant. See Fig. 3 for the dimeric arrangement.

unique photosynthetic apparatus (the primary photoreceptor) is utilized for the sensory transduction. To test the validity of this assumption, dye laser excitation can be used for recording weak action spectra from 550-700 nm, eliminating any contamination of blue light in the action spectral measurement.

b. Can a Carotenoid Function as the Primary Photoreceptor?

The answer to this question can be affirmative only if the excited states can be stabilized to the extent that the excitation energy can be utilized directly (rather than transferring to another primary photoreceptor)' for the primary photoprocess of sensory transduction. One possibility invokes the role of carotenal Schiff's base which functions as a proton uptaker (20). This can be made feasible if the excited state of Schiff's base can be resonance stabilized by exciton coupling with other Schiff's base molecules.

The theory of local heating and rotational reorientation of molecules around the excited molecule which locally releases its energy as heat has been worked out (2). It may be possible that carotenoids absorbing blue light dissipate the energy as heat which activates sensory response systems such as carotenogenesis. In this case, the short excited state lifetime is an advantage as the blue light photoreceptor. However, this conjecture remains highly speculative at the present, since action spectrum based on the local heating would have higher sensitivity in the near UV and UV regions than the blue maximum region.

D. CONCLUDING REMARKS

In this report, we presented arguments that fluorescence life-time can be taken as a reflection of the flavin photoreactivity in the characterization of flavoprotein photoreceptor. We have also shown that the carotenoid excited state can be resonance stabilized to enhance its probability of transferring the excitation energy to primary photoreceptor. Whether carotenoid can function as the primary photoreceptor remains an open question.

ACKNOWLEDGEMENTS

This work was supported by the R. A. Welch Foundation (D-182) and NSF (PCM 7906806). Technical assistance of Drs. R. D. Fugate and P. Koka is greatly appreciated.

REFERENCES

1. P. S. Song, T. A. Moore, & M. Sun, in "The Chemistry of
     Plant Pigments", C. O. Chichester, ed., Academic Press,
     New York, (1972).
2. A. C. Albrecht, J. Chem. Phys. 1413:1414 (1958).
3. P. S. Song, R. D. Fugate, & W. R. Briggs, in "Flavins and Flavo-
     proteins", K. Yagi, ed., Academic Center, Tokyo (1979)
     (in press).
4. A. Danon, & W. Stockenius, Proc. Natl. Acad. Sci. USA 71:1234
     (1979).
5. E. B. Walker, T. Y. Lee, & P. S. Song, Biochim. Biophys. Acta
     587:129 (1979).
6. N. Lasser, & J. Feitelson, J. Phys. Chem. 77:1101 (1973).
7. R. Lovrien, Proc. Natl. Acad. Sci. USA 57:236 (1967).
8. P. S. Song, Q. Chae, & J. Gardner, Biochim. Biophys. Acta
     568:479 (1979).

9. P. S. Song, & T. A. Moore, Photochem. Photobiol. 19:435 (1974).

10. V. Munoz, & W. L. Butler, Plant Physiol. 55:421 (1975).

11. M. Delbrück, A. Katzir, & D. Presti, Proc. Natl. Acad. Sci. USA 73:1969 (1976).

12. D. Presti, W. J. Hsu, & M. Delbrück, Photochem. Photobiol. 26:403 (1977).

13. B. Diehn, & B. Kint, Physiol. Chem. Phys. 2:483 (1970).

14. P. S. Song, J. Agr. Chem. Soc. Korea, C. Y. Lee 60th Birthday Commemorative Issue, 20:10 (1977).

15. W. R. Briggs, Ann. Rev. Plant Physiol. 14:311 (1963).

16. A. C. Brain, J. A. Freeburg, C. F. Weiss, W. R. Briggs, Plant Physiol. 59:948 (1977).

17. A. J. Jesaitis, J. Gen. Physiol. 63:1 (1974).

18. W. Schmidt, & W. L. Butler, Photochem. Photobiol. 24:71 (1976).

19. G. Fairbanks, T. L. Steck, & D. F. H. Wallach, Biochemistry 10:2606 (1971).

20. P. S. Song, J. Agr. Chem. Soc. Korea 20:10 (1977).

21. W. R. Briggs, Carnegie Inst. Yearbook 74:807 (1974).

22. A. W. Galston, S. J. Britz, W. R. Briggs, Carnegie Inst. Yearbook 76:293 (1977).

23. W. E. Kurtin, Dissertation, Texas Tech. University, Lubbock (1969).

24. E. C. De Fabo, R. W. Harding, & W. Shropshire,Jr., Plant Physiol. 57:440 (1976).

25. M. Delbrück, & W. Shropshire,Jr., Plant Physiol.35:194 (1960).

26. K. V. Thimann, & G. M. Curry, in "Light and Life", W. D. McElroy, ed., Johns Hopkins University Press, Baltimore (1961).

27. P. S. Song, P. Koka, B. Prezelin, & F. T. Haxo, Biochemistry 15:4422 (1976).

28. P. Koka, & P. S. Song, Biochim. Biophys. Acta 495:220 (1977).

29. W. Haupt, in"Proc.Biophys. of Photoreceptors and Photobehavior of Microorganisms", G. Colombetti,ed.,Lito Felici,Pisa (1975).

30. R. B. Forward, Planta (Berl) 111:167 (1973).

31. T. Moore, & P. S. Song, J. Mol. Spectrosc. 52:209 (1974).

32. D. B. McCormick, Photochem. Photobiol. 26:169 (1977).

33. A. W. Galston, & R. S. Baker, Am. J. Botany 36:773 (1949).

34. W. R. Briggs, R. D. Tocher, & J. F. Wilson, Science 125:210 (1957).

35. B. G. Pickard, & K. V. Thimann, Plant Physiol.39:341 (1964).

36. M. H. M. Goldsmith, & K. V. Thimann, Plant Physiol. 37:492 (1962).

37. R. Hertel, K. S. Thomson, & V. Russo, Planta 107:325 (1972).

38. W. H. Hagins, Ann. Rev. Biophys.Bioeng. 1:131 (1972).

39. K. V. Thimann, <u>Comprehensive Biochemistry</u> 27:1 (and references therein) (1967).
40. H. von Guttenberg, <u>Planta</u> 53:412 (1959).
41. P. Hemmerich, V. Massey, & G. Weber, <u>Nature</u> 213:728 (1967).

MOLECULAR ASPECTS OF PHOTORECEPTOR FUNCTION: CAROTENOIDS AND RHODOPSINS

René V. Bensasson

Laboratoire de Biophysique, Museum National d'Histoire Naturelle, Paris (France)

ABSTRACT

Among the various biological pigments implicated in photo-reception of solar energy are carotenoids and caroteno-proteins.

The different photochemical properties of the carotenoids, electronically excited in vitro, vary with the number of conjugated double bonds present in the molecule. They can be related to the established or hypothetical function of carotenoids in vivo. These photophysical and photochemical properties are altered when the polyenes are bound to a protein. Two pigments where a C_{20} polyene (retinal) is bound to a protein are the focal point of recent studies rhodopsin, the visual pigment, and bacteriorhodopsin allowing Halobacterium halobium to use light energy to drive its metabolic processes and its movements. The spectral transformations of those two pigments after illumination will be described as well as their functions in the membranes where they are inserted.

INTRODUCTION

Carotenoids are the most important group of natural pigments, they are found in vegetables, flowers, fruits, birds, insects, fishes.....They are brightly coloured and many are used as yellow and red colorants by the food industry. Their chromophore is mainly or entirely a chain of conjugated double bonds. The majority of carotenoids are derived from the acyclic C_{40} H_{56} structure re-

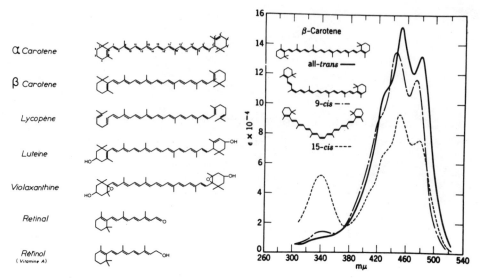

FIGURE 1. Formula of some   FIGURE 2.   Spectra of isomeric
carotenoids   βcarotenes

sulting from the joining together of **eight** isoprenes (1,3 butadiene,
2-methyl) units in the normal head to tail manner, except at the
center where the order is reversed. Fig.1 shows the formula of some
biological important C_{40} and C_{20} carotenoids. The hydrogenation,
dehydrogenation, cyclisation, oxydation (xantophylls) degradation
of this C_{40} skeleton leads to more than 300 natural carotenoids,
Bacteria, algae, and the higher plants are able to synthetize
their carotenoids but animals extract them from their food, mainly
to convert them into C_{20} carotenoids: retinol, the vitamin A necessary for their growth and retinal, the chromophore of their
visual pigment.

The first part of this paper will be devoted to the photophysical and photochemical properties of C_{40} carotenoids in vitro
and to their established or hypothetical role in vivo, in photosynthesis, phototropism and photomovement.....

The second part will deal with the photochemical properties
of rhodopsin, bacteriorhodopsin and of their free chromophore
retinals.

I. CAROTENOIDS C_{40} AND ANALOGS

A) Molecular Properties in Vitro

1. Singlet ground state absorption spectra. The colour of carotenoids is associated with a high intensity band at 25000 to 20000 cm^{-1} (400 to 500 nm) (1,2) usually broken into three sub-maxima. Fig.2 shows the three submaxima found in all trans βcarotene and its 9-cis and 15-cis isomers in hexane. It also shows that change of configuration from trans to cis produces a reduction in extinction coefficient of the main absorption bands and develops or intensifies an additional band in the UV region between 320 and 380 nm, referred to as the "cis" peak. Similar changes are observed between cis and trans polyenes of shorter length as the retinals. An increase in the length of the conjugated system always produce a bathochromic shift with decreasing effects as the number of conjugated double bonds (n) increases. The asymptotic value of the absorption maximum approaches 18000 cm^{-1} (550) and simple π electron theory is able to account for this behaviour(3-4). Shifts of absorptions due to substituents and solvents on the main absorption maxima are well known (1,2). A particularly dramatic blue shift of the ground state absorption to a maximum around 390 nm occurs in certain carotenoproteins due to a formation of aggregates (5-7). A similar shift to the blue is also observed after addition of water to organic solutions of certain carotenoids also due to a stacking of the molecules (8,9).

2. Properties of the first excited singlet state - fluorescence. Weak fluorescence of carotenoids has been reported (10,15), but it is now assigned to impurities and carotenoids are now considered as non fluorescent (16-18), with $(\phi_F) < 5 \times 10^{-4}$ according to Tric and Lejeune (17) and $\phi_F < 10^{-5}$ and $\tau_F = 10^{-14}$ s according to Song and Moore (9).

3. Excited triplet state.

   a) ε_T, ϕ_T and λ_{max} of $T_n \leftarrow T_1$ transition

Triplet excited states of carotenoids are known from triplet energy transfer studies (18-23) or by ion recombination generated in pulse radiolysis (24). Their extinction coefficient is rather high, it reaches for instance $\varepsilon_T = 2.4 \times 10^5$ M^{-1} in hexane at 515 nm for βcarotene (25). However the singlet → triplet intersystem crossing

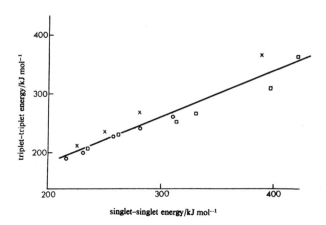

FIGURE 3. Plot of the energies of the $T_n \leftarrow T$ absorption against
the energies of the $S_n \leftarrow S_o$ absorption for series of polyenes

quantum yields are very low $\phi_T < 10^{-3}$ (25). One of the main remarks
on triplet absorption spectra of polyenes is that a regular displacement of the $T_n \leftarrow T_1$ maxima follows the variation of the low energy
absorption of the ground state. The slope and intercept of those
linear relationships are similar for three groups of polyenes (Fig.3):
(i) βcarotene and analog polyenes of different length C_{30}, C_{35},
C_{40}, C_{50} and C_{60} carotenoids (26).   (o) in Fig.3.
(ii) carotenoids from photosynthetic bacteria,15,15'-cis Phytoene,all-
trans phytoene, all-trans ζcarotene, all-trans spheroidene and all-
trans spirilloxanthine (27).   (x) in Fig.3.
(iii) carbonyl containing polyenes from βionone to torularodinal-
drehyde (28).   (□) in Fig.3.
        These slopes and intercept are explained by Hückel theory in
its most simplified form (29).

b) Triplet energy level E_T

An inverse relationship between polyene chain length and the triplet
energy $T_1 \leftarrow S_o$ has been established by a theoretical calculation
based upon the Hückel method (30). Singlet - triplet absorption
bands $T_1 \leftarrow S_o$ induced by oxygen at high pressure have been observed
by Evans (31,32) for ethylene, butadiene, hexatriene and octatetraene.
Using these data Mathis found indeed a straight line by plotting
$1/E_T$ as a funtion of n number of conjugated double bands (26).

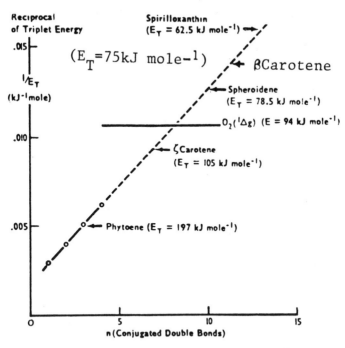

FIGURE 4. Energies of polyenes lowest triplet estimated by extrapolation of Evans data

By extrapolating this straight line, Mathis was able to estimate E_T for βcarotene consistent with its reaction towards $^1O_2$ or chlorophyll lowest triplet. A similar plot allowed an estimation of the triplet energies of bacterial carotenoids (27), phytoene, ζcarotene, spheroidene, spirilloxanthine with respective $E_T \simeq$ 197, 105, 78.5, 62.5 k J mol^{-1} (Fig.4).

   c) Triplet lifetimes

Rate constant for βcarotene triplet decay is 1.7×10^5 s^{-1} (25,26). Plotted on a logarithmic scale, rate constant of triplet decay for carotenoids increase linearly when the energy E_T of the triplet state decrease (26). Flexibility and twisting of these molecules provides an important pathway of the radiationless deactivation of their triplets. This property is well illustrated by Liu and Butt (33) observation that triplet lifetime is greater in a rigid model than in a flexible model of conjugated trienes.

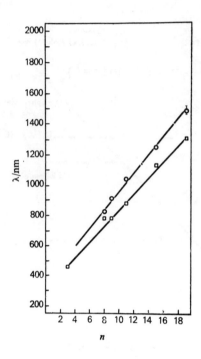

FIGURE 5. λ_{max} of radical anions ($\square$) and radical cations
(o) absorption against the number of conjugated double bands

    4. Anions and cations. The absorption spectra of the radical
anions and cations of nine carotenoid pigments have been determined by pulse radiolysis (34). The cations (o) absorb at a slightly
longer wavelengths than the anions ($\square$) (Fig.5) while Hückel energy
levels predict an identical absorption for both radical ions.
Experimental data show a good linear dependence of λ_{max} on the
number of conjugated double bonds, however the Hückel calculations
predict $\lambda_{max} \sim n^2$ as $n \to \infty$.
    5. Photoisomerization. There are many examples of cis-trans
and trans-cis photoisomerization of ethylenes (stilbenes) dienes
and various polyenes related to retinals (35). But there are no
examples of photoisomerizations of C_{40} carotenoids (17) by direct
excitation.
In conclusion these C_{40} molecules do not fluoresce, they do not
intersystem cross to the triplet manifold by direct excitation,
and they do not photoisomerize. The electronic excited states are

deactivated only into vibrational and thermal energies. How can such molecules with no aptitude for photochemistry play any role in photosynthesis ?

B) Carotenoids in Photosynthesis

We shall now see how by singlet energy transfer, triplet energy transfer and possibly by electron transfer carotenoids are indeed able to accomplish major functions.

1. Singlet energy transfer and light harvesting function. Blinks (36) has reviewed the work performed since 1884 showing the ability of carotenoid pigments to act as accessory light harvesting pigments. This function is due to their ability to transfer singlet energy to chlorophyll and is well shown in vitro and in vivo (37, 40).

The action spectrum of excitation of bacterio-chlorophyll fluorescence in whole cells of Rps. sphaeroides (39) corresponds clearly to the carotenoid absorption spectrum in the visible within the light harvesting antenna. More recently this singlet energy transfer reaction

$$^1CAR^* + BCHL^+ \rightarrow {}^1BCHL^* + CAR$$

$$^1BCHL^* \rightarrow {}^1BCHL + h\nu$$

has also been shown in the reaction centres of Rps. sphaeroides (41). In reconstitution studies with R. rubrum reaction centres the efficiency of energy transfer from the carotenoid to the bacterio-chlorophyll has been estimated: 90 % for spheroidene, 35 % for spheroidenone, 30 % for chloroxanthine and 20 % for spirilloxantine (42).

In the photosynthetic light harvesting complex carotenoid (peridinin) - chlorophyll a-protein complex isolated from dinoflagellates, the energy absorbed by the carotenoid is transferred to chlorophyll with 100 % efficiency (43).

2. Triplet energy transfer and photoprotection. The high carotenoid protective effect against photosensitized oxydation called photodynamic action has been well established (in vitro and in vivo) and reviewed (44).

This photoprotection may occur via the energy transfer mechanism (45):

$$^1O_2 + {}^1CAR \rightarrow {}^3CAR + {}^3O_2$$

but certainly also via the so called "triplet valve" (46):

$$^3CHL + {}^1CAR \rightarrow {}^3CAR + {}^1CHL$$

It is related to n the number of conjugated double bonds (47,48). Thus carotenoids with n > 9 and triplet energy level below that of $^1O_2$ (E $\sim$ 94 k J mol^{-1}) and of the chlorophyll triplet CHL (E $\sim$ 120 k J mol^{-1}) protect efficiently against photodynamic action.

3. Electron transfer. Hypothesis that carotenoids might play a role in electron transfer through a biological membrane has been often reported (49,50) but no experimental evidence has ever been presented.

However efficient electron transfer reactions have been recently observed in vitro between chlorophylls a and b, and carotenoids of varying number of conjugated double bonds (57):

$$CAR^{\bullet +} + CHL \rightarrow CAR + CHL^{\bullet +}$$

$$CAR^{\bullet -} + CHL \rightarrow CAR + CHL^{\bullet -}$$

Positive and negative charge transfer occur with similar rate constants (2 to 12×10^9 dm^3 mol^{-1} s^{-1}) between CHL a and carotenoid ions. But with chl b the positive charge transfer only occurs with the shorter polyenes (septapreno βcarotene and 7,7 dihydroβcarotene) while the negative charge occurs with k between 2 and 8×10^9 dm^3 mol^{-1} s^{-1}. These results of Lafferty et al. (51) support a recent speculation (45) about the CHL a fluorescence quenching by carotene (52) via

$$^1CHL\ a^* + CAR \rightarrow CHL\ a^{\bullet -} + CAR^{\bullet *}$$

which might be followed by

$$CAR^{\bullet +} + CHL\ a \rightarrow CAR + CHL\ a^{\bullet +}$$

such electron processes would produce a charge separated pair,

CHL $a^{\bullet +}$ + CHL $a^{\bullet -}$, which might be involved in the primary process
of PS II. The role of carotenoids as electron carriers is also supported by studies of Mangel et al. (53,54) who found that chlorophyll acting as photosensitizer and βcarotene acting as electron
carrier are both essential components for the photoinduced charge
transport across bilayer lipid membranes. However these results are
in conflict with a photoreduction, irrespective of βcarotene presence,
observed for Copper (II) in liposomes containing chlorophyll a (55).

C) "Blue-light" Effects – C_{40} Carotenoids in Phototropism and Photomotion

The bending of higher plants under light illumination (or phototropism) and the photomotion of microorganisms are processes dependent upon illumination with "blue-light". From the action spectrum of these processes, carotenoids have been suggested as the
photoreceptors. However a resemblance between the absorption spectrum of carotenoids and these action spectra is not sufficient to
assign a role to those pigments. Moreover their photophysical and
photochemical properties do not make them satisfactory candidates
(18,56,57). Nevertheless proteins linked to carotenoids might modify
those photochemical properties as we shall see later in the case of
the C_{20} polyenes retinals and in some cases carotenoproteins,(C_{20} or
C_{40}) might be candidates in the photomotion of microorganisms (57).

II. C_{20} POLYENES AND PROTEINS WHICH BIND C_{20} POLYENES: RHODOPSINS (R) and BACTERIORHODOPSINS (BR)

Until a decade ago rhodopsin visual pigments of vertebrates
and invertebrates were the only proteins known to incorporate a
retinyl C_{20} polyene as a prosthetic chromophore. All consist of an
11-cis, 12 s trans retinal (58,59) covalently bound to a protein
(opsin) through a protonated Schiff base linkage. In the last decade
four other types of protein – polyene complex have been observed:
the yellow vitamin A transporting complex called retinol binding
protein (RBP) (60), the retinochrom found in various kinds of
cephalopod retina absorbing at 490-522 nm and containing the all
trans isomer of the retinyl polyene (61) and two pigments found in
Halobacterium halobium, bacteriorhodopsin (BR) absorbing at 570 nm
(62) and PS 370 absorbing at 370 nm (63). BR converts light into
chemical energy stored in the form of ATP via a light-driven proton
pump and both BR and PS 370 are involved in the light induced motor

response of the microorganism. We shall only discuss the most studied of these proteins: rhodopsins (R) and bacteriorhodopsin (BR).

Photophysical and photochemical properties of (R) and (BR)
were thought to be linked to the properties of the protein-free
form retinal and to the corresponding Schiff bases (RSB). This
stimulated a great number of theoretical, spectroscopic and photochemical investigations on these compounds and on related molecules,
such as retinol, polyenones, polyenals ... We shall review only
some of these studies dealing with retinal isomers and their corresponding Schiff bases (RSB) which are more strictly related to
R and BR.

A) $\underline{C_{20}\ Polyenals}$

In contrast with the C_{40} polyenes, the C_{20} polyenes and their
corresponding RSB have singlet excited states of longer lifetime
able to fluoresce under certain conditions, furthermore they photoisomerize. But these properties are greatly altered as we shall
see later in paragraph B when retinal is linked to a protein.

1. Ground state absorption spectrum. Three bands, a main one
around 365–380 nm and two weaker at 280 nm and 250 nm, characterize
the spectrum of retinals (Fig.6 and 7).
The major band corresponds to an allowed $\pi^* \leftarrow \pi$ transition ($^1Bu^+$).
This intense band is characteristic of all polyenes. It hides a
forbidden π, π^* state, the $^1A_g^-$ state and an n, π^* state (65).

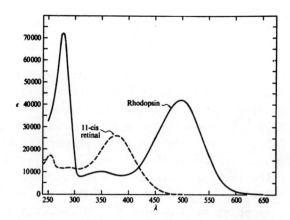

FIGURE 6. Spectra of 11 cis retinal (in hexane at room temperature)
and of rhodopsin in digitonin solution in aqueous glycerol at 4 K).

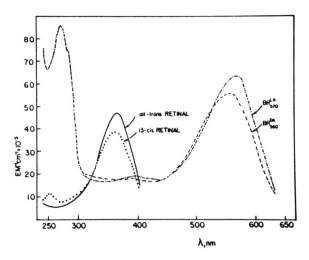

FIGURE 7. Spectra of all trans and 13 cis retinal in hexane at room temperature and of light adapted B_{570}^{LA} and dark adapted B_{560}^{DA} bacteriorhodopsin for aqueous membrane suspension at room temperature

The relative position of the three lowest states $^1B_u^+$, $^1A_g^-$ and $n\pi^*$ plays a major role in the fluorescence, the intersystem crossing and the photoisomerization of retinals isomers (65). Schiff base formation produces a blue shift (approximatively 20 nm) in the position of the main band of retinal isomers absorption. However this band is drastically shifted to the red (440 nm in hydrocarbon solvents) upon protonation and becomes very sensitive to solvent (64).

2. Fluorescence and intersystem crossing. We have reported in Tables I and II some of the data known on the photophysical properties of retinals. Fluorescence of retinal isomers is extremely low at room temperature in alkanes, as reported in Table I, but fluorescence does occur in hydroxylic solvents (EPA, 77K) with $\phi_f \simeq 8 \times 10^{-3}$. This was interpreted as evidence that the $^1n\pi^*$ state is the lowest in dry alkane solvents and that a $^1\pi\pi^*$ ($^1A_g^-$) is the lowest in H bonding environment. This interpretation is consistent also with the relatively high ϕ_T in hexane due to a lowest $^1n\pi^*$ interacting with a $^3\pi\pi$ state. In methanol only inefficient spin orbit coupling will take place between the $^1\pi\pi^*$ which is now the lowest singlet and the underlying $^3\pi\pi^*$ (66). ϕ_T (hexane) /ϕ_T(metha-

nol) varies in the range between 2 (for 13-cis retinal) and 4,5
(for the 11-cis monomer (Table I)).

As retinals,the unprotonated or protonated Schiff bases are
essentially not fluorescent at 295 K (71).

3. Photoisomerization. The main result of light absorption by
a visual pigment is the photochemical isomerization of the chromophore from an 11-cis to an all trans conformation, therefore photoisomerization of model compounds (retinal isomers and Schiff
bases) is of great interest. Table II and Table III report cis-
trans isomerization of retinal isomers and of Schiff bases by direct
excitation and via triplet sensitization. These photoisomerizations
are dependent on wavelength, solvent and the energy of the sensitizer but in all cases the primary products correspond to a one
photon one carbon-carbon double bond isomerization. Direct photoisomerization should involve an excited singlet state or a non re-
laxed vibronic triplet (66).

B) Rhodopsin and Bacteriorhodopsin

1. Absorption spectra. One of the striking changes when retinal
is linked to a protein is the large bathochromic shift of the main
retinal absorption band from about 380 nm in solution to 500 nm in
rhodopsin, 560 or 570 in the dark or light - adapted bacteriorhodopsin (Fig.6 and 7). The generally accepted model for explaining
this spectral shift is that the Schiff base is protonated. This is
mainly based on the observation that protonated alkyl Schiff bases
of retinals absorb well to the red of unprotonated Schiff bases and
on resonance Raman spectroscopy (73-76).

Specific electrostatic interactions between the chromophore
and charges in its protein environment should induce a bathochromic
shift varying with each pigment (77-80). Furthermore an important
charge migration upon vertical excitation of PRSB has been predicted theoretically (77,81-84) but the different calculations do
not agree on the situation in the orthogonal configuration of the
molecule. The theoretical predictions of an important polarization
of the retinyl polyene in its excited state have been confirmed by
an analysis of electrical field effects on the spectrum of 11-cis,
all trans retinal and on the corresponding Schiff bases (85). These
predictions are also consistent with the photochemical behaviour
of retinol (ROH) and retinyl acetate (RAc) in polar solvents (86)
where $^1ROH^* \rightarrow R^+ + OH-$, $^1RAc^* \rightarrow R^+ + Ac^-$.

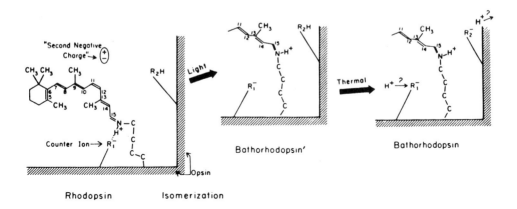

FIGURE 8. The primary events in rhodopsin (ref.93)

2. The primary photochemical event. After illumination rhodopsin transforms itself to batho (prelumi rhodopsin) and then undergoes a series of non photochemical biochemical changes. This first step is the topic of considerable controversy. It has been assumed origi- nally that the primary step is an isomerization of the chromophore from its 11-cis to an all trans conformation (87). A number of re- cent models (88-92) suggests that light initiates a proton transfer and that cis-trans isomerization is a later ground state process. However the simplest and best model in our opinion, providing ex- planations for most of the experimental data dealing with the light energy transduction in visual pigments and bacteriorhodopsin, is the recent model of Honig and al. (93).

This model supports that the cis-trans isomerization is the primary event in vision. Its main argument is that a single common intermediate bathorhodopsin is formed photochemically at room temperature, in the picosecond range from an 11-cis retinal pigment (rhodopsin), a 9-cis retinal pigment (isorhodopsin) and recently also from a 7-cis retinal artificial pigment (T. Yoshizawa un- published).

The model assumes that a negative counter ion presumably the carboxylate group of an aspartic or glutamic acid will be located in front of the Schiff base nitrogen and will form a salt bridge (Fig.8). This interaction with a counter ion accounts for thermal stability, energy storage and spectral shift of the pigment after

isomerization. The primary photochemical step proposed is a cis-
trans isomerization about the 11-12 double bond in visual pigments
and a trans-cis isomerization about the 13-14 double bond in bacteriorhodopsin which disrupts the salt bridge and separates charges
in the protein. This step takes place in less than 6 ps and leads
to the formation of a ground state species "prebatho" or "batho".
Then a deuterium and temperature dependent process takes place and
involves a ground state relaxation of the protein. It may take from
less than 6 ps to 36 ps depending on the temperature and leads to
bathorhodopsin. Raman studies (74-76) have shown that bathorhodopsin
is still a protonated Schiff base, but in the vicinity of the protonated nitrogen, the pK's will be altered and a molecule R_2H might
release a proton (Fig.8). A similar reaction accounts also for the
spectroscopic and proton pumping properties of bacteriorhodopsin
either for the all-trans to 13-cis isomerization in light adapted
BR or for the 13-cis to all trans isomerization in dark adapted
BR.

   3. The dark stages.

   a) The bleaching intermediates of R.

Photoisomerization initiates in rhodopsin a series of dark changes
leading to the formation of all trans retinal separated from the
opsin in vertebrates while it is not detached in invertebrates:

Rhodopsin (λ_{max} = 498 nm) $\xrightarrow{\text{6 - 36 ps}}$ Bathorhodopsin (543 nm)

$\xrightarrow{\text{50 to 170 ns}}$ Lumirhodopsin (497 nm) $\xrightarrow{2 \times 10^{-5} \text{ s}}$ Meta I (480 nm)

$\xrightarrow{10^{-2} \text{s} \div 10^{-3} \text{ s}}$ Meta II (380 nm) $\xrightarrow{10^{3} \text{ s}}$ Meta III (465nm)

$\xrightarrow{10^{5} \text{ s}}$ all trans retinal (380 nm).

This sequence of intermediates has been established (94) in the
photolysis of cattle rhodopsin after illumination at 77 K (95) and
also studied by conventional flash and laser spectrophotometry at
both low (96) and physiological (97-99) temperatures. It is called
bleaching as the final product does not absorb in the visible.
The quantum yield for bleaching in solution is quite high $\sim$ 0.67

(100) much higher than the quantum yield of isomerization for retinals and RSB (Table II and III). This fact shows the important role
of the protein in directing the pathway of isomerization of the
bound chromophore.

b) The photocycles of BR.

In the cases of BR two photocycles have been observed (101). The
main intermediates of these cycles are:
in the case of the light adapted BR_t:

$$BR_t(570) \xrightarrow{h\nu} I-1ps \rightarrow J(625) -10ps \rightarrow K(610) -2\ \mu s \rightarrow L(550)$$

$$-50\ s \rightarrow M(410) \xrightarrow{\rightarrow} N(520) -4\ \underset{\leftarrow}{ms} \rightarrow 0(640) -8\ ms \rightarrow BR_t(570)$$

The time values in the arrows refer to room temperature lifetimes
and wavelength of maximum absorption in nm are indicated in brackets.
In the case of the dark adapted BR:

$$BR\ 13\ cis \longrightarrow K^c\ (610\ nm) \longrightarrow BR_t \text{ with an additional}$$
link between the two BR

$$BR\ 13\ cis \longrightarrow K^c\ (610\ nm) \longrightarrow L^c\ (610) \text{ which can either}$$
regenerate B 13 cis or convert to BR_t.

It seems that a deprotonation occurs in an intermediate between
L 550 and M 410.

c) Molecular aspects of the dark stages.

The intermediates of rhodopsin bleaching are not well defined. The
main hypotheses focussed on conformational changes of the opsin
following photoisomerization. These changes would open ionic pores
in the discal membrane through which Ca^{+2} ions will diffuse and
block pores which permit the passage of Na^+ ions between the extracellular and cytoplasmic spaces. The changes responsible for
this release of the calcium ions is probably related to the MI $\rightarrow$
MII transformation. This modification of the membrane triggers a
signal through the neural network to the brain. Although there is
a decline in favour of the calcium hypotheses of visual transduction,
it is not yet abandoned (102).

The intermediates of the photo cycles of BR are not any better
understood than for rhodopsin. They have been recently linked to
photovoltaic effects which arise from charge displacements in the
bacteriorhodopsin molecules and their photointermediates (103).
Such electrochemical events might control the flagellar response
and the light-driven proton pump of the bacteria.

4. Conclusion. The molecular aspects of the retinal-protein
function which underlie vision or H^+ pumping and light induced
motor responses in Halobacterium are not yet elucidated. Nevertheless great progress has been achieved in undestanding the primary
photochemical step. The recent model presented by Honig et al.
(93) shows in retinal-protein pigments how light might be converted
into chemical energy by a charge separation achieved by a photoisomerization.

GENERAL CONCLUSION

Two remarks can conclude this brief survey:

1. When excited electronically in solution C_{40} carotenoids
and analogs are deactivated mainly via extremely rapid radiationless
relaxation in the singlet manifold. Nevertheless they play a role
of direct photoreceptor in vivo as light harvesting pigment in photosynthetic systems.

2. Photophysical and photochemical properties of retinals,
which are C_{20} carotenoids, are substantially altered when bound
to a protein. Such retinal-protein pigments are photoreceptors responsible for vision in vertebrates and invertebrates and have been
recently found to be responsible for photophosphorylation and photomovement in Halobacterium halobium. These recent findings should
motivate investigations in other microorganisms for carotenoproteins
as the yellow photoreceptors involved in their photobehaviour.

ACKNOWLEDGEMENT

The author is grateful to Prof. M.Ottolenghi for fruitful
discussions and for providing a preprint of reference 101.

TABLE I   FLUORESCENCE AND TRIPLET QUANTUM YIELDS FOR RETINAL ISOMERS

	Temp K	$\phi_F^{(a)}$		$\tau_F^{(a)}$	$\phi_T^{(b)}$			
SOLVENT →		3MP	EPA	3MP	HEXANE		METHANOL	
λ_{exc} (nm) →		380	380	380	265	353	265	353
RETINAL								
All trans	295	10^{-5}			0.40	0.43	0.08	0.12
	77	10^{-4}	8×10^{-3}	1.0				
9 cis	295	10^{-4}			0.70	0.61	0.07	0.20
	77	5×10^{-4}		1.6				
11 cis	295	5×10^{-5}			0.13	0.51	0.06	0.11
	77	5×10^{-4}		2.0				
13 cis	295	10^{-4}			0.18	0.39	0.10	0.19
	77	10^{-4}		2.5				

(a) ref 65

(b) ref 66

228   R. V. BENSASSON

TABLE II   QUANTUM YIELDS OF ISOMERIZATION[c] via direct excitation ϕ_{ISO}^D
and via triplet energy transfer ϕ_{ISO}^T for retinal isomers

	SOLVENT	ϕ_{ISO}^D	(PRODUCTS)^D	ϕ_{ISO}^T	(PRODUCTS)^T
All trans	hydrocarbon	0.08	13 cis : 9 cis 4 : 1		
	methanol	0.06	13 cis:11cis:9cis:7cis 60 : 30 : 10 : 1		
	acetonitrile			0,03	13 cis
9 cis	hydrocarbon	0,18	trans : 9,13 dicis 2 : 1		
	methanol	0.04	trans : 9,13 dicis 4 : 1		
	acetonitrile			0,20	trans: 9 cis,13 cis 5 : 1
11 cis	hydrocarbon	0,24	trans : 11 cis,13 cis 5 : 1	0,16-1(d)	
	methanol	0,04	trans		
	acetonitrile			0,17	trans
13 cis	hydrocarbon	0,21	trans		
	methanol	0,05	trans		
	acetonitrile			0,15	trans

(c) ref 67   (d) ref 72 values vary monotonically with the energy of the sensitizer

TABLE III

ISOMERIZATION QUANTUM YIELDS

of unprotonated and protonated n-Butylamine Retinal Schiff Bases

SOLVENT	ϕ^D_{ISO}	Ref	ϕ^T_{ISO}	Ref
All trans				
hydrocarbon	2×10^{-3}	68	0.03	68
ethanol (H^+)	$5 \times 10^{-4} - 0,27$*	70	< 0.05	69
9 cis				
hydrocarbon			0.06	68
ethanol (H^+)			0.5	69
11 cis				
hydrocarbon	4×10^{-3}	60	0.45	68
ethanol (H^+)	$5 \times 10^{-3} - 0.25$*	70	1.0	69
13 cis				
hydrocarbon			0.08	68
ethanol (H^+)			0.2	69

* values depend on the excitation wavelength

REFERENCES

1. L. Zeichmeister, in "Cis-trans Isomeric. Carotenoids, Vitamins A & Arylpolyenes", Springer, Vienna (1962).
2. W. Vetter, G. Englert, N. Rigassi & U. Schwietter, in "Carotenoids", O. Isler, ed., Birkhauser Verlag (1971).
3. H. Kuhn, J. Chem. Phys. 17:1198 (1949).
4. J. R. Platt, J. Chem. Phys. 25:80 (1956).
5. M. Buckwald & W. P. Jencks, Biochemistry 7:834 (1958).
6. V. R. Salares, N. M. Young, P. R. Carey and H. J. Bernstein, J. Raman Spectroscopy 6:282 (1977).
7. M. L. Mackentum, R. D. Tom & T. A. Moore, Nature 279:265 (1979).
8. A. Hager, Planta 91:38 (1970).
9. P. S. Song & T. A. Moore, Photochem. Photobiol. 19:435 (1974).
10. G. Klein and H. Linser, Osterr. Bot. Z. 79:125 (1930).
11. D. I. Sapozhnikov,The Physico-chemical Basis for the Evolution of a Phototropic Type of Feeding. Thesis. Leningrad (1955).
12. G. N. Lyalin, G. I. Kobyshev & A. N. Terenin, Doklady Akad. Nauk, SSSR 150:407 (1963).
13. F. T. Wolff & M. V. Stevens, Photochem. Photobiol. 6:597 (1967).
14. R. J. Cherry, D. Chapman & J. Langelaar, Trans. Faraday Soc. 64:2304 (1968).
15. J. Langelaar, R. P. H. Rettschnick, A. M. F. Lambooy & G. T. Hoytinck, Chem. Phys. Letters 1:609 (1968).
16. E. I. Rabinovitch, Photosynthesis 2:798 (1951).
17. C. Tric & V. Lejeune, Photochem. Photobiol. 12:339 (1970).
18. P. S. Song & T. A. Moore, Photochem. Photobiol. 19:435 (1974).
19. M. Chessin, R. Lingstone & T. G. Truscott, Trans. Faraday Soc. 62:1519 (1966).
20. P. Mathis, Photochem. Photobiol. 9:55 (1969).
21. C. S. Foote & R. W. Denny, J.A.C.S. 90:6233 (1968).
22. A. Sykes & T. G. Truscott, Chem. Commun. D274 (1969).
23. C. Wolff & H. T. Witt, Z. Naturforsch. 24b:1031 (1969).
24. E. J. Land, A. Sykes & T. G. Truscott, Photochem. Photobiol. 13:311 (1971).
25. R. Bensasson, E. A. Dawe, D. A. Long & E. J. Land, J. Chem. Soc. Faraday Trans. I, 73:1319 (1977).
26. P. Mathis & J. Kleo, Photochem. Photobiol. 18:343 (1973).
27. R. Bensasson, E. J. Land & B. Maudinas, Photochem. Photobiol. 23:189 (1976).

28. R. S. Becker, R. V. Bensasson, J. Lafferty, T.G. Truscott & E. J. Land, J. Chem. Soc. Faraday Trans. II, 74:2246 (1978).

29. O. L. J. Gijzeman & A. Sykes, Photochem. Photobiol. 18:339 (1973).

30. N. C. Baird (personal communication).

31. D. F. Evans, J. Chem. Soc. 1735 (1960).

32. D. F. Evans, J. Chem. Soc. 2566 (1961).

33. R. S. H. Liu & Y. Butt, J. Chem. Soc. Chem. Comm. 799 (1973).

34. J. Lafferty, A. C. Roach, R. S. Sinclair & . T. G. Truscott, J. Chem. Soc. Faraday I, 73:416 (1977).

35. N. J. Turro, in "Modern Molecular Photochemistry", the Benjamin/cummings Publishing Company (1978).

36. L. R. Blinks, in "Photophysiology", A. C. Giese, ed., 1:199 Academic Press (1964).

37. H. J. Dutton, W. M. Manning & D. M. Duggar, J. Phys. Chem. 47:308 (1943).

38. F. W. J. Teale, Nature 181:415 (1958).

39. J. C. Goedher, Biochim. Biophys. Acta 35:1 (1959).

40. F. T. Haxo, H. H. Kylia, H. W. Siegelman   G. F. Somers, Abstr. of Com. Intern. Symp. on Carotenoids, Cluj, Romania (1972).

41. R.J. Gogdell, W. W. Parsons & M. A. Kerr, Biochim. Biophys. Acta 430:83 (1976).

42. F. Boucher, M. Van Der Rest & G. Gingras, Biochim. Biophys. Acta 461:339 (1977).

43. P. S. Song, P. Koda, B. B. Prezelin & F. T. Haxo, Biochemistry 15:4422 (1976).

44. N. I. Krinsky, in "Carotenoids", O. Isler, ed., Birkhauser Verlag (1971).

45. C. S. Foote & R. W. Denny, J. Am. Chem. Soc. 90:6233 (1968).

46. K. Witt & Ch. Wolff, Z. Naturforsch. 25b:387 (1970).

47. H. Claes & T. O. Nakayama, Z. Naturforsch. 14b:746 (1959).

48. M. M. Mathews-Roth, T. Wilson, E. Fujimori & N. I. Krinsky, Photochem. Photobiol. 19:217 (1974).

49. J. T. Dingle & J. A. Lucy, Biol. Rev. (1965).

50. B. Pullman & A. Pullman, Quantum Chemistry,Interscience,New York (1963).

51. J. Lafferty, T. G. Truscott & E. J. Land, J. Chem. Soc. Faraday I 74:2760 (1978).

52. C. S. Beddard, R. S. Davidson & K. R. Trethewey, Nature 267:373 (1977).

53. M. Mangel, D. S. Berns & A. Ilani, J. Membrane Biol. 20:171 (1975).

54. M. Mangel, Biochim. Biophys. Acta 430:459 (1976).
55. K. Kurihara, M. Sukigara & Y. Toyoshima, Biochim. Biophys. Acta 547:117 (1979).
56. P. S. Song, T. A. Moore & M. Sun, in "Chemistry of Plant Pigments", C. O. Chichester, ed., Academic Press (1972).
57. R. Bensasson, in "Research in Photobiology", A. Castellani, ed., Plenum Pub. Corp., London (1977).
58. T. Ebrey, R. Govindjee, B. Honig, W. Pollak, W. Chan, R. Crouch, A. Yudd & K. Nakanishi, Biochemistry 14:3933 (1975).
59. R. H. Callender, A. Doukas, R. Crouch & K. Nakanishi, Biochemistry 15:1621 (1976).
60. Kanai, Raz & Goodman, J. Clin. Invest. 47:2025 (1968).
61. T. Hara & R. Hara, Nature 206:1331 (1965).
62. D. Oesterhelt & W. Stoeckenius, Nature New Biol. 233:149 (1971).
63. N. A. Dencher, in "Energetics and Structure of Halophilic Microorganisms", S. R. Caplan & M. Ginsburg, eds., Elsevier/North Holland (1978).
64. P. E. Blatz, Photochem. Photobiol. 15:1 (1972).
65. R. S. Becker, G. Hug, P. K. Das, A. M. Schaffer, T. Takemura, N. Yamamoto & W. Waddell, J. Phys. Chem. 80:2265 (1976).
66. R. V. Bensasson & E. J. Land, Nouv. J. Chimie 2:503 (1978).
67a. W. H. Waddell, R. Crouch, K. Nakanishi & N. J. Turro, J. Am. Chem. Soc. 98:4189 (1976).
67b. W. H. Waddell & D. L. Hopkins, J. Am. Chem. Soc. 99:6457 (1977).
68. T. Rosenfeld, A. Alchalel & M. Ottolenghi, Photochem. Photobiol. 20:121 (1974).
69. A. Alchalel, B. Honig, M. Ottolenghi & T. Rosenfeld, J. Am. Chem. Soc. 97:2161 (1975).
70. T. Rosenfeld, B. Honig, M. Ottolenghi, J. Hurley & T. G. Ebrey, Pure Appl. Chem. 49:341 (1977).
71. W. H. Waddell, A. M. Schaffer & R. S. Becker, J. Am. Chem. Soc. 99: 8456 (1977).
72. T. Rosenfeld, O. Kalisky & M. Ottolenghi, J. Phys. Chem. 81:1496 (1976).
73. A. Lewis, R. Fager & E. Abrahamson, J. Raman Spect. 1:465 (1973).
74. A. Oseroff & R. Callender, Biochemistry N. Y. 13:4243 (1974).
75. A. Lewis, J. Spoonhower, R. A. Bogomolni, R. H. Lozier W. Stockenius Proc. Nat. Acad. Sci. USA 71:4462 (1974).

76. G. Eyring & R. Mathies, Proc. Nat. Acad. Sci USA 76:33 (1979).

77. H. Suzuki, T. Komatsu & H. Kitajima, J. Phys. Soc. Jpn. 37: 177 (1974).

78. M. Mautione & B. Pullman, Int. J. Quant. Chem. 5:349 (1971).

79. A. Kropf & R. Hubbard, Ann. N.Y. Acad. Sci. 74:266 (1958).

80. T. G. E. Brey & B. Honig, Quart. Rev. Biophys. 8:129 (1975).

81. B. Honig, A. D. Greenberg, U. Dinur & T. G. Ebrey, Biochemistry 15:4593 (1976).

82. A. M. Schaffer, T. Yamaoka & R. S. Becher, Photochem. Photobiol. 21:297 (1975).

83. L. Salem & P. Bruckmann, Nature 258:526 (1975).

84. L. Salem, Accounts of Chem. Res. 12:87 (1978).

85. R. Mathies & L. Stryer, Proc. Nat. Acad. Sci. USA 73:2169 (1976).

86. T. Rosenfeld, A. Alchalel & M. Ottolenghi, in "Excited States of Biological Molecules", J. B. Birks, ed., Wiley, London, (1976).

87. T. Yoshizawa & G. Wald, Nature 197:1279 (1963).

88. K. Peters, M. Applebury & P. Rentzefis, Proc. Nat. Acad. Sci. USA 74:3119 (1977).

89. K. Van Der Meer, J. C. C. Mulder & J. Lugtenberg, Photochem. Photobiol. 24:363 (1976).

90. A. Lewis, Proc. Nat. Acad. Sci. USA 75:549 (1978).

91. A. Warshel, Proc. Nat. Acad. Sci. USA 75:2558 (1978).

92. J. Favrot, J. M. Leclerq, R. Roberge, C. Sandorfy & D. Vocelle, Photochem. Photobiol. 29:99 (1979).

93. B. Honig, T. Ebrey, R. H. Callender, U. Dinur & M. Ottolenghi, Proc. Nat. Acad. Sci. USA 76:2503 (1979).

94. S. E. Ostroy, Biochim. Biophys. Acta 463:91 (1977).

95. T. Yoshizawa & G. Wald, Nature 197:1279 (1963).

96. K. H. Grellmann, R. Livingston & D. C. Pratt, Photochem. Photobiol. 3:121 (1964).

97. C. R. Goldschmidt, T. Rosenfeld & M. Ottolenghi, Nature 263: 169 (1976).

98. R. A. Cone, Nature (New Biol.) 236:39 (1972).

99. R. Bensasson, E. J. Land & T. G. Truscott, Photochem. Photobiol. 26:601 (1977).

100. H. Dartnall, in "Handbook of Sensory Physiology", W. R. Lowen-Lowenstein, ed., Springer Verlag, Berlin (1972).

101. M. Ottolenghi, in "Advances in Photochemistry", Vol. 12
       (in press) (1980).
102. P. Fatt, Nature 280:355 (1979).
103. S. B. Hwang, J. I. Korenbrot & W. Stoeckenius, Biophys.
       Biochim. Acta 509:300 (1979).

MOLECULAR ASPECTS OF PHOTORECEPTOR FUNCTION: PHYTOCHROME

Pill-Soon Song

Chemistry Department, Texas Teach University

Lubbock, Texas 79409 (U.S.A.)

ABSTRACT

The major factor in the molecular affinity of the P_{fr} form of phytochrome for its "receptor" site is the availability of a large, hydrophobic area on the apoprotein which becomes available as a result of the $P_r \rightarrow P_{fr}$ transformation by red light (660 nm).

INTRODUCTION

Although phytochrome controls photomorphogenic responses in higher plants, its involvement in the photosensory behaviors of motile algae is an open question. However, several findings suggest that phytochrome is present in certain species of algae (1) and that the photoresponses of Gyrodinium dorsum (phototactic) (2) and Mougeotia (chloroplast photo-orientation) (3) are indeed regulated by phytochrome. In this lecture, we summarize results which lead to a working model for the molecular basis of the physiologically active form of P_{fr}, relative to the inactive P_r form:

$$P_r \underset{\text{in dark}}{\overset{660 \text{ nm}}{\rightleftharpoons}} P_{fr} \overset{x}{\longrightarrow} P_{fr} \cdot X \longrightarrow \longrightarrow \text{Photoresponses}$$

For more details, the readers may refer to our original publication (4).

235

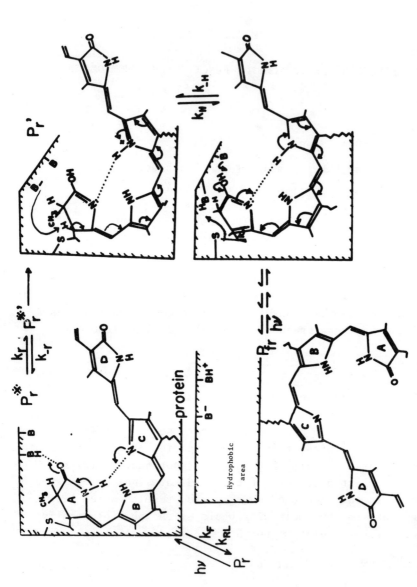

FIGURE 1. A possible mechanism of the phototransformation of phytochrome. Asterisk represents the excited state. The apoprotein is represented by enclosing lines. The thioether linkage is at C-3¹ position (10). Similar mechanisms can be written for other structures having the thio-ether linkage at C-3(4). Evidence supports the thioether linkage at C-3¹ (10).

SUMMARY OF RESULTS

    (i) The P_r and P_{fr} chromophores resume essentially identical
conformation as shown in Fig. 1. This conclusion is based on the
fact that the oscillator strength ratios $f_{Qx,y}/f_{Bx,y}$, are approximately unity for both P_r and P_{fr} form. The red shift of λ_{max} for
the latter is interpreted in terms of the phototautomeric shift of
a ring A proton, as illustrated in Fig. 1 (4).

    (ii) The P_r chromophore is more tightly bound to the apoprotein
than the P_{fr} chromophore, as a result of the phototransformation
which involves release of one of the covalent linkages (i.e., -SH
group; Fig. 1). This conclusion is based on the fact that the
P_r form shows much stronger induced optical activity (CD) than P_{fr}.
The CD intensity of P_r is virtually quenched upon denaturation of
the protein, suggesting that the induced optical activity due to
the P_r chromophore-protein interaction is stronger than the intrinsic optical activity of the chromophore (i.e., chirality at
ring A and helicity of the chromophore, if any) (4).

    (iii) The P_r chromophore binding site includes one or more Trp
residues (4), making the binding site hydrophobic in nature. The
Trp residue(s) is either lost or its juxtaposition with respect to
the chromophore is removed upon proteolytic digestion of large P_r
(120,000 daltons) to small P_r (60,000 daltons). These conclusions
are based on the fact that an efficient energy transfer from Trp
to the P_r chromophore occurs only with large P_r, but not with
small P_r (4).

    (iv) There is no change in the protein conformation in going
from P_r to P_{fr} as judged by the CD spectrum at 220 nm (4,5) and
by fluorescence anisotropy of the pyrene-maleimide labeled P_r and
P_{fr} (unpublished).

PROPOSED MODEL

    On the basis of the above results and additional lines of in-
direct evidence reviewed in our original publication (4), we
propose a working model for the physiologically active form of
P_{fr} as shown in Fig. 2.

    The P_r model depicts a tightly bound chromophore at the hydro-
phobic site, consistent with CD results described above (ii, iii).
The P_{fr} form losses one of the two covalent linkages, releasing
-SH groups, upon phototransformation from the P_r form (cf. (i) and
Fig. 1). At the same time, the chromophore of P_{fr} becomes flexible

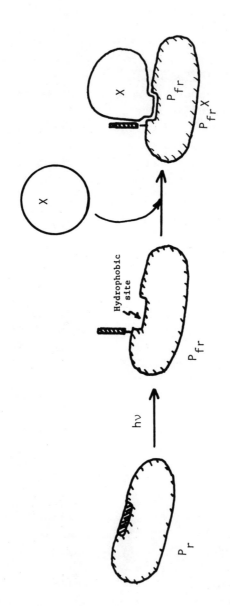

FIGURE 2. Model for Pr and Pfr; X represents Pfr receptor which is protein and/or membrane.

relative to its binding site, thus causing the loss of induced CD which was strong in P_r.

The major consequence of the $P_r \rightarrow P_{fr}$ phototransformation is, therefore, the availability of a large hydrophobic surface area on the protein in P_{fr} form.

The P_{fr} model shown in Fig. 2 can accomodate the following characteristic properties of P_{fr}:

(a) Pelletability of P_{fr} in the presence of cations (6); this may be due to the hydrophobic interaction of P_{fr} in its chromophore-vacated region.

(b) Lack of protein conformation changes in going from P_r to P_{fr} (vide supra).

(c) Association of P_{fr} to membrane fragments and membraneous structures (7).

(d) Release of the bound charged species, acetylcholine, from mung bean root slices upon phototransformation of $P_r \rightarrow P_{fr}$ (8).

(e) P_{fr} is more reactive to metal ions, N-ethylmaleimide, and P-mercuribenzoate than P_r (see references in (4)), probably due to the exposure of the P_{fr} chromophore, and:

(f) P_{fr} is bleached more rapidly than P_r in the presence of 5M urea (9), probably due to the exposure of the P_{fr} chromophore.

In conclusion, we note that the proposed model provides a satisfactory account of the physiological reactivity of P_{fr} in terms of strong hydrophobic interactions between the chromophore-vacated surface on the protein and as yet unidentified phytochrome receptor, probably membrane. Work is now in progress to further examine this model in vitro and in vivo.

ACKNOWLEDGEMENTS

This work was supported by the Robert A. Welch Foundation (D-182) and the National Science Foundation (PCM 7906806).

REFERENCES

1. H. H. van der Velde, & A. M. Hemrika-Wagner, Plant Sci. Lett. 11:145 (1978).

2. R. Forward, Planta (Berl.) 111:167 (1973).

3. W. Haupt (this volume).

4. P. S. Song, Q. Chae, & J. Gardner, Biochim. Biophys. Acta 576:479 (1979).

5. E. Tobin, & W. R. Briggs, Photochem. Photobiol. 8:487 (1973).

6. P. H. Quail, Photochem. Photobiol. 27:147 (1978).

7. L. H. Pratt, Photochem. Photobiol. 27:81 (1978).

8. M. Jaffe, Plant Physiol. 46:768 (1970).

9. W. L. Butler, H. W. Siegelman, & C. O. Miller, Biochemistry
   3:851 (1964).

10. G. Klein, S. Grombein, & W. Rüdiger, Hoppe-Seyler's Z. Physiol.
    Chem. 358:S.1077 (1977).

MOLECULAR ASPECTS OF PHOTORECEPTOR FUNCTION IN STENTOR COERULEUS

Pill-Soon Song, Edward B. Walker and Min Joong Yoon

Chemistry Departement, Texas Tech University
Lubbock, TX 79409 (USA)

ABSTRACT

Stentor coeruleus shows both step-up photophobic and negative
phototactic responses with maximum response to red light. Videotape recordings of these photoresponses have been made for illustration (in collaboration with D. Häder and K. L. Poff). The photoreceptor has been isolated and characterized in terms of its
chromophore structure and subunit organization of the protein. On
the basis of preliminary results obtained from light-induced pH
measurements and effects of ionophores on the Stentor photoresponse, we propose a proton flux event as the primary mechanism
of the Stentor photoreceptor.

INTRODUCTION

Stentor coeruleus, a non-photosynthetic protozoan ciliate, exhibits photophobic response to visible light (1-3). Fig. 1 shows
action spectra for the step-up photophobic and negative phototactic
responses of Stentor coeruleus (3,4). However, no systematic study
of the photobiology of Stentor at phenomenological and mechanistic
levels has been reported until recently (3-5). In this lecture, we
will review the recent progress toward understanding of the Stentor
photoreceptor and its function.

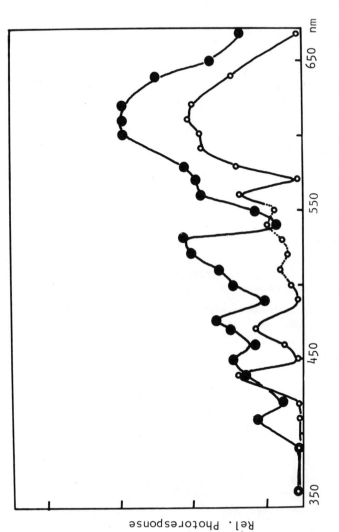

FIGURE 1. The action spectra for step-up photophobic (solid circles) and negative phototactic (open circles) responses of *Stentor coeruleus*. The uncertain portion (dotted line) of the phototactic action spectrum represents averages of two sets of the fluence response data, one of which showed very little responses. (After Song et al. (3,4)).

PHOTORECEPTOR

The chromophore of Stentor photoreceptor has been suggested
as a derivative of the naphthoanthraquinone type of compounds (6).
Recently, we have identified the chromophore as hypericin covalently linked to protein, on the basis of spectroscopic and chro-
matographic characterization (5). Fig. 2 shows two possible structures. The action spectra shown in Fig. 1 are consistent with the
absorption spectra of Stentor coeruleus and its isolated pigment
protein (Fig. 3).

Isoeletric focussing of a sucrose-density gradient fraction
of the protein exhibits a single chromoprotein band (Fig. 4),
which is composed of four protein subunits (mol wts 13,000,
16,000, 65,000 and 130,000) which can be resolved on an SDS-poly-
acrylamide gel electrophoresis (5). The chromo-protein is membrane-
bound, and the 65,000 mol wt subunit is soluble in acetone, ethanol and n-pentane.

MECHANISM

From the absorption spectrum of Stentor (Fig. 3), it can be
inferred that the concentration of photoreceptor protein is rather
high ($\sim 10^{13}$ -10^{14} molecules per cell). The chromophore structure
is rigid. There were no observable absorbance changes in Stentor
in situ at room temperature and 77 K upon red light irradiation,
indicating that light-induced reactions such as the photoreduction
of chromophore did not occur to any significant extent. On the
other hand, excitation of Stentor resulted in fluorescence from
anionic form(s) of the photoreceptor chromophore (Fig. 5). Furthermore, the photoreceptor protein consisted of 4 subunits exhibiting similar anionic fluorescence, but the identical excitation spectra were obtained with respect to emission maxima
for the neutral and anionic fluorescence bands (E.B. Walker and
P.S. Song, unpublished results). Based on these observations, we
suggested that the photoreceptor itself functions as a proton
source in a light-induced proton flux across the membrane as the
primary transducing mechanism in Stentor, since there are six
hydroxyl protons whose excited state pK_a^{*}'s are significantly reduced upon excitation by light (5).

Fig. 6 shows red light-induced pH change in the Stentor suspension, showing that a proton flux triggered by light eventually

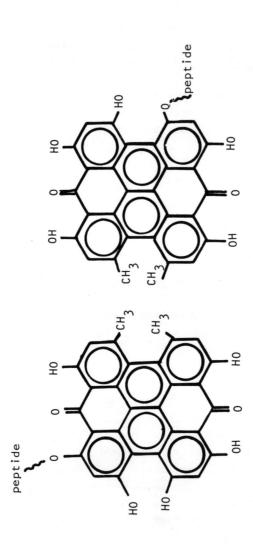

FIGURE 2. Two of the possible structures for the stentorin chromophore. Other structures having peptide linkage elsewhere and involving more than –OH group are also possible. A cyclic peptide linking two –OH or –CH$_3$ groups cannot be ruled out at this stage. We have ruled out a peptide linkage through the –CH$_3$ groups, however, on the basis that the peptide is very easily hydro-lyzed, releasing the chromophore identified as hypericin. The most likely hydrolyzable linkage is the ester linkage with a carboxylic group of the peptide.

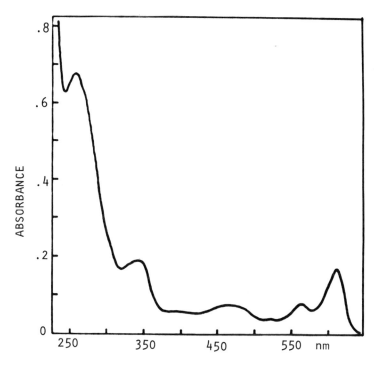

FIGURE 3/a. The absorption spectrum of the photoreceptor protein isolated from <u>Stentor coeruleus</u> in the phosphate buffer (10 mM), pH 7.4. (After Walker et al. (5)).

resulted in the leakage of protons to the medium. Similar pH drops can also be observed with the photoreceptor-embedded liposome suspension. Fig. 7 shows that the interior pH of the liposome does indeed decrease upon light excitation.   Fig. 8 presents a working model for the light-induced proton flux across the membrane.

It is noterworthy that the 65,000 dalton subunit and hypericin itself exhibit long fluorescence lifetime (τ_F > 5 nsec), while the native photoreceptor protein shows about 3 times shorter lifetime (Table 1). The shortening of τ_F and the predominance of anionic fluorescence of the native photoreceptor <u>in vitro</u> and <u>in vivo</u> are suggestive of the possibility that proton dissociation effectively

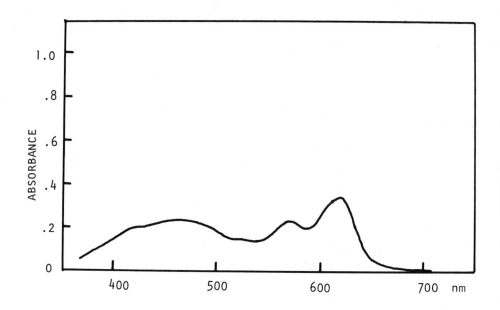

FIGURE 3/b. The absorption spectrum of Stentor coeruleus in vivo
at room temperature.   (After Song et al. (3)).

competes with other radiative and radiationless relaxation processes in the excited photoreceptor.

To further support the light-induced proton flux mechanism
in Stentor, we have carried out preliminary experiments (3,4),
which demonstrated that light-induced spike potential ($\sim$10 mV,
rise-time $\sim$5 nsec) across the Stentor cell wall and membrane
was completely abolished by proton uncouplers, 1 μM FCCP and TPMP[+].

These protonophores also strongly inhibited photophobic and
phototactic responses of Stentor, while nigericin, valinomycin,
gramicidin and A23187 did not inhibit the photoresponses (unpublished).

CONCLUSION

Our results suggest that Stentor coeruleus utilizes a light-
induced proton flux across membrane as the primary source of driving force for the photosensory transduction. The photoreceptor

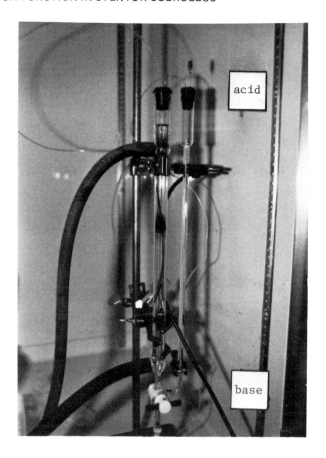

FIGURE 4. The isoelectric focussing of the native photoreceptor
of <u>Stentor</u>. Approximately 10 mg of total protein was applied to
a 25 ml electrofocusing column (5-50 % sucrose gradient, 20 %
Ampholines). After electrofocusing for 20 hrs at 4°C and 500 V,
the column was drained and assayed by absorbance at 612 nm and
Bio-Rad protein assay reagent.

itself seems to provide as the initial proton source which is conducted across the membrane through a proton-conduction network
consisting of juxtaposed acid-base groups of proteins in the membrane. Further work is warranted to definitely establish the mechanism of photosensory machinery of <u>Stentor coeruleus.</u>

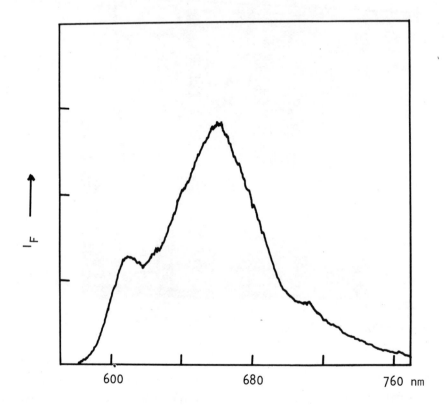

FIGURE 5. The fluorescence emission spectra of <u>Stentor</u> <u>in vivo</u> at room temperature. Emission bandpass 5.4 nm, λ_{ex} 460 nm.

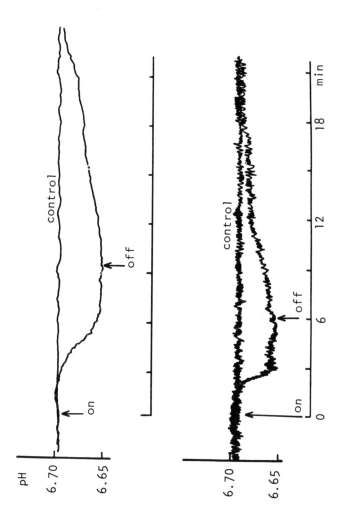

FIGURE 6. Light-induced pH drop due to H$^+$ release from Stentor coeruleus. The cell population, ca. 5x10^4/ml. Top: pH measured by pH electrode. (600 nm, Bausch & Lomb high intensity monochromator, 150 W Xe bandpass = 100 nm, 78 W/m^2), IR filter. Bottom: pH monitored by beta-methyl-umberiferone (fluorescence). (600 nm, Bausch & Lomb plus CS3-66 filter, 63 W/m^2), IR filter. Irradiation from the top of a Stentor-containing cuvette, and the fluorescence of the probe detected simultaneously.

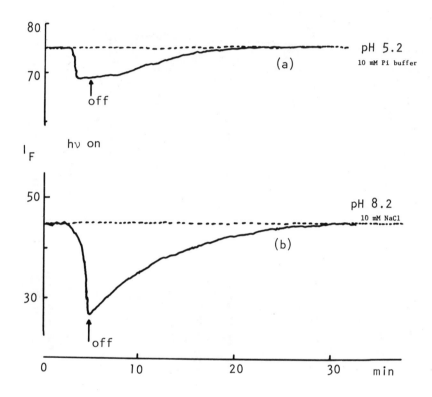

FIGURE 7. pH drop induced by light excitation of stentorin-liposomes:

(a) DMPC liposome entrapped with stentorin (inside), as measured by the quenching of 9-amino-acridin (outside → inside).

(b) Stentorin-embedded liposome (DMPC), as measured by the pH quenching of the fluorescence of umberiferone (outside).

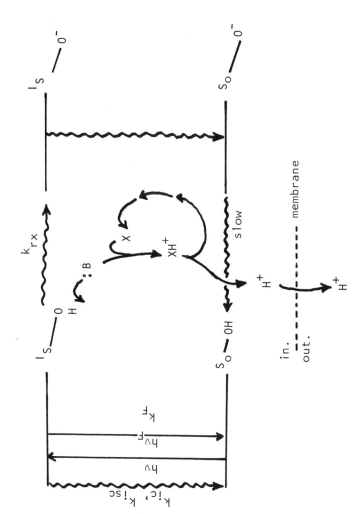

FIGURE 8. Relationship between fluorescence (left, k_F) and the proton transfer (k_{rx}) in the excited state of stentorin, ^1S-OH, of membrane bound photoreceptor. It is hypothesized that the slow rate of re-protonation in the ground state, S_o-O^-, is controlled by the topographic conditions of membrane as the result of the excited state proton transfer. A possible model for the proton release across the membrane is also shown schematically.

ACKNOWLEDGEMENTS

    This work was supported by the PHS NIH grant NS 15426 and
the Robert A. Welch Foundation (D-182). A grant from the Institute
of College Research, Texas Tech University (No. 12-C666-200000),
enabled one (EBW) of us to conduct some preliminary experiments
during the early stages of this work. We also thank Drs. Häder and
K.L. Poff for valuable collaborations.

REFERENCES

1. H. S. Jennings, Pub. Carnegie Inst. Washington, N.16:1 (1904).
2. D. C. Wood, Photochem. Photobiol. 24:261 (1976).
3. P. S. Song, D. Häder & K. L. Poff, Photochem. Photobiol.
   in press.
4. P. S. Song, D. Häder & K. L. Poff, Photochem. Photobiol.
   in press.
5. E. B. Walker, T. Y. Lee & P. S. Song, Biochim. Biophys. Acta
   587:129 (1979).
6. K. M. Møller, Comp. Rend. Trav. Lab. Carlsberg, Ser. Chim. 32:
   472 (1962).

POSSIBLE PHOTOREGULATION BY FLAVOPROTEINS

Vincent Massey

Department of Biological Chemistry

University of Michigan, Ann Arbor, Mi. 48109 (USA)

A wide variety of biological phenomena are blue light-dependent, responding to light of wavelengths below 500 nm. This topic has been the subject of many review articles; a recent one by Schmidt (1) is particularly useful, both in its description of the phenomena and in exploring the basis of the possible molecular mechanisms involved. Many individual examples of blue light photoreception will be dealt with in detail at this meeting; I will simply list here those examples where action spectra permit a reasonable conclusion about the nature of the photoreceptor. There now seems to be general agreement that flavin is the blue light photoreceptor (1). This conclusion is based on several lines of evidence. The action spectra of most blue light dependent biological phenomena show a prominent band around 450 nm, often with a distinct shoulder in the region of 470-480 nm. This spectral characteristic is shared by two classes of widely spread biological compounds, flavins and carotenes. In addition most action spectra also exhibit a near-UV band, in the region of 350 nm. There is however much less consistency about the shape, position and "intensity" of this band, a fact chiefly responsible for the past uncertainty about whether flavin or carotene is the blue light photoreceptor. Flavins do possess such an absorption band, as does cis-β-carotene. However trans-β-carotene does not, and the biological occurrence of cis-β-carotene is apparently quite rare. Particularly important is the failure to detect this form in one of the archetypal organisms exhibiting blue light photoresponse,

Phycomyces (2). Another important piece of evidence is that caro-
tene-deficient mutants of various biological species generally exhibit the same blue light responses as the wild types (3-5).
Finally while carotene is present in high concentrations in the
growing avena coleoptile, it has been shown by microchemical techniques to be absent from the apex, which is the most photosensitive
zone of the coleoptile (6).

Table I is a partial listing of the blue light photoreceptor
systems where evidence suggestive of the involvement of flavin has
been obtained (see reference 1).

TABLE I
BLUE LIGHT RECEPTOR SYSTEMS PROBABLY INVOLVING FLAVIN

Biological Phenomenon	System	Reference
Phototropism	Phycomyces sporangiophores	7-9
	Philobolus sporangiophores	10
	Avena coleoptile	11,12
Phototaxis	Euglena	13
Metabolic control	Neurospora carotenogenesis	14
	Mycobacterium carotenogenesis	15
	Fusarium carotenogenesis	16
	Chlorella O_2 uptake	17,18
	Scenedesmus metabolism	19
Photomorphogenesis	Dyropteris filix-mas	20,21
Circadian Rhythm	Drosophila	4,5
	Pectinophora gossypiella	22
Structure	Funaria chloroplast rearrangement	23

    Attemps at identifying particular photoreceptors have of neces-
sity relied heavily on the differential response of the particular
biological phenomenon to different wavelengths. It is therefore important to appreciate the variations in spectral properties that
flavins may show under different conditions. For example as illustrated in Fig. 1, the polarity of the medium in which the flavin
is dissolved has a very pronounced influence on the absorption spectrum. Thus in the biological context, if free flavin were the photoreceptor, considerable differences would be possible in action spectra depending on whether the flavin was localized in a polar or non
polar environment. In fact there is good reason to believe that
blue light photoreceptors would be membrane bound, probably in the
form of flavoproteins (24-30). This enlarges considerably the range
of spectral shapes possible for a flavoprotein photoreceptor, as illustrated in Figs. 2-5. Firstly, there may be quite extensive
changes in extinction coefficients and spectrum on binding flavin
to a particular apoprotein (Fig.2). Secondly the spectral differences between various flavoproteins can also be quite marked, as
shown in Fig. 3. Thirdly the spectra of flavoproteins may also be
changed quite dramatically depending on the nature of associated
ligands. This is illustrated in Figures 4 and 5 for the case of the

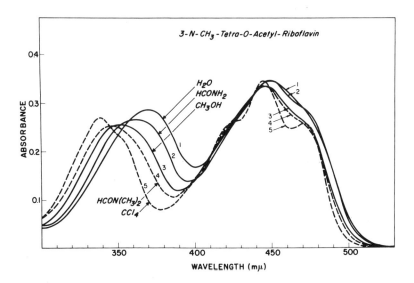

FIGURE 1. Effect of solvents on the spectrum of 3-methyl-tetra-
acetyl-riboflavin

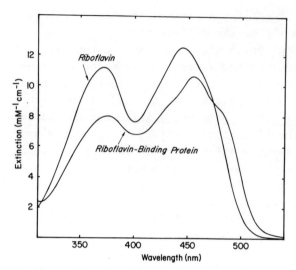

FIGURE 2. Effect of riboflavin binding protein on the spectrum of riboflavin. Conditions, 0.1M phosphate, pH 7, 25°

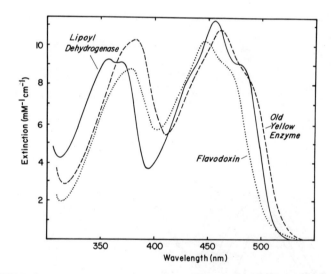

FIGURE 3. Representative spectra of flavoproteins

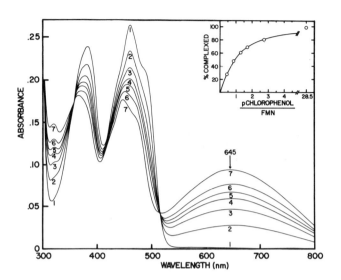

FIGURE 4. Formation of a charge transfer complex between p-chloro-phenol and Old Yellow Enzyme. The enzyme (curve 1) in 0.1M phosphate, pH 6.5, 25°, was titrated with p-chlorophenol as shown. From ref. 32.

Yellow Enzyme of brewers yeast. This enzyme has been found to be associated in nature with p-hydroxybenzaldehyde (31) in the form of a charge transfer complex between the phenolate as donor and oxidized flavin of the enzyme as acceptor (31,32). Old Yellow Enzyme has been found to give such complexes with a wide variety of aromatic and heteroaromatic compounds containing an ionizable hydroxyl group. Fig. 4 shows a typical example, with p-chlorophenol as ligand. The phenolate group is not necessary for binding, only for the charge transfer interaction. This is illustrated in Fig.5, which shows a large perturbation of spectrum, but no charge transfer band, on binding of p-methoxybenzaldehyde.

    The susceptibility of the free flavin coenzymes to photochemical reduction has been known for may years, with a wide variety of naturally occurring compounds, such as pyridine nucleotides, amino acids, hydroxy acids and simple carboxylic acids acting as photodonors (33-35). A significant milestone in the understanding of this photochemistry was the discovery of Hemmerich et al. (36)

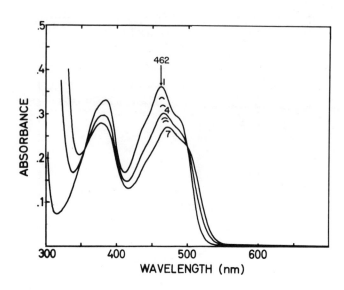

FIGURE 5. Perturbation of the spectrum of Old Yellow Enzyme (curve 1) on complex formation with p-methoxybenzaldehyde (curves 2-7). From ref. 32.

that photoreduction with carboxylic photodonors proceeded by way of covalent derivatives of the flavin. A comprehensive account of this work has appeared recently (37) and will be described also in the present volume.

The available evidence suggests that most physiological blue light responses involve a redox reaction. Oxygen is often necessary for the response to be elicited (15,38,39), suggesting multiple turnover of the receptor. In fact, surprisingly, the blue light carotenogenesis in Fusarium is mimicked by H_2O_2 (40). The significance of these findings is however unclear, since Rau et al. (41) also found mercuribenzoate to replace the blue light effect.

There are several different ways in which flavin photochemistry could be applied to advantage. Flavoproteins are readily converted to their semiquinoid forms photochemically; this process is mediated very efficiently by trace amounts of free flavin in a catalytic fashion (42). The semiquinoid flavin so produced could react rapidly with a variety of natural acceptors such as quinones, heme proteins or with molecular O_2. In the latter case the product is O_2^- (43), which by subsequent metal-catalysed Fenton chemistry could yield the highly reactive hydroxyl radical, $OH^{\bullet}$ (44). However, because of the indiscriminate reactivity of this species, the involvement of $OH^{\bullet}$ does not appear very likely.

Electron transfer from a photo-reduced flavoprotein to another redox carrier such as a heme protein would offer much more possibility of providing a specific reaction which could mediate the primary photo-event. Indeed in the last few years considerable attention has been paid to "light induced absorption changes" which give some support to such a possibility. The activity in this area was sparked by studies of Phycomyces (45,46) where reduction of a heme protein was found to be induced by blue light. In several cases the action spectra for photoreduction of heme proteins are closely similar to flavin spectra (47,48). The subject is reviewed in detail by Schmidt (1). Unfortunately the finding of a flavin action spectrum for heme protein reduction by no means proves that this is the basis of the photoreception phenomenon. As pointed out by Schmidt (1) the widespread occurrence of flavins and the ability of their reduced forms to react with heme proteins is sufficiently general that one might expect such correlations even if the photoreception system did not involve heme proteins or flavoproteins. In this case one might expect that added artificial redox dyes could replace the natural flavin, and give a completely different action spectrum. Indeed this has been done with methylene blue and a membrane fraction of corn coleoptiles by Widell et al. (49). The dangers of drawing false conclusions are obvious. Fortunately in the above case the heme protein reduced appeared to be the same (as judged by wavelength maxima and extinction) whether the photochemically active species was the endogenous flavin or the added methylene blue, and represented less than 20% of the heme protein which could be reduced by dithionite (50).

A potentially valuable method for substantiating evidence for flavin involvement is the inhibition of the photoresponse by phenyl-acetate. This compound forms stable N(5)- and C(4a)- flavin adducts on photoreaction with flavins (36,51) and a variety of flavoproteins (52). It has been found recently to be a specific inhibitor of phototropism in maize seedlings (53), thereby adding considerably to the evidence that the flavin is the photoreceptor in this system. Another potential inhibitor of flavin photoreceptors is sulfite. Oxidized flavins are susceptible to nucleophilic attack by sulfite to form an N(5)-sulfite adduct (54). The ability to form sulfite adducts varies considerably among the different classes of

flavoproteins, with oxidases, in general, forming rather stable adducts, especially α-hydroxyacid oxidases such as glycollate oxidase
and L-lactate oxidase (55).

Recent studies on the photochemical reactions of the bacterial
flavoenzyme, L-lactate oxidase (56,57) offer other potential inhibitors of photoreception, and possibly an alternative chemical explanation of the biological phenomena. This enzyme, which is closely
related to the glycollate oxidase of plant sources, undergoes remarkably facile photoreactions with a variety of naturally occurring
or closely related carboxylic acids. Thus stable N(5)-flavin adducts
are formed by photodecarboxylation of oxalate, oxamate, pyruvate,
malonate, tartronate, succinate, etc.

As remarkable as the facility of the photoreaction is the enhanced stability of the protein bound adduct. For example, the protein bound -N(5)-CHOH-COOH adduct from the light reaction with
tartronate has a half life of 20 min at 25°, decaying to oxidized
flavoenzyme and glycollate. This should be compared to the extreme
lability of the protein-free form; on liberation from the protein
stopped flow techniques have to be employed to monitor the decay
process. Even more stable adducts are the N(5)-carboxymethyl and
carboxyethyl derivatives formed by light reaction with malonate and
succinate. These appear to be indefinitely stable to O_2 in the dark,
but are photooxidized to the stable blue radical form. By comparison,
the flavin adducts on release from the proteins, are converted fairly
rapidly to the radical state, even in the dark ! The stability of
the adducts from these small molecules thus offers a wider choice
of potential inhibitors, which could prove useful when the more
bulky phenylacetate is not able to penetrate to the photoreceptor
site.

The studies with lactate oxidase also raise the possibility of

photo-generated flavin adducts actually being involved in the photoreception process. The conversion of oxidized to reduced flavin
results in the transformation of the planar oxidized ring system
to the "butterfly-shaped" reduced state (58). The possibility has
been raised that the directional effects of the light reaction are
connected in some way with such a conformational change (1). In
general, N(5)- and C(4a)-substituted flavins are more non-planar
than the unsubstituted reduced flavin (58); hence such adducts
would be even better candidates for triggering a conformation-based
response. The adduct might be either a spontaneously unstable one,
or could be reconverted to oxidized flavin in a secondary photoreaction. For example, it has been found that the stable N(5)-car-
bonate or glycollyl adducts formed photochemically with lactate
oxidase, and oxalate or tartronate, are rapidly reconverted to oxidized enzyme when irradiated with near-UV light corresponding to
their absorption spectra (57). As such spectra (Fig.6) overlap with
the near-UV band of oxidized flavin, detection of such a contribution in action spectra would be difficult. The occurrence of such
a reaction could however account for the occasional inference that
the two banded action spectrum is due to two separate and separable
pigments, one containing only a near UV-peak (59).

While the reactions outlined above appear to offer a reasonable
basis on which to build a biological photoresponse, there are a
number of experimental results which suggest that the phenomena are
triggered from a secondary reaction with reduced flavin, such as
the reduction of heme proteins discussed above. Particularly important are the results of Lang-Feulner and Rau (60) who showed
that mycelia of _Fusarium_ infused with artificial redox dyes exhibited
photodependent carotenogenesis with action spectra typical of the
artificial dye. Those which were able to replace the natural receptor all had redox potentials such that they would be capable of reduction of b-type cytochromes (methylene blue, E_o' 11mV, toluidine
blue, 34mV and neutral red, -325mV) while those that were inactive
had higher potentials (such as for instance dichlorophenol indo-
phenol 219 mV). Coupled with the work from the laboratory of Winslow
Briggs already discussed, these studies support the concept of electron transfer to a b-type cytochrome being an important step in the
light transduction process. An attractive idea is that the flavin-
heme interaction takes place across a membrane, with a proton gradient being established. Such a gradient could be established with
flavin alone, because of the pK near neutrality of the dihydroflavin.

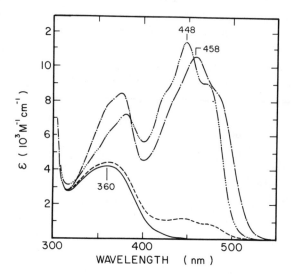

FIGURE 6. Photochemical adduct formation from the lactate oxidase-tartronate complex. The enzyme, 3.1×10^{-5}M, in 0.01 M imidazole HCl buffer pH 7.0 (-·-) was complexed with tartronate, 5×10^{-3}M (-·-·-). The complex was irradiated at 4° through a Corning CS 372 filter cutting off light at wavelengths shorter than 445 nm. The dashed curve shows the spectrum obtained, which still contains some oxidized enzyme contribution from photodecomposition of the glycollyl adduct. If unfiltered white light is used only about 50 % adduct is obtained at steady state. The solid line shows the calculated spectrum of the flavin -N(5)-glycollyl adduct.

$$AH_2 + Fl_{ox} \xrightarrow{h\nu} A + Fl_{red} H_2$$

$$Fl_{red} H_2 \underset{\longleftarrow}{\overset{pK \sim 6.5}{\rightleftharpoons}} Fl_{red} H^- + H^+$$

In the case of electron transfer from reduced flavin to heme protein (via the flavin radical state) one or two protons would be released depending on whether the reduced flavin was in its anion or neutral form. Thus the overall stoichiometry catalysed by flavin would be:

$$AH_2 + 2\ Cyt^{3+} \longrightarrow A + 2\ Cyt^{2+} + 2H^+$$

If the flavoprotein were localized in a membrane and the photodonor
on one side and the cythochrome on the other, a proton gradient
would thus be set up on irradiation with blue light.

Finally, attention should be drawn to the discovery in recent
years of various naturally occurring flavins with optical properties
considerably different from those of riboflavin and its derivatives.
Covalent attachment of flavin to protein through the C-8α linkage
has been known for many years, but results in only minor modification of the flavin absorption spectrum (see ref. 61 for a recent
review). Another side of covalent linkage between flavin and protein which does have a somewhat greater influence on the spectral
properties is the flavin 6-cysteinyl linkage found in the bacterial
enzyme trimethylamine dihydrogenase (62).

Much more dramatic alterations in spectral properties are displayed by three new flavin structures discovered in the last few
years. These all involve modifications in the benzene ring portion
of the flavin structure. Roseoflavin is a red colored flavin which
is produced by Streptomyces sp and whose structure has been shown
to be 8-dimethylaminoriboflavin (63).

In this case, in contrast to the other derivatives to be discussed
below, roseoflavin would appear to exist in solution predominantly
as the neutral form, rather than the Zwitterion. The riboflavin
binding protein of egg whites binds preferentially neutral, non
polar flavins (64). Fig. 7 shows the very small spectral change
which occurs when roseoflavin binds to the riboflavin binding protein. If there had been any significant contribution of the Zwit-
terionic species to the structure, much more extensive spectral
changes would have been expected.

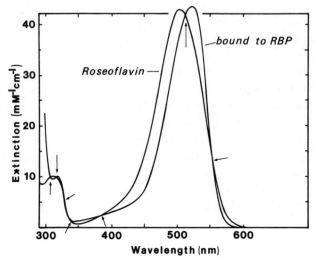

FIGURE 7. Effect of riboflavin binding protein on the spectrum of roseoflavin. Conditions, 0.05 M phosphate, pH 6.8, 25°. The roseoflavin was a generous gift from Dr. K. Matsui.

To illustrate this point further, Fig. 8 shows the spectral changes which occur when riboflavin binding protein binds to 6-hydroxyriboflavin. This flavin was discovered in enzyme preparations from the anaerobic microorganism, <u>Megasphera elsdenii</u> and shown to have the following properties (65):

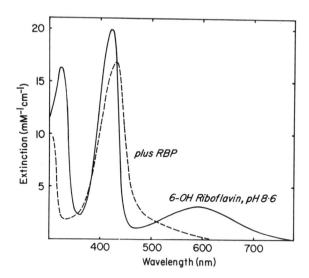

FIGURE 8. Effect of riboflavin binding protein on the spectrum of 6-hydroxyriboflavin. The latter, in 0.1 M pyrophosphate, pH 8.6, 25°, and therefore in its anionic form, was titrated with ribo-flavin binding protein. The flavin was a generous gift from Dr. Peter Hemmerich.

At pH values above its pK, the monoanion contains substantial contri-bution from the benzoquinoid structure with its characteristic long wavelength contribution to the spectrum (65). Riboflavin binding protein completely eliminates this contribution, in keeping with its vastly preferential binding to neutral flavin molecules.

    The spectral characteristics of 8-hydroxy riboflavin, free and in complex with riboflavin binding protein, are shown in Fig.9. This flavin was also isolated from M.elsdenii and shown to have similar properties of those of 6-hydroxyflavins with the predominant ionized form having the benzoquinoid structure (66):

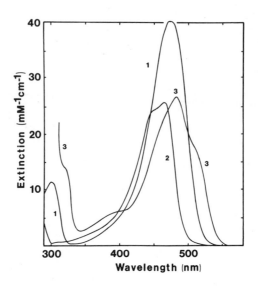

FIGURE 9. Effect of riboflavin binding protein on the spectrum of
8-hydroxyriboflavin. The flavin, in 0.1 M phosphate, pH 7.0, 25°,
(curve 1) was titrated with riboflavin binding protein, to give
the spectrum of curve 2. The anionic form bound to RBP (curve 3)
was obtained in a separate experiment by titration to pH 10.5, in
the presence of a five fold excess of RBP. The apparent pK was
skifted from 4.8 for the free flavin to 9.0 for the protein-bound
form. The flavin was a generous gift from Dr. Sandro Ghisla.

Again the spectral changes at pH 7 are consistent with the preferential binding of RBP to neutral flavins. At high pH values, in
the presence of a large excess of apo-RBP, the spectrum of the protein-bound anion can be recorded. The affinity for flavin binding
at pH 11 can be estimated to be 4-5 orders of magnitude weaker than
at pH 5; illustrating again the vastly preferential binding of neutral flavins by this protein.
    The spectral characteristics of 8-hydroxyflavin are particularly
interesting in view of some of the published action spectra of blue
light biological phenomena. Either in its neutral or anionic form,
8-hydroxyflavin has only one strong absorption band in the visible,
with a wavelength maximum varying from 430-490 nm depending on the

protein to which it is bound (67). Thus the possibility exists
that action spectra ascribed to carotenes (see examples in reference 1) could in fact to be due to 8-hydroxyflavins.

REFERENCES

1. W. Schmidt, Structure and Bonding (in press) (1979).
2. D. Presti, W. -J. & M. Delbrück, Photochem. Photobiol. 26:403
     (1977).
3. M. L. Sargent & W. R. Briggs, Plant Physiol. 42:1504 (1967).
4. W. E. Zimmerman & T. H. Goldsmith, Science 171:1167 (1971).
5. E. Klemm & H. Ninnemann, Photochem. Photobiol. 24:369 (1976).
6. E. Bünning, Planta 26:719 (1937).
7. G. M. Curry & H. E. Gruen,Proc.Natl.Acad.Sci.USA 45:797 (1959).
8. M. Delbrück & W. Shrophire Jr. Plant Physiol. 35:194 (1960).
9. M. Delbrück, A. Katzir & D. Presti, Proc. Natl.Acad. Sci. 37:
     1969 (1976).
10. R. M. Page & G. M. Curry, Photochem. Photobiol. 5:31 (1966).
11. M. Everett & K. V. Thimann, Plant Physiol. 43:1786 (1968).
12. G. M. Curry, Ph. D. Thesis, Harvard University, Cambridge, Mass.
     (USA) (1957).
13. B. Diehn, Biochim. Biophys. Acta 177:136 (1969).
14. E. C. De Fabo, R. W. Harding & W. Shrophire Jr., Plant Physiol.
     57:440 (1976).
15. C. D. Howes & P. P. Batra, Arch. Biochem. Biophys. 137:175
     (1970).
16. W. Rau, Planta 72:14 (1967).
17. J. Zurzycki, Acta Soc. Bot. Pol. 39:483 (1970).
18. J. M. Pickett & C. S. French, Proc.Natl.Acad.Sci.US.57:1587 (1967).
19. G. Brinkman & H. Senger, Plant and Cell Physiol. 19:1427
     (1978).
20. H. Mohr, Planta 47:127 (1956).
21. A. M. Steiner, Naturwissensch. 18:497 (1967).
22. K. G. Bruce & D. H. Minis, Science 163:583 (1969).
23. J. Zurzycki, Acta Protozoologica 11:189 (1972).
24. P. S. Song & T. A. Moore, Photochem. Photobiol. 19:435 (1974).
25. L. F. Jaffe, Exp. Cell Research 15:282 (1958).
26. E. S. Castle, J. Gen. Physiol. 17:751 (1934).
27. A. J. Jesaitis, J. Gen. Physiol. 63:1 (1974).
28. F. Mayer, Z. Bot. 52:346 (1964).
29. W. Shropshire, Jr., Science 130:336 (1959).

30. L. Jaffe & H. Etzold, J. Cell Biol. 13:13 (1962).

31. R. G. Matthews, V. Massey & C. C. Sweeley, J. Biol. Chem. 250:9294 (1975).

32. A. S. Abramovitz & V. Massey, J. Biol. Chem. 251:5329 (1976).

33. L. P. Vernon, Biochim. Biophys. Acta 36:177 (1959).

34. W. R. Frisell, C. W. Chung & C. G. Mackenzie, J. Biol. Chem. 234:1297 (1959).

35. G. R. Penzer & G. K. Radda, Biochem. J. 109:259 (1968).

36. P. Hemmerich, V. Massey & G. Weber, Nature 213:728 (1967).

37. W. Haas & P. Hemmerich, Biochem. J. 181:95 (1979).

38. M. Zalonkar, Arch. Biochem. Biophys. 50:71 (1974).

39. H. C. Rilling, Biochim. Biophys. Acta 60:548 (1962).

40. R. R. Theimer   W. Rau, Planta 92:129 (1970).

41. W. Rau, B. Feuser & A. Rau-Hund, Biochim. Biophys. Acta 136: 589 (1967).

42. V. Massey, M. Stankovich & P. Hemmerich, Biochemistry 17:1 (1978).

43. V. Massey, S. Strickland, S. G. Mayhew, L. G. Howell, P. C. Engel, R. G. Matthews, M. Schuman & P. A. Sullivan, Biochem. Biophys. Res. Comm. 36:891 (1969).

44. J. A. Fee, in "ISOX III",T. E. King, H. S. Mason & M. Morrison, eds., University Park Press, Baltimore (in press) (1979).

45. D. S. Berns & J. R. Vaughn, Biochem. Biophys. Res. Comm. 39: 1094 (1970).

46. K. L. Poff & W. L. Butler, Nature 248:799 (1974).

47. E. D. Lipson & D. Presti, Photochem. Photobiol. 25:203 (1977).

48. V. Muñoz & W. L. Butler, Plant Physiol. 55:421 (1975).

49. S. Widell, J. Britz & W. R. Briggs, Carnegie Institution Year Book 77:344 (1978).

50. M. H. M. Goldsmith & W. R. Briggs, Carnegie Institution Year Book 77:347 (1978).

51. W. H. Walker, P. Hemmerich & V. Massey, Eur. J. Biochem. 13: 258 (1970).

52. S. Ghisla, V. Massey, J. -M. Lloste & S. G. Mayhew, Biochemistry 13:589 (1974).

53. W. Schmidt, J. Hart, P. Filner & K. L. Poff, Plant Physiol. 60:736 (1977).

54. F. Müller & V. Massey, J. Biol.Chem. 244:4007 (1969).

55. V. Massey, F. Müller, R. Feldberg, M. Schuman, P. A. Sullivan, L. G. Howell, S. G. Mayhew, R. G. Matthews & G. P. Foust, J. Biol. Chem. 244:3999 (1969).

56. S. Ghisla & V. Massey, J. Biol. Chem. 250:577 (1975).

57. S. Ghisla, V. Massey & Y. S. Choong, J. Biol. Chem. (in press) (1979).

58. P.Hemmerich, in "Progress in the Chemistry of Organic Natural Products", W. Herz, H. Grisbach & G. W. Kirby, eds., Springer Verlag, Vienna (1976).

59. W. Kowallik, Plant Physiol. 42:672 (1967).

60. J. Lang-Feulner & W. Rau, Photochem. Photobiol. 21:179 (1975).

61. D. E. Edmondson & T. P. Singer, FEBS Lett. 64:255 (1976).

62. D. J. Steenkamp, W. McIntire & W. C. Kenney, J. Biol. Chem. 253:2818 (1978).

63. S. Kasai, R. Miura & K. Matsui, Bull. Chem. Soc. Japan 48: 2877 (1975).

64. G. Blankenhorn, Eur. J. Biochem. 82:155 (1978).

65. S. G. Mayhew, C. D. Whitfiled, S. Ghisla & M. Schuman-Jorns, Eur. J. Biochem. 44:579 (1974).

66. S. Ghisla & S. G. Mayhew, Eur. J. Biochem. 63:373 (1976).

67. S. Ghisla, V. Massey & S. G. Mayhew, in "Flavins and Flavoproteins", T. P. Singer, ed., Elsevier, Amsterdam (1976).

BLUELIGHT RECEPTION AND FLAVIN PHOTOCHEMISTRY

Peter Hemmerich and Werner Schmidt

Faculty of Biology, University of Konstanz

D-7750 Konstanz (W. Germany)

Support is accumulating for the idea that biological blue-light reception is mediated by a flavin derivative. Recently the available evidence has been reviewed comprehensively (1). With respect to these data, it is the aim of the present lecture: 1) to give a survey of the pertinent photochemical data known for the flavin chromophore; 2) to show most recent, preliminary chemical data about a flavin known to accumulate in phototropically active avena coleoptiles, the so-called "FX" first observed by Zenk (2).

In order to clarify the complex picture of chemical and biological activities inherent in the flavin chromophore, Hemmerich and Massey (3) have advanced a classification scheme, which allows one to differentiate mechanistically the three principal flavin activities:
1) de - or transhydrogenation
2) single electron ("1 e$^-$") transfer
3) dioxygen activation

It is the postulate of these authors, that flavin action is regulated by regiospecific (hydrogen) bonds blocking lone pairs either in the N(1)/O(2α) - or in the N(5)-region (Scheme 1): 1/2α-blocking promotes dehydrogenation, 5-blocking promotes 1e$^-$ - -transfer, while dioxygen activation requires both tautomeric conformations of the flavo-holoprotein in a sequence.

In the present context of "flavin-dependent sensory physiology" the discussion of mechanisms has been confined, hitherto, to 1e$^-$ - -transfer. The photochemistry of oxidized flavin (Fl_{ox}) involved

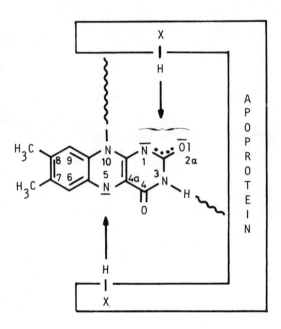

SCHEME 1. Regulation of flavoprotein action by H-bridges from
apoprotein. Note that (e.g. text) - in the free state - Fl_{ox} and
$^1Fl_{ox}^*$ are protonated at 1/2α, while $^3Fl_{ox}^*$ and Fl^- are protonated
at $N(5)$.

here, however, is by no means confined to one-electron reactions.
On the contrary, quite like in ground state biochemistry, we must
differentiate in flavin photochemistry between: a) true $1e^-$-transfer
and b) $2e^-$-transfer reactions. A $2e^-$-reaction requires, of course,
a carrier for the electron pair. If we denote the $2e^-$-donor "substrate" as RH, the pair-carrier is shown to be (4) R^- (carbanion,
mercaptide ion, cyanide) rather than H^- (hydride). Only in the
enforced case of borohydride (BH_4^- or, more elegantly, BH_3CN^-) hydride moieties will truly be transferred photochemically (but not in
the dark!) towards Fl_{ox} (5). In all other cases, we find a competition of single electron ($1e^-$) and carbanion ($2e^-$) transfer, as
reviewed in the following sentences:
1) The most specific $2e^-$-donor we found is HCN (6) reacting according to eq. (1):

$$(1) \quad Fl_{ox} + HCN \xrightarrow{h\nu} HFl_{red}CN$$

The product 6- and 9-cyanodihydroflavin results from a typical
"$2e^-$-transfer", the electron pair being carried by CN^-. The reaction
is identical with a nucleophilic addition in the excited state. This
prototypical case is, of course, irreversible.
2) Another highly specific, but reversible $2e^-$-reduction is by
α-phenylpropionate, which is photodecarboxylated by flavin according
to eqs.(2), (7):

$$(2) \quad C_6H_5CH(CH_3)COO^- + {}^3Fl^*_{ox} \longrightarrow C_6H_5CH(CH_3)\text{-}5\text{-}Fl^-_{red} + CO_2$$

In this case, the "adduct formation" can be reversed with acid
yielding starting Fl_{ox} and ethylbenzene, $C_6H_5CH_2CH_3$. It follows
from these examples that, obviously, cases of photo-substrates can
be achieved, in which the photochemical addition of RH to Fl_{ox} to
yield $Rfl_{red}H$, is thermally - even if slowly - reversible. It must
be emphasized that the suffix "red" denotes not a net reduction of
the system, but only the formation of the dihydroflavin chromophore,
i.e. a bleaching induced by illumination at 450 nm.
      In general, $2e^-$-transfer upon Fl_{ox} obeys reaction (1):

Reversed bleaching

$$(1) \quad {}^3Fl_{ox} + RH \xrightarrow{\text{Bleaching}} Rfl_{red}H \left\{ \begin{array}{l} \longrightarrow Fl_{ox} + RH \\[2ex] \overset{\longleftarrow}{\longrightarrow} HFl^-_{red} + R^+ \end{array} \right.$$

Oxidoreduction

It depends, of course, on the nature of R as well as on the environment, whether reversal of bleaching or oxidoreduction will take
place. If, for example, R = Cys-S, both possibilities may physiologically compete with each other.
      So much in order to show that flavin-dependent biological
bluelight reception would not result mandatorily in $1e^-$-transfer,
i.e. radical formation, but in a reversible σ-bond formation with
a substrate or an environmental prosthetic group. The picture of
possible $1e^-$-reactions is outlined in Scheme 2 (8).
      In experimental terms, we must say: A $1e^-$-transfer reaction
is defined by the observed formation of flavosemiquinone HFl. In
the protein-free chemical flavin system, the semiquinone has the
structure 5-HFl, λ_{max} 570-610 nm, depending on the environment (4),
and is moderately stable as defined by the dismutation equilibrium

$$^3Fl^*_{ox} + RH \longrightarrow \begin{cases} \dot{F}l^- + R\dot{H}^+ \longrightarrow Fl_{ox} + RH \\ \quad\quad pK\ 8.5 \qquad\quad \text{"Back donation":} \\ \quad\quad\quad\quad\quad\quad\quad\quad \text{zero bleaching} \\ H\dot{F}l + \dot{R} \longrightarrow HFl^-_{red} + R^+ \end{cases}$$

"Double le^--transfer":
full bleaching

$$\longrightarrow 1/2\ (Fl_{ox} + H_2Fl_{red} + R_2)$$

Dismutation: half bleaching

SCHEME 2

according to equation (3)

(3) $H\dot{F}l \xrightleftharpoons[\hspace{2cm}]{} 1/2\ (Fl_{ox} + H_2Fl_{red})$

It dissociates at pK $\sim$ 8.5 (4) to yield the red anion $\dot{F}l^-$. We thus
find: if ever the le^--uptake by the flavin triplet is accompanied
by a (diffusion controlled) proton transfer, resulting in the
easily observable formation of the blue 5-H$\dot{F}$l, the subsequent radical decay may change from back donation towards dismutation/dime-
rization, as shown by the permanent bleaching extent of $\sim$ 50%, but
not towards "double le^--transfer". In terms of biology, however,
this decay is generally improbable, since it requires rapid interaction of two flavin molecules. In certain enzymes, however, this
case is provided for by nature (4). Presently, we want to neglect
this complication in the context of flavin-dependent bluelight
reception.

Summarizing, we can achieve mechanistic verification for the
case of biological triplet decay only by: 1) reversible (bleached)
adduct formation ($2e^-$-transfer and back donation or oxidoreduction);
2) reversible (colored) radical formation (le^--transfer and back
donation).

On the other hand, the chemically frequent case of decay by
dismutation/dimerization is biologically improbable, while the case
of radical formation and subsequent full bleaching (double le^--
-transfer) turns out to be chemically unsound for the flavin chromophore.

Thus, if bluelight reception is connected with a net redox
change (not involving back donation) mediated by flavin, it must
be a non-radical reaction, involving group transport and major
conformational changes. This again is improbable. Thus, we might
confine ourselves to the consideration of $1e^-$- and $2e^-$-back donation
mechanisms. Differentiation between these two mechanisms is, in the
model experiment, easy, since the $1e^-$-mechanism requires at neutral
pH the intermediate formation of blue HFl. If a 5-protonation is
hindered at the specific active site (thermodynamically or kineti-
cally) we are left with the red radical species $\dot{F}l^-$ (4) or 1-HFl
(9), respectively. While such radicals can be generated artificially
in certain flavoenzymes (3), they could not be verified hitherto in
any case as true biological intermediates. In fact, if 5-protonation
is hindered, the flavin system is "bound for $2e^-$-transfer", as stated
before (cf. (10)).

In addition to flavin-dependent mechanisms mentioned before,
we can, however, devise a third one which is much less complex:
it might well be that flavin-dependent bluelight reception would
not even involve a redox-reaction at all, and even if it were a
reversible one, we have already seen, that we are forced to consider electron transfer together with proton transfer in the flavin
system - and even with regiospecific proton transfer towards N(5)
for $1e^-$- and towards N(1)/O(2) for $2e^-$-mechanisms. But even if we
might not find any flavin photoreduction connected with bluelight
reception considering only triplet excitation and (physical) deactivation, we are confronted with the fact, that the triplet $^3Fl^*_{ox}$
accepts a proton at N(5), while the ground state $^1Fl^o_{ox}$ does so at
N(1). A mere triplet excitation could thus involve a cyclic or even
linear proton transfer, (Scheme 3) (4).

The energy of the light could thus be converted into conformational energy or "proticity", i.e. flavin could be a light driven
proton pump, if suitably arranged in a membrane. Scheme 3 is based
on the known structure of 1-HFl^+_{ox}, which we derived from comparison
with 1-CH_3Fl_{ox} (11) and from the triplet pK, which is five units

$$Fl_{ox} \quad \xleftarrow{\;\;pK \sim 0\;\;} \quad 1\text{-HFl}^+_{ox}$$

$$h\nu \;\Big\downarrow\Big\uparrow\; \text{dark}$$

$$^3Fl^*_{ox} \quad \xrightarrow{\;\;pK \sim 5\;\;} \quad 5\text{-HFl}^{+*}_{ox}$$

SCHEME 3.

higher than the singlet-pK and can most likely be even more increased by a membrane or protein to become physiologically relevant.
The final evidence that this difference is due to the difference
in protonation site is shown in Fig. 1.

If we replace the flavin N(5) by CH ("5-dezaflavin" (12)), the
characteristic pH-dependence of triplet photochemically vanishes.
The three possible triplet decay cycles are, in summary, visualized in Scheme 4.

If we want to apply these considerations to intact organisms –
as we are forced to do in bluelight receptor research – we run
above all, risk of hitting by light such flavins, which show a
photochemical, but artificial response. This can give rise to
phenomena such as photoreduction of cytochromes or even reversal
of respiratory electron flow. Thus, all observed "light-induced
absorption changes" should be taken with greatest caution in such
systems. The main conclusion from the foregoing should be kept in
mind: absorption changes relevant in the context of bluelight reception should be reversible in reasonably short time, leaving some
kind of energized proton rather than stored electron.

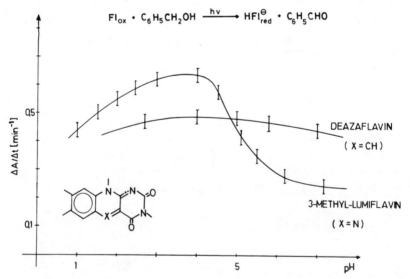

FIGURE 1. Anaerobic reaction rates of photoreduction of flavin and
deazaflavin (dFl) by benzyl alcohol as measured by the decrease
of Fl_{ox} (16) or dFl_{ox} absorption, as a function of pH. Clearly,
there is no break in the curve for dFl near pH 4.5.

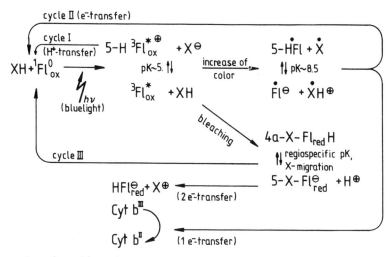

Scheme 4. Visualization of the three possible triplet decay cycles (1).

Before a rigorous investigation of the photo-events of physiological bluelight action becomes feasible, it appears to be an essential prerequisite to elucidate the chemical nature of the photoreceptor pigment itself. At present we have only indirect evidence
(1) that it is "a flavin". The Cholodny-Went-Theory is now generally
accepted as explanation for the so-called "tip-phototropism" of the
grass coleoptile (avena) and for light intensities which include
the "first positive curvature" (1).

Higher light intensities induce the so-called "second phototropic curvature" which develops from the coleoptile base rather
than the tip (such light condition is probably given under natural
conditions). For this latter type of bending the old "photolysis
theory", as introduced by Galston (13) (which in turn, is a special
case of the still older theory of Blaauw (14)) is still under
discussion. This is based on the fact that naturally occurring
flavins are capable of destroying the plant growth hormone auxin
(indoleacetic acid), thereby depleting the light exposed side of

the coleoptile from the hormone and inhibiting growth. Since the
shadowed side continues to grow normally, positive phototropism
develops.

In addition to the well known flavin-mediated photodecarboxylation of auxin in vitro as shown by Galston (13) and others before,
Zenk (2) obtained clear-cut evidence for this process occurring in
vivo as well. He isolated the photoreceptor flavin involved in
this decarboxylation and designated it "FX". Although nearly all
flavin of the etiolated oat coleoptile is FX, from the chemist's
point of view only extremely small amounts could be purified preventing a chemical identification of this hitherto unknown flavin.
In addition, FX hydrolyses under mild alkaline and acidic conditions
to riboflavin. It does not contain phosphate. Zenk assumes it to be
a riboflavin ester (2). It is interesting to note that only 10% of
FX are protein-bound, the greater part is in the free state (wall
fraction and cytoplasma). However, Zenk found only FX in solution
(not the protein-bound FX) to be capable of mediating the photodecarboxylation of only about 1% of the hormone present in vivo.
For these reasons he concluded that this reaction cannot be the
cause of phototropic induction. He rather hypothesized that the
FX-triplet (iodide is capable to quench photodecarboxylation) oxi-
dizes an enzyme system involved in cellwall biosynthesis.

Whatever the actual mechanism of photoinduction of phototropism
of avena coleoptiles will turn out to be, it appears to be indispensable to know what the chemical nature of the blue-light-receptor-
flavin is. In order to characterize FX chemically, we are in the
process of growing large amounts of avena seedlings, and we have
developed an efficient three step purification and separation
procedure, starting from 1 to 1.5 cm long coleoptiles, including
the primary leaves: after alcohol extraction we separate all flavins
from other chromophores by affinity chromatography on a column bearing immobilized "egg white flavin binding protein" (15) and separate afterwards FX from normal riboflavin by chromatography on
silica gel. We confirmed Zenk's observations, that pure FX is
extremely light, acid and base sensitive. Hydrolysis transforms it
into normal riboflavin. The FX absorption and fluorescence spectra
are shown in Figs. 2 and 3 together with riboflavin blanks.

The small absorption difference is probably not characteristic.
Fig. 4 represents a comparison by thin layer chromatography of FX
with riboflavin and 5'-monoacetylriboflavin. We prepared this
compound, as documented by NMR in Fig. 5, by very mild acetylation
of riboflavin, since 5'-monoacetylriboflavin (M. Delbrück, personal

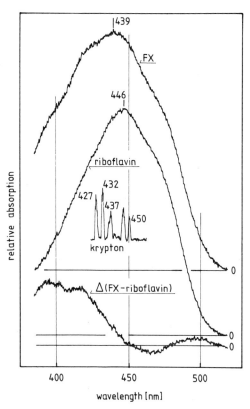

FIGURE 2. Absorption spectra in the blue of FX and riboflavin in
ethanol as taken with a single beam spectrophotometer on-line
with a small computer. Both concentrations have been roughly
adjusted to 2μM and the final spectra were normalized at 450 nm.
The FX-preparation contains a contaminant (not removable by freeze-
-drying) causing strong absorption below 380 nm (not shown). The
increase of the difference spectrum (bottom) at 450 nm possibly
reflects this contaminant as well, since the fluorescence excitation spectra (Fig. 3, left), of FX and riboflavin are practically
identical. For the purpose of wavelength calibration the "blue"
krypton lines are included.

comm.) was recently suspected to be the structure of FX.

Fig. 4 a shows unequivocably, that this is not the case. Since the sequence of R_f-values 5' -Ac-RF RF FX, which occurs in a less polar medium, is reversed in aqueous environment, we can conclude that the acid residue to be suspected in position of FX, must be greater and more hydrophylic than acetyl.

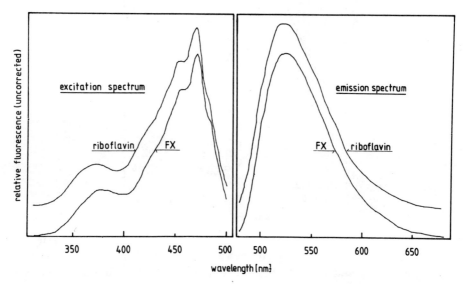

FIGURE 3. Uncorrected fluorescence excitation and emission spectra of FX and riboflavin of the same samples used in Fig. 2.
Excitation spectrum: emission at 520 nm.
Emission spectrum: excitation at 470 nm.
Again, the spectra were adjusted to the same height and separated for clearer illustration. The fluorescence quantum efficiencies of FX and riboflavin cannot be distinguished ($\sim$28%).

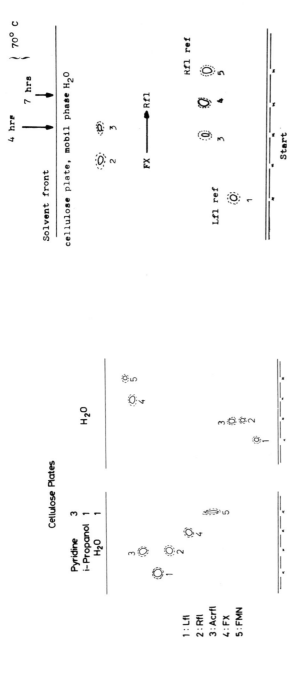

FIGURE 4a. Thin layer comparison (cellulose plates) of FX, riboflavin, FMN, lumiflavin and 5'-monoacetylriboflavin run in more or less polar medium.

FIGURE 4b. Thin layer control of the hydrolysis of FX to yield riboflavin in hot water saturated n-butanol (70°). Addition of acid or base dramatically increases the rate.

Tetraacetylriboflavin

5'-Monoacetylriboflavin
(cont. 50% Riboflavin)

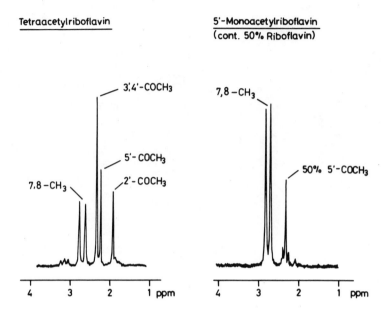

FIGURE 5. Comparison of tetraacetylriboflavin and 5'-monoacetyl-riboflavin by NMR spectroscopy. Riboflavin was 50% monoacetylated by keeping it in a 1:1 v/v mixture of glacial acetic acid and acetic anhydride for 12 hrs. After thorough removal of the solvent in vacuo, the solid showed in CF_3 COOH 50% 5'-acetyl per Fl_{ox}, as compared with authentic fully acetylated riboflavin. Only the acetyl and $ArCH_3$ region of the NMR-spectrum is shown.

ACKNOWLEDGEMENTS

     Mr. R. Mack is thanked for excellent technical assistance. This work was financially supported by the Deutsche Forschungs-gemeinschaft (Donderforschungsbereich 138, Projekt A1).

REFERENCES

1. W. Schmidt, in "Structure and Bonding",(in press) (1979).
2. M. H. Zenk, Z. Pflanzenphysiol. 56:57 (1967).

3. P. Hemmerich & V. Massey, in "Oxidases and Related Redox
   Systems III", T. E. King and H. S. Mason, eds., Pergamon
   Press, Oxford (1979).

4. P. Hemmerich, in "Progr. Chem. Org. Natl. Prod.", H.Griesebach,
   ed., Springer Verlag, Wien (1976).

5. P. Hemmerich, W. R. Knappe, R. Traber & H. E. A. Kramer,
   Eur. J. Biochem. (in press) (1979).

6. R. Traber & W. R. Knappe, Eur. J. Biochem. (in press) (1979).

7. P. Hemmerich & W. Haas, in "Recent Developments in the Study
   of "Fully Reduced Flavin" ", K. Yagi, ed., University of
   Tokyo Press, Tokyo (1975).

8. P. Hemmerich, R. Traber & H. E. A. Kramer, Eur. J. Biochem.
   (in press) (1979).

9. H. J. Duchstein, H. Fenner, P. Hemmerich, & W. R. Knappe,
   Eur. J. Biochem. 95:167 (1979).

10. P. Hemmerich, in "Transport by Proteins, FEBS Symp. N° 58",
    G. Blauer, & H. Sund, eds., Walter de Gruyer, Berlin
    (1979).

11. K. H. Dudley, A. Ehrenberg, P. Hemmerich & F. Müller, Helv.
    Chim. Acta 47:1354 (1964).

12. P. Hemmerich, V. Massey & H. Fenner, FEBS Letters 84:5 (1977).

13. A. W. Galston, Proc. Natl. Acad. Sci. 35:10 (1949).

14. A. H. Blaauw, Licht und Wachstum III, Mededeelingen, Landbou-
    whoogeschool, Wageningen 15:89 (1918).

15. G. Blankenhorn, D. T. Osuga, H. S. Lee, & R. E. Feeney, Bio-
    chim. Biophys. Acta 386:470 (1975).

16. A. Haas, & P. Hemmerich, Biochem. J. 181:95 (1979).

BASIC CONCEPTS IN BIOENERGETICS

Bruno A. Melandri and Giovanni Venturoli

Institute of Botany, University of Bologna

Bologna, ITALY

INTRODUCTION

A living cell is an open thermodynamic system which can exchange heat and matter with the surrounding world; the surrounding for a unicellular organism is represented by the environment in which it lives and for a multicellular organism by the adjacent cells with which a functional interrelation can exist. Cells are however also multicomponent systems, often subdivided into separate compartments, the intracellular organelles; this situation can be thermodynamically described as a multiphase system, in which the water phase, the predominant phase in a cell, is divided into multiple compartments by a different phase, the biological membranes. The thermodynamic equilibration within a phase implies that all intensive properties are constant throughout the volume of the phase itself (temperature, pressure, electric potential and chemical potentials). This constancy of the thermodynamic parameters does not apply, strictly speaking, to a narrow region adjacent to the boundary between phases; this interphase region is an independent part of the system, characterized by its own properties, different from those of the bulk phase.

The intracellular compartmentation hampers severely the theoretical and practical utilization of thermodynamics for biological systems; in fact it is often impossible to measure exactly the quantities useful for the complete thermodynamic description of a single compartment. As an example, the concentrations of specific chemical components, or the electric potential difference between compartments are indispensable for the correct determination of the thermodynamic state of a cell, but their experimental determination is often impossible.

A second severe limitation consists in the fact that classic thermodynamics deals, strictly speaking, only with reversible processes, taking place at an infinitely slow rate; only this extreme sluggishness can in fact satisfy the macroscopic homogeneity of the internal parameters within a system, which is the very essence of thermodynamic itself. This necessity is however in sharp contrast with the actual needs of a cell, which requires often very high rates of physiological transformations, which occur therefore in an irreversible way. This situation can be described, at least with a good approximation, by the thermodynamics of irreversible processes.

In spite of these limitations the thermodynamic approach is nevertheless for the students of biological sciences a valuable tool, which offers a rational framework for the description of cellular bioenergetics.

THE GIBBS' FREE ENERGY

The fundamental law for the description of an open system, a cell or an intracellular organelle, has been proposed by Gibbs, and links the internal energy of the system with its entropy, its volume and its composition. In its integrated form the Gibbs' equation takes the form:

$$(1) \quad U = TS - PV + \sum_i \mu_i n_i$$

where U is the internal energy, P and T the pressure and the temperature respectively, S the entropy, V the volume and μ_i and n_i the chemical potential and the number of moles of the component i. Eq. (1) holds for any single compartment of a composite system; therefore it holds also for the system as a whole, as a summation of the contributions of the different compartments.

Usually in a cell all compartments are thermally equilibrated and the pressure difference between them is negligible (P = cost.). Under these conditions a different function of state, the free energy, which can be obtained from a linear combination of other functions of state, is particularly useful. The free energy (G) is defined as:

(2) $G = U + PV - TS$

and its change in value measures the energy available to perform useful work (at T and P constant), or more precisely the maximal work obtainable during a transformation taking place in a reversible, quasi static fashion.

At constant T and P, the change in free energy corresponding to an infinitely small change in composition of the system can be obtained from eqs. (1) and (2), in their differential form:

(3) $dG_{T,P} = \sum_i \mu_i \, dn_i$

and therefore:

(4) $\mu_i = \left(\dfrac{\partial G}{\partial n_i} \right)_{T,P,n_{j \neq i}}$

The chemical potential μ_i is therefore an intensive property of the component i, which measures the change in free energy occurring upon addition of an infinitesimal amount of i to a system of well defined composition and at a given T and P. The actual value of the chemical potential is a function of the composition of the system; more specifically it depends upon the concentration of the component i, according to the relation:

(5) $\mu_i = \mu_i^0 \, (T,P) + RT \ln f_i C_i$

where μ_i^0 is the standard chemical potential (i.e. the chemical potential at $f_1 C_1 = 1$, namely of an ideal solution at unit concentration) and f_i is a parameter (the activity coefficient) depending on C_i and measuring the deviation of the component i from an ideal behavior.

When the component i is electrically charged, its chemical potential is conveniently substituted by the electrochemical potential $\tilde{\mu}_i$. This quantity takes into account the electrical work performed if charged molecules are transferred between two compartments which are at different electrostatic potentials. The explicit form of the electrochemical potential is:

(6) $\tilde{\mu}_i = \mu_i^0 \, (T,P) + RT \ln f_i C_i + Z_i F \psi$

where F is the Faraday constant, Z_i the electrical charge of i and ψ the electric potential of the compartment. It has been recognized,

however, that there is no physical way for measuring the chemical
part (concentration dependent) of the electrochemical potential
separately from the electric component; $\tilde{\mu}_i$ is therefore the primary
thermodynamic quantity and is only formally separable into a chemical and an electrostatic component. This situation arises from the
very large amount of energy necessary for a macroscopic separation
of positively and negatively charged species (electroneutrality
principle) and the consequent impossibility of specifying the electrostatic potential of phase in terms of its composition (owing to
the exceedingly small change in composition following the formation
of a difference in electrostatic potential caused by the transfer
of charged species between two adjacent compartments).

On the basis of the first and second principles of thermodynamics, from which eq. (1) is derived, special states of an open
system, characterized by a minimum of free energy, can be identified. These special states, states of equilibrium, can be recognized
from the constancy in time of their thermodynamic parameters. Any
other state of the system, for which the value of the free energy
is higher than the minimum, will evolve spontaneously (irreversibly)
towards an equilibrium condition, with a rate depending upon the
difference in free energy between the initial non-equilibrium and
the final equilibrium state. The absolute value of the minimum free
energy attainable at equilibrium is dictated by the kinetic internal
constraints of the system; these constraints do not allow the system
to reach equilibrium states for which, in principle, the free energy
could be still lower. If, for example, a biological membrane is completely impermeable to a solute, a difference in electrochemical
potential across this membrane will not drive the diffusion of that
solute and the system will be characterized by a free energy higher
than the theoretical minimum, in respect to the distribution of
that solute. The induction of a specific carrier (i.e. the removal
of kinetic constraints will allow the equilibration of the solute
and a lower value of the free energy will be reached by the system.

If the system is multiphasic and/or with more than one compartment, the overall equilibrium condition will be obtained if every
component i will be equally distributed in all phases, i.e. when
$\tilde{\mu}_i$ for all components will be constant throughout the system; this,
of course, will be possible only in the absence of diffusion barriers between the various compartments.

ISOTHERMAL CHEMICAL EQUILIBRIUM IN A MONOPHASIC SYSTEM

The utilization of the concepts outlined above in the case of isothermal chemical reactions is part of the elementary chemical thermodynamics. When the chemical interconversion of some components into others is the only energetic transformation taking place in an isothermal system, the equilibrium state can be studied considering the change that this reaction brings to the overall free energy. From eqs. (3-6), and applying the equilibrium condition ($\Delta G_{T,P}$=0), we obtain:

$$(7) \quad \Delta G_{T,P} = \sum_i \nu_i \tilde{\mu}_i = \Delta G^O_{T,P} + RT \sum_i \ln \left[f_i C_i \right]^{\nu_i}_{eq.} = 0$$

where the summation is extended to all products and reagents of the reaction, ν_i are the stoichiometric coefficients (positive for products and negative for reagents) and subscription eq. indicates the activity ratios, which determine an equilibrium situation. The value of $\Delta G^O_{T,P} \sum_i \nu_i \tilde{\mu}^O_i$, the standard free energy change, defines the equilibrium constant of the reaction at the temperature T.

If, as an example, a cell is considered as a monophasic system, the equilibrium condition for the metabolic reactions would correspond to an absolute minimum of free energy for all metabolites, and consequently to the impossibility to perform any kind of work. Of course this is not allowed by kinetic barriers on the reaction rates among metabolites, rates which are regulated by the concentration and activities of specific enzymes. This situation causes a steady metabolic flow continuously controlled and capable of performing useful work.

It must be underlined therefore that a correct quantitative study of the thermodynamic aspects of metabolism can be carried on, only if it is realized that the system is far from equilibrium and is not undergoing quasi static (reversible) transformations. In the following we will deal with some examples of this situation; in these examples, however, emphasis will be put on phenomena involving more than one compartment (mass transport phenomena and energy transductions associated with biological membranes).

FLOWS AND FORCES IN IRREVERSIBLE PROCESSES

A multicompartment thermodynamic system, far from equilibrium, but thermally and mechanically equilibrated (T = cost.; P = cost.), will evolve irreversibly towards a state of lower free energy. The

rates of the processes taking place during this transition will be
a function of the electrochemical potentials of all species in all
compartments, which are causing the disequilibrium situation. These
rates will be described, as a first approximation, by a set of linear phenomenological equations of the form:

$$(8) \quad J_i = L_{ii} X_i + \sum_{K \neq i} L_{iK} X_K \quad (i = 1,n)$$

In eq. (8) J_i represent generalized flows (rates of processes) and
X_i generalized forces (i.e. intensive quantities driving the flows).
There are no special restrictions in the definition of J_i and X_i;
the only condition is that the product $J_i X_i$ must have the dimensions
of a power (work performed per unit of time). In the special cases
discussed in this paper, however, we will deal only with mass fluxes
(mass transferred across a boundary per unit of time) and with rates
of chemical reactions (mass of product formed per unit of time):
the related forces will be respectively differences in electrochem-
ical potentials between two compartments $(-\Delta \tilde{\mu}_i = \tilde{\mu}_i^I - \tilde{\mu}_i^{II})$ and af-
finities of chemical reactions ($A_{Ch} = -\Delta G_{Ch}$, negative free energy
changes for chemical reactions).

Equations of the type (8) must be written for all processes
involved in the evolvement of the system towards equilibrium, and
will form a set of linear phenomenological equations describing
completely the overall spontaneous transformation. Two important
points need to be commented in regards to eq. (8): the various flows
are linearly correlated to the forces directly driving the process
(conjugated forces, e.g. X_i for J_i) through "straight" coefficients
L_{ii}; they are however related, also linearly, to other forces of
the system, X_K, by "cross" coefficients L_{iK}. Thus in eq. (8) the
coupling between different energetic transformations is explicitly
represented.

It must be understood that linear equations of the type (8)
are a simplification of the general interactions between flows and
forces, since the rate of a process is not always linear with a
thermodynamic force. Eq. (8) strictly hols in fact, only for conditions not too far from equilibrium, when the processes are relatively slow. However, in biological systems the linearity between
flows and forces has been tested experimentally in a surprisingly
large variety of situations and eqs. like (8) can be utilized for
the description of many physiological processes, for which a steady
state kinetics has been attained (steady state kinetics of enzymatic
reactions or of carrier mediated transport, etc).

In writing eq. (8) no "a priori" assumption has been made on the structure of the system which can be composed of several components and of several compartments. Phenomenological equations can be written therefore for the description of the coupling of two flows of mass (secondary active transport), for the coupling between chemical reactions in a homogeneous phase, or for the coupling between a chemical reaction and a flow of mass (primary active transport).

The complete experimental determination of all phenomenological coefficients L_{iK}(n x n constants) would be a formidable experimental task, should the number of independent constants not be drastically reduced by the simmetry condition of the coefficient matrix (demonstrated by Onsager), for which $L_{iK} = L_{Ki}$. This notation means that when an interaction between a flow and a non conjugated force occurs (energy coupling between two processes), the effect of the force K on the flow i, measured by the cross coefficient L_{iK}, is equal to the effect of force i on the flow K (measured by L_{Ki}).

THE DISSIPATION FUNCTION

When a system evolves to equilibrium at a finite rate a net increase of entropy of the system plus that of the surrounding takes place, owing to the irreversibility of the process. The rate of entropy production, multiplied by the absolute temperature of the isothermal transformation, defines the so called dissipation function ϕ and measures the rate of loss in free energy due to irreversibility. If all flows and forces are chosen correctly, so that the overall transformation of the system is represented completely, the dissipation function is given by the summation of all products of flow and conjugated forces, and has the dimension of a power:

$$(9) \quad \phi = \sum_{i}^{n} J_i X_i$$

For its own definition the dissipation function must always be a positive entity and will become equal to zero when the system is at equilibrium and no net production of entropy occurs. When external constraints are imposed on the system (e.g. a force is maintained constant by an external action), so that the number of the degrees of freedom is decreased, the flows and forces will eventually be stabilized during time and a stationary state will be attained. In these conditions the dissipation function is at a minimum, but is non-zero, and the rate of entropy production is

minimal. The constancy in time of the parameters of the system
should not be considered erroneously an equilibrium state: in fact
such a state would be characterized by $\phi = 0$ and by a complete stability of the intensive parameters without external constraint
$(\phi = - \dot{G}_{total} = 0$, i.e. $- G_{total} = $ cost. or $\Delta G = 0)$.

THE COUPLING BETWEEN TWO FLOWS OF MASS: SECONDARY ACTIVE TRANSPORT

Let us consider the interaction between flows of two different
chemical species which are present at different electromechanical
potentials in the aqueous phases of two cellular compartments
separated by a membrane; this rather common situation in a living
cell occurs when the flow of a metabolite is coupled to that of
another chemical species. This process is defined as secondary active
transport and involves generally an ion, in most cases Na^+ or H^+,
and ionic or non-ionic metabolites, such as aminoacids, carboxylic
acids and sugars.

The phenomenological equations of this simple process can be
quite nicely utilized as an example of irreversible thermodynamic
treatment of a simple energy transduction and to define parameters
not yet mentioned in earlier paragraphs. The phenomenological equations for the flows of the chemical species A and B between the two
compartments I and II will be:

$$(10a) \quad J_A = L_A \, (- \Delta \tilde{\mu}_A) + L_{AB} \, (- \Delta \tilde{\mu}_B)$$

$$(10b) \quad J_B = L_{AB} \, (- \Delta \tilde{\mu}_A) + L_B \, (- \Delta \tilde{\mu}_B)$$

In eqs. (10) the conventions used for the positive sign of flows
and forces are explicitly set: a flow $J_i^{I \to II} > 0$ will correspond to
a coniugate force $-\Delta \tilde{\mu}_i = -(\tilde{\mu}_i^{II} - \tilde{\mu}_i^I) > 0$. $(-\Delta \tilde{\mu}_i)$ will represent therefore the negative free energy, i.e. the positive affinity, for the
spontaneous diffusion of the component i from compartment I to compartment II. Thus a straight phenomenological coefficient is always
positive; this is not always true for the cross coefficient (see
below).

The three phenomenological coefficient L_A, L_B and L_{AB}, can be
substituted by other, resulting from combinations of the former.
If we define:

$$(11) \quad Z \equiv \sqrt{\frac{L_A}{L_B}} \qquad \text{and} \quad (12) \quad q \equiv \frac{L_{AB}}{\sqrt{L_A L_B}}$$

eqs. (10) will take the form:

$$(13a) \quad J_A = L_A \left[(-\Delta\tilde{\mu}_A) + \frac{q}{Z} (-\Delta\tilde{\mu}_B) \right]$$

$$(13b) \quad J_B = L_B \left[qZ (-\Delta\tilde{\mu}_A) + (-\Delta\tilde{\mu}_B) \right]$$

The coefficient q, the degree of coupling, is of fundamental importance since it defines the extent of interaction between the flows, and determines the maximal efficiency of the energetic transduction between $\Delta\tilde{\mu}_A$ and $\Delta\tilde{\mu}_B$. In fact, when q = 0 ($L_{AB} = 0$), the two flows are completely independent and their intensities will be determined only by the extent of their conjugate forces; the two flows are totally uncoupled. When q ≠ 0, however, the intensity of J_A will depend on that of J_B and "viceversa". Let us assume that for example $\tilde{\mu}_A^I > \tilde{\mu}_A^{II}$ and $\tilde{\mu}_B^I = \tilde{\mu}_B^{II}$. If q > 0, also in the absence of a conjugate force, a flow $J_B > 0$ will take place (since L_A, L_B and Z are positive coefficients). J_B will therefore be parallel to J_A, and a phenomenon of cotransport (symport) will occur, i.e. the accumulation of solute B at the expenses of an electrochemical potential difference of A.

If, on the other hand, q<0, a flow J_B will still be present, but in an opposite direction; this situation will be defined as a process of antitransport (antiport), in which the two flows are coupled in an antiparallel fashion.

The degree of coupling q has an upper limit of $|q| = 1$ (this limit derives from the positive sign of the dissipation function, for which must always be $L_{ii} L_{KK} \geq L_{iK}^2$, i.e. $|q| \leq 1$). In the limit ideal case in which q = ± 1 the two flows are completely coupled and $J_A/J_B = \pm Z$ (where the flow ratio is positive for symport and negative for antiport); only under these conditions Z represents the mechanicistic stoichiometry of the coupling, i.e. the stoichiometry of the molecular catalyst of the coupling (in this instance the carrier). This is however an ideal situation; in real systems (for which is $|q| < 1$), the flow ratio measures merely the "phenomenological" stoichiometry; in the real case, in fact, when no gradient of B exists, will be:

$$(14) \quad \frac{J_B}{J_A} = \frac{q}{Z} \quad (\text{for } \Delta\tilde{\mu}_B = 0)$$

namely a constant containing implicitly three different phenomenological coefficients and therefore ambiguous in its interpretation.

From this example the limitation of the thermodynamic approach

stems up clearly: this approach is scarcely able to supply direct
information about the molecular mechanisms of the coupling machinery (in this instance the carrier), but can only describe phenomeno-
logically the process of energy transduction. This limitation is
a direct consequence of the fact that thermodynamics is the study
of macroscopic transformations and cannot therefore discriminate
among different molecular mechanisms, which were all thermodynamically consistent. The undetermination of the flow ratio of real
processes is due to the fact that L_A, L_B and L_{AB} contain implicitly phenomena decreasing the efficiency of energy conversion in
the coupling of two flows. These include all passive leaks of A
and B across the membrane, all possible mediated flows (uniports
of A or B) and also the failures of the carrier in coupling the
two flows during a single catalytic turnover. These two different
ways of uncoupling the two flows have been defined as "external"
and "internal" leaks respectively by Heinz; this author has suggested that the straight coefficients can be considered as the sum
of the coefficients of coupled and uncoupled flows.

Lastly let us consider another property of a co-transport (or
antiport) system: if $(-\Delta\tilde{\mu}_A)$ is kept constant in a system of limited
capacity, a flow J_B will produce a force $(-\Delta\tilde{\mu}_B)$, which will be positive for symport or negative for antiport and which will eventual-
ly reach an upper limit, when J_B will become zero. This situation
corresponds to a stationary state (Φ is at a minimum) defined as
static head for solute B. Since the coupling is not complete,
however, $J_B = 0$ will not correspond to $J_A = 0$; at static head of
B the phenomenological equations (13) will become:

$$(15a) \qquad J_A = (1 - q^2)L_A \ (-\Delta\tilde{\mu}_A)$$
$$\text{(for } J_B=0)$$
$$(15b) \qquad (-\Delta\tilde{\mu}_B)_{max} = - \ qZ \ (-\Delta\tilde{\mu}_A)$$

The flow of A, which is dissipated in order to maintain solute
B at static head, is not only a function of $(-\Delta\tilde{\mu}_A)$, but also of the
degree of coupling (since in eq.(15a) the term q^2 appears, it
is irrelevant if we are dealing with symport or antiport). The
limit value of $(-\Delta\tilde{\mu}_B)$ will also be a function of the degree of
coupling. The force ratio at static head, often considered as a
measure of the stoichiometry of cotransport, contains therefore
again three undeterminated phenomenological coefficients. The me-
chanicistic stoichiometry of the carrier can be obtained at static
head only if $q = \pm 1$; in this case for B at static head J_A would

be zero, the dissipation function will also be zero and the stationary state would correspond to a real electrochemical equilibration of solutes A and B, mediated by the carrier, with a stoichiometry Z ($(-\Delta\tilde{\mu}_B) = \mp Z \ (-\Delta\tilde{\mu}_A)$).

THE COUPLING BETWEEN FLOW OF MASS AND CHEMICAL REACTIONS: PRIMARY
ACTIVE TRANSPORT

Primary active transport, i.e. direct coupling between an exoergonic chemical reaction (usually ATP hydrolysis) and the transmembrane translocation of certain ionic species is a very general process in living cells. The coupling is actuated by enzymes strictly integrated in the membrane structure (intrinsic proteins), which catalyze the translocation of ions across the membrane itself in parallel with the hydrolysis of ATP; for the latter process substrates are assumed and products are released on the same face of the membrane. These two processes appear mediated by two different parts of the enzyme: a first one, which catalyzes the chemical reaction and a second portion forming a ion-conducting gated channel. Examples of such enzymes are the Na^+, K^+-ATPase of plasma membranes in animal cells, the Ca^{++}-ATPase of sarcoplasmic reticulum and the H^+-ATPase of mitochondria, of chloroplasts and of bacterial plasmalemma *.

The phenomenological treatment of these processes is not fundamentally different from that of secondary active transport. Formally now the forces involved are the electrochemical potential differences across the membrane of the ion(s) translocated and the affinity of ATP hydrolysis in one compartment of the system. For the simple instance of the translocation of a single ionic species (e.g. the H^+-ATPase) the two phenomenological equations will be:

$$(16a) \qquad J_H = L_H(-\Delta\tilde{\mu}_H) + L_{HP}A_P = L_H \left[(-\Delta\tilde{\mu}_H) + \frac{q_{HP}}{Z_{HP}} A_P \right]$$

---

*   The phosphoenolpyruvate-dependent vectorial translocation of sugars, present in many bacteria (cfr. e.g. Kaback (1970)) is not discussed here since, strictly speaking, it cannot be considered as an example of primary active transport: in fact the sugar translocated is phosphorylated during its passage across the membrane and no difference in $\Delta\tilde{\mu}$ of the sugar is formed during the process.

$$(16b) \quad J_P = L_{HP}(-\Delta\tilde{\mu}_H) + L_P A_P = L_P \left[q_{HP} Z_{HP}(-\Delta\tilde{\mu}_H) + A_P \right]$$

where J_H and J_P are the flows of protons and the rate of ATP-hydrolysis respectively and A_P is the affinity of the reaction of ATP
decomposition (often indicated as "phosphate potential"):

$$(17) \quad A_P = -\Delta G_P = \tilde{\mu}_{ATP} - \tilde{\mu}_{ADP} - \tilde{\mu}_{Pi}$$

Analogously to eqs. (11) and (12):

$$(18) \quad Z_{HP} = \sqrt{L_H/L_P} \quad \text{and} \quad q_{HP} = \frac{L_{HP}}{\sqrt{L_H L_P}}$$

since A_P and $(-\Delta\tilde{\mu}_H)$ are the negative free energy for an exoergonic
chemical reaction and for proton diffusion respectively, L_H, L_P
and Z_{HP} are positive quantities. The sign of q_{HP} (or of L_{HP}) is
somewhat arbitrary and will be discussed below.

When Ap in compartment I is positive, as it is nearly always
the case, ATP-hydrolysis will cause a positive J_H and, if the
system has a finite capacity, a difference in $(-\Delta\tilde{\mu}_H)$, ($(-\Delta\tilde{\mu}_H) < 0$,
i.e. $\tilde{\mu}_H^{II} > \tilde{\mu}_H^{I}$). If the protonic current is measured when $-\Delta\tilde{\mu}_H = 0$,
however, the flow ratio will be again a measure of an ambiguous
parameter, including the stoichiometry and the degree of coupling
$(J_H/J_P = q_{HP}/Z_{HP})$. On the other hand, the maximal protonic electrochemical potential difference attainable at static head for protons
at a given phosphate potential will also be a function of the degree
of coupling ($(-\Delta\tilde{\mu}_H)_{max} = -q_{HP}/Z_{HP} A_P$ for $J_H = 0$). Again this undetermination derives from the incomplete coupling of the system
$(|q_{HP}| < 1)$ and reflects both the passive proton leaks of the membrane and the statistical failures in the interaction between the
gated channel and the active site for ATP-ase.

Equations (16) deserve some special comments: J_P, although
indicated as a generalized flow, is actually the rate of a reaction
taking place in an isotropic phase. J_P is therefore a scalar quantity and the same holds for A_P. This situation introduces the necessity to utilize a vectorial parameter in the phenomenological
description of the coupling of A_P with the vectorial proton flux
(Curie-Prigogine principle); in this case the vectorial parameter
is the cross coefficient L_{HP}, which includes not only a quantitation of the interaction between J_H and Ap (and between Jp and
$(-\Delta\tilde{\mu}_H)$ as well), but also the unidirectional nature of the structure
and function of the H^+-ATPase. Since in all cases ATP-hydrolysis
causes a flow of protons away from ATP active site, L_{HP} (and q_{HP} as

well) will be positive for an A_P measured in compartment I
$(J_H^{I \to II} > 0)$.

If the translocating system is reversible eqs. (16) will be
valid also with opposite signs: a negative $(-\Delta\tilde{\mu}_H)$ (i.e. $\tilde{\mu}_H^{II} \gg \tilde{\mu}_H^{I}$),
which will drive a negative J_H, will be able to drive also a nega-
tive J_P, i.e. ATP-synthesis, if A_P is sufficiently small, although
positive $(|-\Delta\tilde{\mu}_H| > |A_P/q_{HP}Z_{HP}|)$. Phosphorylation of ADP driven by
artificially formed ion gradients has been indeed demonstrated for
all the three systems mentioned above (Na^+, K^+-ATPase, Ca^{++}-ATPase
and H^+-ATPase).

ATP SYNTHESIS COUPLED TO REDOX REACTIONS

The synthesis of ATP coupled to redox reactions, a fundamental
process in respiration and photosynthesis, is the main, and in some
instances, the only primary source of chemical energy for a living
cell. This process is always associated with biomembranes and is
extremely sensitive to damages of the impermeability of the membrane to ions. According to the well known chemiosmotic hypothesis by
Mitchell, the mechanism of energy transduction between redox reactions and ATP synthesis is mediated by a circulation of protons
across the membrane, being the protons translocated by several
primary active transport systems, one of which the H^+-ATPase, operating in parallel and reversibly between two compartments. The
thermodynamic treatment of this complex system can evidentiate the
distinctive characters of chemiosmotic coupling and can offer some
experimental criteria for testing the validity of Mitchell's views.
Thermodynamics, however, will never be able, for its own nature,
to contribute to our knowledge of the molecular mechanisms of the
primary active transport processes.

The first non-equilibrium thermodynamic treatment of oxidative
phosphorylation was presented by Caplan and Essig and was further
discussed and studied experimentally by several authors. Caplan
and Essig proposed phenomenological equations for oxidative ATP-
-synthesis which are valid for any coupling mechanism and which are
taking into account also the flow of protons, since active proton
translocation is invariably linked to oxidoreductive reactions of
respiration or photosynthesis and to the hydrolysis of ATP catalyzed
by phosphorylating membranes. These general phenomenological equations take the form:

$$(18a) \quad J_P = L_P A_P + L_{HP} (-\Delta\tilde{\mu}_H) + L_{OP} A_O$$

(18b) $J_H = L_{HP}A_P + L_H (-\Delta\tilde{\mu}_H) + L_{OH}A_O$

(18c) $J_O = L_{OP}A_P + L_{OH}(-\Delta\tilde{\mu}_H) + L_O A_O$

where J_O and A_O are respectively the rate and the affinity of a
redox reaction and the other quantities have the same meaning as
in eqs. (16).

If the chemiosmotic mechanism holds, however, the translocation of protons is driven by two enzymatically independent processes; the system can be therefore subdivided into two subsystems
(α and β), in which only a redox reaction of ATP hydrolysis drives
the proton current. Eqs.(18) can be rewritten accordingly as follows:
for subsystems α, ATP-driven:

(19a) $J_P = L_P A_P + L_{HP} (-\Delta\tilde{\mu}_H)$

(19b) $J_H^{\alpha} = L_{HP}A_P + L_H^{\alpha}(-\Delta\tilde{\mu}_H)$

and for subsystem β, driven by redox reactions:

(19c) $J_H^{\beta} = L_H^{\beta}(-\Delta\tilde{\mu}_H) + L_{OH}A_O$

(19d) $J_O = L_{OH}(-\Delta\tilde{\mu}_H) + L_O A_O$

In eqs. (19) the subsystem has to be specified only for the flows
of protons and for the straight coefficients L_H, since all forces
are determined in the same homogeneous compartment and the coupling
between protonic current and oxidative or ATPase reaction is specific for a given subsystem. The net proton flow between the two
compartments will be obtained from the summation of the partial
flows in the two subsystems:

(19e) $J_H = J_H^{\alpha} + J_H^{\beta} = L_{HP}A_P + (L_H^{\alpha} + L_H^{\beta})(-\Delta\tilde{\mu}_H) + L_{OH}A_O$

and the four equations (19a-d) will be again reduced to a set of
three equations (eq. 19a, 19d and 19e) which will describe pheno-
menologically the three basic flows of the process.

From a comparison of eqs. (18) and eqs. (19) a distinctive
characteristic of the chemiosmotic mechanism is readily apparent,
namely the absence of a cross coefficient between redox system
and ATP synthesis, i.e.:

(20) $L_{OP} = 0$

This conclusion means that when $(-\Delta\tilde{\mu}_H)$ is maintained constant $(J_H=$ = 0, for a system of finite capacity) the rates of the redox and ATPase reactions are independent from each other and will depend only on the extent of the respective conjugate force (plus a constant contribution given by the protonic force). In particular, when $(-\Delta\tilde{\mu}_H)=0$ J_P and J_O are totally uncoupled; it has been indeed amply demonstrated that all uncouplers of oxidative or photosynthetic phosphorylation can act as transmembrane proton carriers, capable of dissipating the electrochemical potential difference of protons through passive protonic flows across the membrane.

In an intact, unperturbated system of finite capacity, however, an exoergonic reaction, such as the aerobic oxidation of a respiratory coenzyme or the oxidation of an electron acceptor photoreduced during a photosynthetic event, will be able to generate a $J_H>0$ and a negative $(-\Delta\tilde{\mu}_H)(\tilde{\mu}_H^{II} > \tilde{\mu}_H^{I})$; the upper limit for $(-\Delta\tilde{\mu}_H)$ will correspond to the condition of static head for protons of the redox proton pump (eq. 19c). If this $(-\Delta\tilde{\mu}_H)$ is large enough as compared to A_P $(|(-\Delta\tilde{\mu}_H)| > |L_P/L_{HP}A_P|)$ the protonic force will drive also a negative J_P, i.e. synthesis of ATP produced by the operation in reverse of the H^+-ATPase. This mechanism represents in synthesis the essential concept of chemiosmotic coupling, whose basic validity is now generally accepted.

Chemiosmotic coupling can occur several times along a respiratory enzymatic system, while electrons are delivered to more and more positive electron acceptors. Actually a typical respiratory chain is subdivided stepwise into three (or sometimes only two) sequential oxidoreductive segments, of which every one is coupled to the translocation of protons and therefore to ATP synthesis (phosphorylation sites). The J_P/J_O flow ratio (the so called P/O ratio, a measurement routinely performed experimentally) cannot however in any way be considered as a stoichiometric ratio, since again it represents a phenomenological quantity including the degree of coupling of the oxidoreductive and of the ATPase proton translocators. Since these degrees of coupling in real systems are certainly less than one (although in mammalian mitochondria the overall degree of coupling can be as high as 0.92), no integral number for the J_P/J_O ratio must be expected.

A situation opposite to the one described above can also occur: when the phosphate potential A_P is high enough to generate a large $(-\Delta\tilde{\mu}_H)$, an inversion of the oxidoreductive proton translocator and

an endoergonic reduction of an electron carrier by a more positive
electron donor can take place. This phenomenon is the so called
reverse electron flow driven by ATP hydrolysis, and, especially
for bacteria, represents a major physiological pathway for the reduction of low potential coenzymes (e.g. NADH, $E_{m,7}$=-0.320 V),
extracting electrons from substrates at relatively high redox potential (e.g. succinate, $E_{m,7}$ = + 0.005 V). The same kind of interaction, i.e. the coupling of exoergonic and endoergonic reactions
can also take place between different segments of the respiratory
chain (oxidation driven reverse electron flow) or between respiratory and photosynthetic enzyme system (light-induced reverse electron flow in facultative photosynthetic bacteria).

Thus, as a consequence of chemiosmotic coupling, the mechanism
of energy transduction between oxidoreductive reactions and group
transfer reactions,both scalar in nature, is reconducted to the
more general principle of primary active transport of ions (protons).
For this function biological membranes play obviously a fundamental
role, since they represent the diffusion barrier indispensable for
conserving the free energy of ion gradients; in addition they supply
the necessary structural support for the unidirectional function of
the enzyme catalyzing the coupling between scalar reactions and
vectorial proton translocation.

A last comment may be concerned with the role of transmembrane
electrochemical potential difference of protons as primary free
energy source for living cells. A large body of evidence indicates
in fact that in bacteria and in intracellular organelles the diffusion of protons is coupled to the secondary active transport of
many ions and metabolites, by the action of symport and antiport
carriers. Energy for the accumulation of useful substrates can
therefore be supplied directly by the photosynthetic or respiratory
reactions, avoiding the utilization of ATP for performing this
kind of work. On the other hand, in anaerobic conditions or in
organisms lacking a photosynthetic or respiratory metabolism, the
same function can be performed by the H^{+}-ATPase, which exploits
as energy source the hydrolysis of the ATP produced by fermentation.

REFERENCES

Introductory Reviews

R.S. Caplan, in "Current Topics in Bioenergetics", D. R. Sanadi, ed.,
    Vol. 1, Academic Press, New York (1971).

H. Rottenberg, R. S. Caplan & A. Essig, "Membranes and Ion Trans-
    port", E.E. Bittar Ed., Vol. 1, Wiley, New York (1970).

H. R. Kaback, "Current Topics in Membranes and Transport", F. Bron-
    ner and A. Kleinzeller Eds., Vol. 1, Academic Press, New York
    (1970).

E. Heinz, "Current Topics in Membranes and Transport", F. Bronner
    and A. Kleinzeller Eds., Vol. 5, Academic Press, New York
    (1974).

D. Walz, Biochim. Biophys. Acta 505:279 (1979).

Advanced Textbooks

K. Denbigh, "The Principles of Chemical Equilibrium", Cambridge
    University Press, London (1964).

E. A. Guggenheim, "Thermodynamics", North Holland Publishing Co.,
    Amsterdam (1950).

E. Katchalsky & P. Curran, "Non-equilibrium Thermodynamics in
    Biophysics", Harvard University Press, Cambridge, Mass. (1965).

I. Prigogine, "Introduction to Thermodynamics of Irreversible Pro-
    cesses", Interscience Pub., New York (1967).

PROPERTIES OF THE PHOTOSYNTHETIC MEMBRANE

Jan Amesz

Dept. of Biophysics, Huygens Laboratory
University of Leiden
Leiden, The Netherlands

INTRODUCTION

The chloroplast is the major biological energy converter on earth, and the amount of pigment involved in photosynthesis is many orders of magnitude larger than involved in all other photobiological processes taken together. Under optimal conditions the conversion of light energy occurs with an efficiency of 30-35 percent, nearly half the maximum efficiency thermodynamically obtainable (1-3). For reasons that need not concern us here, in practice the efficiency is usually considerably lower. Nevertheless, the amount of energy fixed by photosynthesis exceeds that expended by modern technology by at least an order of magnitude.

All of the basic processes of photosynthesis occur in specialized membranes. The primary reaction of photosynthesis consists of the transfer of an electron from an excited bacteriochlorophyll or chlorophyll molecule to an acceptor. Only a small fraction (in the order of one percent) of the chlorophyll or bacteriochlorophyll present in the membrane takes part in the primary reaction; the remainder serves, together with other pigments, as "light-harvesting" pigments to collect the light energy and to transfer it to the photoactive pigment in the reaction center.

The most widely used method to study primary and secondary electron transport in photosynthesis is by measurement of the small changes of absorbance due to changes in the redox state of reactants that occur upon illumination. This method is well suited to

study the photooxidation of the reaction center pigment. In purple
bacteria (with the exception of those species that contain bacteriochlorophyll b instead of bacteriochlorophyll a) oxidation results
in the disappearance of an absorption band near 870-890 nm (4).
For this reason the reaction center bacteriochlorophyll is called
P870. By use of ultra-rapid flash spectroscopy it has recently been
demonstrated that the photooxidation of P870 is a very fast reaction, and that the oxidized form (P870$^+$) can be observed within a
few ps after the onset of the flash (5,6), which confirms the hypothesis that P870 is the primary electron donor in purple bacteria.
Another method which has become quite important for studying the
mechanism of electron transport is measurement of electron spin
resonance (ESR). The free radical signal of P870$^+$ was among the first
to be observed upon illumination of photosynthetic material. The
width of the ESR line as well as the results of ENDOR studies indicate that P870$^+$ is a dimer, the free radical being delocalized
over both bacteriochlorophylls (7-9). Other substances that can
be studied by ESR are metallo-proteins, such as the iron-sulfur
proteins that serve as electron acceptors in higher plant photosynthesis (9,10).

A basic difference between photosynthesis in bacteria and in
higher plants is that the latter contain two different types of
reaction center with different primary electron donors and acceptors. Each type of reaction center is associated with its own set
of light-harvesting pigment molecules. Fig. 1 gives a scheme of
electron transport in green plants. The vertical arrows in this
scheme denote the primary electron transfer reactions. All other
arrows denote "dark" reactions; they serve to bring about the oxidation of water to oxygen, to connect the two photosystems and to
produce NADPH which is used in the Calvin cycle for the reduction
of CO_2. The primary electron donors of photosystem 1 and 2 are
called P700 and P680, respectively, after the wavelength of maximum
bleaching upon oxidation (11). Both compounds are probably chlorophyll a dimers. From the midpoint potentials of the primary reactants it can be seen that for each photosystem about 1.0 eV is
produced as "chemical" energy per electron transferred. Since a
little more than one light quantum is needed to transfer one
electron, this corresponds to an energy conversion yield of 1.0/
/1.82 $\approx$ 55 percent in red light (680 nm). The overall yield for
the transfer of electrons from water to NADP is only about 30 percent because of the losses occurring in the dark reactions. However,
when electron transport is coupled to phosphorylation, like in

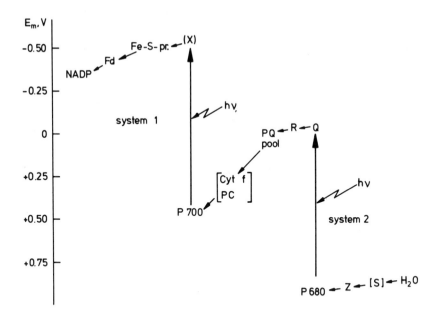

FIGURE 1. Simplified scheme of electron transport in oxygen evolving organisms. The arrows indicate the direction of electron transport. The vertical scale indicates the approximate or estimated redox potentials of the intermediates shown. The primary reactions of systems 1 and 2 are denoted by vertical arrows. The meaning of the various symbols is discussed in the text.

intact cells, an additional amount of energy is stored in ATP and the yield may be increased to 35 per cent or perhaps more, depending on the stoichiometry of ATP production (12).

THE PRIMARY REACTION

The nature of the primary reaction has been most extensively studied in purple bacteria. The main reason is that from many species of purple bacteria it is possible to obtain so-called reaction center particles, i.e. relatively small pigment-protein complexes that in their purest form are devoid of light-harvesting pigments and contain only the pigment and reaction components that are essential for the primary photochemical reaction (13). These

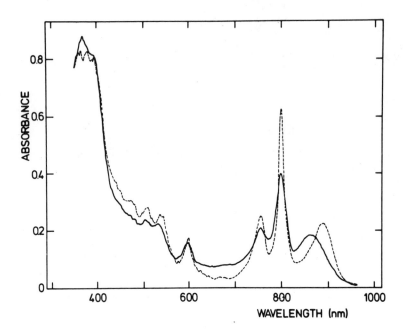

FIGURE 2. Absorption spectra of a reaction center preparation obtained from the purple bacterium <u>Chromatium vinosum</u> (15), measured at room temperature (solid line) and at 100 K (broken line). The bands near 600, 800 and 870 nm are due to bacteriochlorophyll a. The latter band disappears upon oxidation of the primary electron donor P870. The band near 750 nm is due to bacteriopheophytin.

complexes can be prepared by use of various detergents; their molecular weights vary between 85 and 200 kD or more, depending on source and method of preparation. They contain four bacteriochlorophyll and two bacteriopheophytin molecules per reaction center. Fig. 2 shows the absorption spectrum of such a preparation. Studies of light-induced ESR signals and absorbance changes in the blue and near-ultraviolet region of the spectrum indicate that the so-called primary electron acceptor in purple bacteria is a quinone--iron complex, which upon reduction is converted to the semiquinone anion. The quinone is either ubiquinone, as in <u>Rhodopseudomonas sphaeroides</u> and <u>Rhodospirillum rubrum</u>, or a naphtoquinone, presumably menaquinone (vitamin K_2), as in <u>Chromatium vinosum</u> (14,15). By use of flash spectroscopy it has been shown that the quinone is not the very first electron acceptor: it was observed that the

bleaching of P870, induced by a very short laser flash, occurs
within a few ps (5,6,16,17), but the absorption difference spectrum
measured shortly after the flash showed additional bands that could
be explained by the photoreduction of bacteriopheophytin (18),
which is now usually denoted as the "intermediary acceptor", I. The
intermediary charge pair P870$^+$ I$^-$ is quite unstable, and decays in
about 10 ns if a flash is given when the quinone is in the reduced
state (19). Normally, I$^-$ reduces the quinone in about 200 ps. The
charge stabilization obtained in this way occurs at the expense of
an energy loss of at least 0.5 eV, since the midpoint potentials
of the quinone and I are about - 100 and less than - 600 mV, respectively (20). Further stabilization is brought about by secondary
electron transport, such as the reduction of P870$^+$ by cytochrome c
(21). It is possible to accumulate I$^-$ by prolonged illumination in
reducing conditions in preparations where the reaction between
cytochrome c and P870$^+$ does occur. By this method the difference
spectrum of pheophytin reduction can be obtained free of interfering absorbance changes due to P870 (20).

Relatively little is known about the primary reaction in green
bacteria. The primary electron donor, P840, is probably bacteriochlorophyll a, and, as in purple bacteria, its photooxidation has
been shown to occur at temperatures down to a few K (22,23). The
redox potentials of P840 and of the primary electron acceptor are
lower than in purple bacteria. For this reason it is unlikely that
the primary acceptor is a quinone; it may be an iron-sulfur protein
as in system 1 of algae and higher plants.

Signals due to photooxidation of P700, the primary electron
donor of photosystem 1 of algae and higher plants (11,16,17) were
first observed more than 20 years ago by means of optical and ESR
measurements. The electron acceptors of system 1 (X and Fe-S-pr.
in Fig. 1) can be distinguished on the basis of their redox potentials by measurement of the ESR signals induced by illumination at
low temperature and low redox potential (10,24,25), and also on the
basis of their kinetics at room temperature by means of optical
measurements (26). The results indicate that there are two iron-
sulfur centers with fairly similar ESR spectra that can be reduced
by light. Optical measurements suggest that X, the primary acceptor,
may be an iron-sulfur protein also (27), but the ESR spectrum of
X is somewhat unusual for such a compound. Measurements of absorbance changes induced by a flash have shown that the photooxidation
of P700 occurs in less than 10 ps (28), and indicate that this
reaction is coupled to the reduction of an intermediary acceptor,

as in purple bacteria (27,28). In contrast to purple bacteria, however, the acceptor appears to be chlorophyll a rather than pheophytin as indicated by the absorption difference spectrum.

The photooxidation of P680, the electron donor of system 2 is more difficult to observe than that of P700, because the reaction of $P680^+$ with the secondary electron donor Z (Fig. 1) is very fast (16,29-31). However, the life time of P680 can be increased by applying special conditions, such as low temperature, low pH, detergent treatment and the addition of oxidants, and in these conditions light-induced absorbance and ESR signals that can be attributed to the formation of $P680^+$ are more easily observed (16,31). The acceptor side of system 2 bears resemblance to the photosystem of purple bacteria: measurements of light-induced absorbance changes under reducing conditions suggest that the intermediary acceptor is pheophytin a (32). The primary acceptor Q (Fig. 1) is probably plastoquinone (31,33) a substance related to ubiquinone that occurs only in green plant material.

SECONDARY ELECTRON TRANSPORT

On the reducing side of system 1 the electron transport chain contains, besides the iron-sulfur proteins mentioned already, a soluble ferredoxin (Fd in the scheme of Fig. 1), a flavoprotein (ferredoxin-NADP-reductase, not shown in the figure) and NADP. NADP, together with ATP is used to drive the Calvin cycle for the conversion of CO_2. Evidence for this portion of the electron transport pathway comes mainly from biochemical studies with isolated chloroplasts (34). Electron transport components between system 1 and system 2 include the copper-protein plastocyanin, cytochrome f, a relatively large pool of plastoquinone and an intermediate R which, like Q, may be a "special" plastoquinone molecule, distinct from the main plastoquinone pool (31,34). There are still many details that require clarification. Studies of the reaction rates of P700 and plastocyanin suggest that there is an as yet unidentified intermediate between these two components (35) and that cytochrome f may be located in a side path (36). Moreover, the reaction sequence between these compounds and plastoquinone may be more complicated than suggested by the scheme of Fig. 1 (37). Secondary electron transport in bacteria is reviewed in ref. 21.

A small fraction of the light energy absorbed by the photosynthetic pigments is re-emitted as fluorescence by chlorophyll or bacteriochlorophyll. Fluorescence studies have yielded much infor-

mation, not only about the properties of the pigments and transfer
of excitation energy between them, but also about primary and secondary photosynthetic electron transport. In oxygen evolving organisms most of the chlorophyll fluorescence is emitted by chlorophyll a molecules associated with system 2. As the yield of this
fluorescence is strongly affected by the oxidation-reduction level
of the primary reactants Q and P680, fluorescence measurements have
traditionally provided a useful tool to study primary and secondary electron transport near system 2 (31,38). A phenomenon that is
peculiar to photosynthetic material, and does not occur in solutions
of chlorophyll is that of delayed fluorescence. The emission spectrum of delayed fluorescence is essentially the same as that of
"prompt" fluorescence, but the emission occurs in the μs and ms,
instead of the ns and sub-ns time range. The effect is generally
assumed to arise from a reversal of the primary photochemical reaction (in algae and higher plants mainly of system 2), causing re-
excitation of chlorophyll or bacteriochlorophyll. Since the rate
of back reaction of the primary products is a function of many
parameters, delayed fluorescence has been studied extensively in
order to obtain information about the mechanism of photosynthesis.
Recent reviews are given by refs. 39 and 40.

Little is known about the identity of the intermediates
between water and P680, and most of the information about this
section of the electron transport chain comes either from measurements of oxygen evolution or from studies of the yield of fluorescence and of delayed fluorescence of chlorophyll a, often combined with the use of inhibitors or treatments to bring about more
or less specific inactivation of components of the chain (31,34).
The results of these studies indicate that the terminal enzyme (S)
may be a manganese containing protein that is able to accomodate
four oxidation equivalents ("positive charges") in a sequential
set of reactions with the oxidized intermediate Z. The chemical
nature of Z is not known; its existence has been mainly inferred
from the kinetic of P680 (16,31).

PIGMENT-PROTEIN COMPLEXES

With a few exceptions, the light-harvesting and reaction center pigments are located in the photosynthetic membrane. From work
in several laboratories it has gradually became apparent that these
pigments are not incorporated in the lipid bilayer, but that they
are contained in protein complexes. As a rule, these are hydrophobic

proteins, that are partially submerged in the membrane, and therefore much of the available information is based on detergent fractionation and solubilization of the membrane components (41,43).

Two types of pigment protein complexes can be discerned: the
light-harvesting and the reaction center complexes. The reaction
center complexes from purple bacteria were mentioned already in an
earlier section. In addition to these two types of light-harvesting
complexes have been obtained from purple bacteria. The first type
contains bacteriochlorophylls absorbing at 850 and 800 nm; the
second type shows an absorption maximum at 870-890 nm (42). The
basic units of these complexes appear to contain two or three
bacteriochlorophylls and one carotenoid molecule; the complexes
show no reaction center activity and are apparently only involved
in light absorption.

A somewhat different situation appears to exist in green bacteria and in algae and higher plants. From chloroplasts of higher
plants and green algae a light-harvesting pigment-protein complex
has been obtained that contains approximately equal amounts of
chlorophyll a and b (41,43). In addition to this a complex that
contains the reaction center of system 1 has been isolated. However,
in contrast to what was observed in purple bacteria, this "complex
I" is not a "pure" reaction center complex, but it contains about
40 chlorophyll a molecules per P700 (41,43). The complex has been
isolated from a large variety of organisms, including red and blue-
green algae. Its characteristics are remarkably independent of the
source of material, which suggests that it originated early in the
evolution of photosynthesis. A similar complex may exist in system
2 (43).

The major bacteriochlorophyll of green bacteria is chlorobium
chlorophyll (bacteriochlorophyll c, d or e). Fractionation experiments have shown that chlorobium chlorophyll is present in the
so-called chlorobium vesicles, which are situated adjacent to the
cytoplasmic membrane. In this membrane, bacteriochlorophyll a and
the reaction center appear to be located (22). The amount of
antenna pigment in green bacteria relative to the number of reaction
centers is much larger than in other photosynthetic organisms,
presumably as an adaptation to the natural habitat of these bacteria, which is in the deeper anaerobic layers of fresh water ponds,
where little light penetrates. A typical bacteriochlorophyll to
reaction center ratio is 1500.

After disruption of the cells and repeated gradient centrifugation it is possible to obtain a membrane fraction devoid of

chlorobium chlorophyll and containing about 80 bacteriochlorophyll
a molecules per reaction center (22,44). Treatment of this preparation with the chaotropic agent guanidine-HCl released a water
soluble light harvesting bacteriochlorophyll a protein complex
accounting for about half of the total bacteriochlorophyll a present,
but the reaction center activity in the remaining membrane preparation appeared to be largely destroyed (23). The light-harvesting
complex was found to be identical to one earlier obtained by extraction of broken cells at high pH. This complex so far is the only
one that has been fully characterized by X-ray crystallography (45).
It has a molecular weight of about 145 kD and consists of three
identical subunits, each containing 7 bacteriochlorophyll a molecules.

Swarthoff (unpublished experiments) recently succeeded in obtaining photochemically active pigment-protein complexes from a
green bacterium. Starting from a membrane fraction a pigment protein
complex containing about 75 bacteriochlorophyll a molecules per
reaction center was obtained by detergent treatment. The complex,
which apparently contained all or nearby all the bacteriochlorophyll
present in the membrane has a molecular weight of about 600 kD.
Subsequent treatment with guanidine-HCl yielded a complex of about
350 kD, which contained about 35 bacteriochlorophylls per reaction
center. The light-induced difference spectrum of this complex
(Fig. 3) was similar to that of the membrane preparation used as
starting material, indicating that the structure of the reaction
center is not basically affected by the isolation procedure. It
can be seen that the spectrum is quite complicated. The bands at
610 and 842 nm are usually ascribed to oxidation of P840; the band
at 553 nm is due to oxidation of cytochrome c553. The other bands
that are observed in the spectrum are probably in part due to
changes in exciton interaction between P840 and neighboring pigment
molecules and to the electric field resulting from the light-induced electron transfer. At present it is not clear if there is a
significant contribution to the spectrum caused by the reduction
of one or more electron acceptors.

ELECTROCHROMISM

A phenomenon that has been studied extensively in various
laboratories is the electrochromism of photosynthetic pigments.
It is caused by the interaction between an electric field and
a permanent or induced electrical dipole. In such a field the

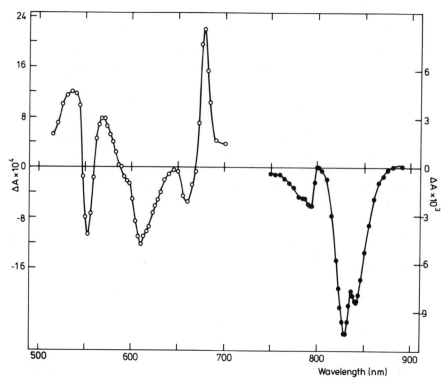

FIGURE 3. Spectrum of the absorbance changes induced in a reaction center pigment complex of the green photosynthetic bacterium <u>Prosthecochloris aestuarii</u> by illumination. The left-hand scale refers to measurements in the region 500-700 nm, the right-hand scale applies to the region 750-900 nm.

energy levels of the molecule will be changed by an amount equal to the scalar product of the dipole moment ($\vec{\mu}$) and the electrical field. As the dipole moments and polarizabilities are different for different electronic states, the energy differences between two states, and thus the absorption spectrum will be affected by the field. When, for an ensemble of molecules, the dipoles are more or less aligned, onset of the electrical field will result in (small) shifts of the absorption bands; if the dipoles are randomly oriented broadening of the absorption bands will result. The extent of the shift is proportional to the scalar product of $\Delta \vec{\mu}$ and field. Therefore, if the molecule has a dipole, either permanent or induced by charges on neighboring molecules, the shift is to first approximation linear with the field. Otherwise the extent of the

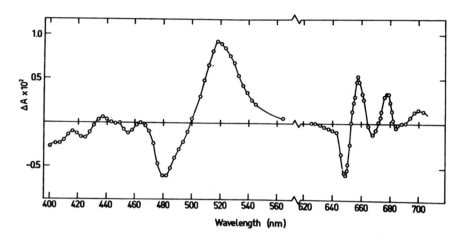

FIGURE 4. Spectrum of the electrochromic absorbance changes observed in spinach chlorophlasts upon illumination (48).

shift is quadratic with the field and proportional to the difference in polarizability in the ground and excited states. Carotenoids have a high polarizability, but some are symmetrical and have
no dipole moment of their own.

    Light-induced shifts of the absorption spectra of photosyn-
thetic pigments have been observed in a large variety of organisms.
The best known example of such an effect is the well-known absorbance increase at 515-520 nm, caused by a red shift of the long-
wave band of a carotenoid (see reviews 31, 46, 47). The complete
spectrum of the electrochromic absorbance changes of spinach chloroplasts is shown in Fig. 4. It can be seen that the spectrum contains not only contributions of carotenoid, but also of chlorophyll
b (e.g. near 650 nm) and of chlorophyll a. This spectrum was recorded in such conditions that contributions due to changes in the
redox state of electron carriers, including the primary photoreact-
ants, were minimized (48). It is now generally accepted that these
absorbance changes are due to the generation of an electrical potential across the photosynthetic membrane, caused by light-induced electron transport. There is various evidence that, as formu-
lated by Mitchell, the electron donors and acceptors of photosynthesis are located near opposite sides of the membrane (47,49) so
that the electron transport acts as an "electrogenic" pump by
transfer of electrons from one side of the membrane to the other.

Ionic movements in the aqueous phases on both sides of the membrane bring about a rapid delocalization of the potential. Measurements at -30 to -40°C, where the leakage of charge through the
membrane due to diffusion of ions is very small, so that it is
possible to obtain large membrane potentials, provided a quantitative analysis of the electrochromic absorbance changes in membrane
preparations of the purple bacterium Rhodopseudomonas sphaeroides
(50). It was found that a membrane potential of 100 mV, corresponding to a field of about 2×10^5 V cm^{-1}, gave a shift of a few
tenths of a nm of the absorption bands of the carotenoid spheroidene; for bacteriochlorophyll the shift was three times smaller.
These values are in good agreement with the theory of electrochromism. Interesting, only about one-third of the spheroidene showed
electrochromism. This pool apparently has a large induced dipole,
caused by neighboring molecules, perhaps bacteriochlorophyll (51),
and is present in an ordered array; the remaining two-third, although chemically identical, shows no or little electrochromism.
This is presumably due to the absence of a permanently induced
dipole, as indicated by the absorption spectrum of this pool. We
shall see that there is evidence for similar pools in spinach
chloroplasts also.

   De Grooth et al. (52) recently studied electrochromism in
photosynthetic membranes induced by an externally applied electric
field. Osmotically swollen chloroplasts were used for these experiments. These consist of spherical vesicles, called "blebs",
obtained by dilution of a chloroplast suspension by a large volume
of distilled water, which are aften more than 10 μm in diameter
and are formed by unfolding of the thylakoid membrane. It can be
shown that, if such a suspension is brought between two electrodes,
the field in the membrane, which has a much lower conductivity
than the surrounding and internal media, is strongly enhanced as
compared to the average field in the system. The field, and thus
the potential is proportional to the diameter of the bleb, and
e.g. for a bleb of 10 μm diameter, the maximum enhancement (in
the regions opposite the electrodes) is 1500-fold. Thus, potential
differences across the membrane comparable to those obtained by
illumination, and even larger, are easily obtained. There is,
however, one fundamental difference: upon illumination a membrane
potential is generated with the same polarity in the whole thylakoid system, whereas the externally applied field induces potential
differences that are inverted in one half of the bleb as compared
to the other half. Therefore, effects that are linearly proportio-

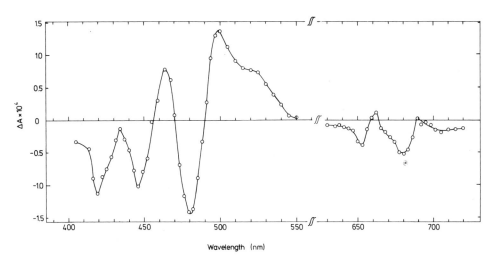

FIGURE 5. Spectrum of the electrochromic absorbance changes induced in the thylakoid membrane of spinach chloroplasts by an external el-electric field (52).

nal to the field and are due to existing dipoles will tend to cancel out, and only quadratic effect will contribute significantly to the difference spectrum. As it can be seen by comparing Figs. 4 and 5, the difference spectrum is indeed quite different from that obtained by illumination; in the blue-green region the spectrum is now dominated by carotenoids that do not have a (permanent) dipole and which absorbs at shorter wavelength than those responding to a light-induced field, and the absorbance increase at 520 nm is replaced by one at 500 nm. The generation of the field across the membrane is not instantaneous, when an electrical pulse is applied, but its rate depends on the capacity of the membrane and the resistance of the medium. By measurement of the rate of the electrochromic changes, which follow the field in the membrane practically instantaneously, the specific capacitance of the membrane was found to be 3 μF cm^{-2}.

Measurements of electrochromism, especially when combined with those of linear dichroism thus provide evidence about the structure of the membrane and the location of the pigments. If combined with measurement of electron transport reactions, they also give information about the location of the electron transfer components (see e.g. ref. 21) and, in some cases, about their reaction sequences (e.g. 37).

Research performed in this laboratory was supported by the Netherlands Foundation for Chemical Research (SON), financed by the Netherlands Organization for the Advancement of Pure Research (ZWO).

REFERENCES

1. L. N. M. Duysens, & J. Amesz, Plant Physiol.34:210 (1979).
2. R. T. Ross, & M. Calvin, Biophys. J. 7:595 (1967).
3. R. S. Knox, in "Primary Processes of Photosynthesis", S. Barber, ed., Elsevier, Amsterdam (1977).
4. W. W. Parson, in "The Photosynthetic Bacteria", R. K. Clayton, and W. R. Sistrom, eds., Plenum Press, New York-London (1978).
5. W. W. Parson, R. K. Clayton, & J. J. Cogdell, Biochim. Biophys. Acta 387:265 (1975).
6. P. L. Dutton, K. J. Kaufmann, B. Chance, & P. M. Rentzepis, FEBS Letters 60:275 (1975).
7. J. R. Norris, M. E. Druyan, & J. J. Katz, J. Am. Chem. Soc. 95:1680 (1974).
8. G. Feher, A. J. Hoff, R. A. Isaacson, & L. C. Ackerson, Ann. New York Acad. Sci. 244:239 (1975).
9. A. J. Hoff, Physics Reports (in press) (1979).
10. R. Malkin, & A. J. Bearden, Biochim. Biophys. Acta 505:147 (1978).
11. A. J. Bearden, & R. Malkin, Quart. Rev. Biophys. 7:131 (1975).
12. M. Avron, Ann. Rev. Biochem. 46:143 (1977).
13. G. Gingras, in "The Photosynthetic Bacteria", R. K. Clayton, and W. R. Sistrom, eds., Plenum Press, New York-London (1978).
14. J. R. Bolton, in "The Photosynthetic Bacteria", R. K. Clayton, and W. R. Sistrom, eds., Plenum Press, New York-London (1978).
15. J. C. Romijn, & J. Amesz, Biochim. Biophys. Acta 461:327 (1977).
16. K. Sauer, Ann. Rev. Phys. Chem. (in press) (1979).
17. A. J. Koff, in "Light Reactions in Photosynthesis", F. K. Fong, ed., Springer, Berlin (1979).
18. J. Fajer, D. C. Brune, M. S. Davis, A. Forman, & L. D. Spaulding, Proc. Natl. Acad. Sci. USA 72:4956 (1975).
19. R. J. Cogdell, T. G. Monger, & W. W. Parson, Biochim. Biophys. Acta 408:189 (1975).
20. V. V. Klimov, V. A. Shuvalov, I. N. Krakmaleva, N. V. Karapetyan, A. A. Krasnovskii, Biochem. USSR (Engl.Transl.) 41:1169

(1976).

21. P. L. Dutton, & R. C. Prince, in "The Photosynthetic Bacteria",
    R. K. Clayton, & W. R. Sistrom, eds., Plenum Press, New
    York-London (1978).

22. J. M. Olson, R. C. Prince, & D. C. Brune, Brookhaven Symp.
    Biol. 28:238 (1977).

23. W. B. Whitten, R. M. Pearlstein, & J. M. Olson, Photochem.
    Photobiol. 29:823 (1979).

24. A. R. McIntosh, M. Chu, & J. R. Bolton, Biochim. Biophys.
    Acta 376:308 (1975).

25. P. Heathcote, D. L. Williams-Smith, C. K. Sihra, & M. C. W.
    Evans, Biochim. Biophys. Acta 503:333 (1978).

26. K. Sauer, P. Mathis, S. Acker, & J. A. van Best, Biochim.
    Biophys. Acta 503:120 (1979).

27. V. A. Shuvalov, E. Dolan, & B. Ke, Proc. Natl. Acad. Sci.USA
    76:770 (1979).

28. J. M. Fenton, M. J. Pelling, K. J. Govindjee, & K. J. Kauf-
    mann, FEBS Letters 100:1 (1979).

29. J. A. van Best, & P. Mathis, Biochim. Biophys. Acta 503:178
    (1978).

30. G. Renger, H. J. Eckert, & H. E. Buchwald, FEBS Letters
    90:10 (1978).

31. J. Amesz, & L. N. M. Duysens, in "Primary Processes of Photo-
    synthesis", J. Barber, ed., Elsevier, Amsterdam (1977).

32. V. V. Klimov, A. V. Klevanik, V. A. Shuvalov, & A. A. Krasno-
    vskii, FEBS Letters 82:183 (1977).

33. H. J. Van Gorkom, Biochim. Biophys. Acta 347:439 (1974).

34. J. H. Golbeck, S. Lien, & A. San Pietro, in "Photosynthesis
    I, Photosynthetic Electron Transport and Phosphorylation",
    A. Trebst, & M. Avron, eds., Springer, Berlin (1977).

35. B. Bouges-Boucquet, & R. Delosme, FEBS Letters 94:100 (1978).

36. W. Haehnel, Biochim. Biophys. Acta 459:418 (1977).

37. B. R. Velthuys, Proc. Natl. Acad. Sci. USA 75:6031 (1979).

38. J. Lavorel, & A. L. Etienne, in "Primary Processes of Photo-
    synthesis", J. Barber, ed., Elsevier, Amsterdam (1977).

39. S. Malkin, in "Primary Processes of Photosynthesis", J. Barber,
    ed., Elsevier, Amsterdam (1977).

40. J. Amesz, & H. J. van Gorkom, Ann. Rev. Plant Physiol. 29:47
    (1978).

41. J. P. Thornber, J. P. Markwell, & S. Reinman, Photochem.
    Photobiol. 29:1205 (1979).

42. R. J. Cogdell, & J. P. Thornber, in "Ciba Foundation Symp.,

61, New Series, Chlorophyll Organization and Energy Transfer in Photosynthesis", Excerpta Medica, Amsterdam (1979).

43. J. Amesz, Progr. Botany 41:55 (1979).

44. C. F. Fowler, N. A. Nugent, & R. C. Fuller, Proc. Natl. Acad. Sci. USA 68:2278 (1971).

45. R. E. Fenna, & B. W. Matthews, Brookhaven Symp. Biol. 28:170 (1977).

46. J. Amesz, Progr. Botany 39:48 (1977).

47. H. T. Witt, Biochim. Biophys. Acta 440:301 (1976).

48. J. Amesz, & B. G. de Grooth, Biochim. Biophys. Acta 440:301 (1976).

49. A. Trebst, Ann. Rev. Plant Physiol. 25:423 (1974).

50. B. G. de Grooth, & J. Amesz, Biochim. Biophys. Acta 462:247 (1977).

51. K. U. Sewe, & R. Reich, FEBS Letters 80:30 (1977).

52. B. G. de Grooth, H. J. van Gorkom, & R. F. Meiburg, Biochim. Biophys. Acta (in press) (1979).

COMPARATIVE DISCUSSION OF PHOTORECEPTION IN LOWER AND HIGHER ORGANISMS. STRUCTURAL AND FUNCTIONAL ASPECTS

Eilo Hildebrand

Institut fuer Neurobiologie Kernforschungsanlage
D-5170 Juelich 1
Federal Republic of Germany

1. INTRODUCTION

There are at least three reasons for a comparative view. The first is identical with the central goal of science, namely to discover general rules and relationships. The second is the search for similarity and diversity which derive from genetic constancy and adaptation to the environment, and is therefore concerned with our interest in evolution. The third is a more practical one. Comparative considerations often uncover new problems, which may be tackled experimentally and hence stimulate further research.

Sensory photoreceptors which are presently best studied are the visual cells of certain arthropods, especially of _Limulus_, and the vertebrate rod. Nevertheless, our knowledge of the molecular process of vision is incomplete and the mechanism of photosensory transduction _sensu strictu_ is not understood. All metazoan photoreceptors so far investigated contain rhodopsin as the receptor pigment (1). It was rather surprising when some years ago in _Halobacterium halobium_ a similar retinylidene protein, bacteriorhodopsin, was found which has, besides its light-harvesting property, a photosensory function (2). It seems tempting to concentrate in our comparative survey mainly on these rhodopsin systems and to discuss certain aspects of general interest. "Photoreception" will be used here as a term covering the entire chain of events which is displayed in a single cell.

The following considerations are based on a great number of

contributions from various fields of research. To simplify matters
mainly review articles or recently appeared papers are cited, from
which previously published literature can be obtained.

2. MOLECULAR ORGANIZATION OF PHOTORECEPTIVE STRUCTURES. LOCATION AND ORIENTATION OF PHOTOPIGMENTS AND OF THEIR CHROMOPHORES

Photoreceptors are specialized organelles designed to maximize
the absorption of photons and to translate the light stimulus into
an appropriate signal which can be processed by the cell. This requires a close relation of photosensitive parts to excitable structures or to effector organelles. According to these general relationships photoreceptive structures are expected to have some basic
features in common.

2.1. Chemical Nature of Photoreceptor Pigments

The visual pigment, rhodopsin, consists of one molecule of
retinal, the aldehyde of vitamin A, and a polypeptide chain of
about 350 amino acid residues. Retinal is covalently attached to
the protein part via the ε-amino group of a lysine residue thereby
forming a protonated Schiff base. Retinal exists in several stereo-
isomeric configurations among them only the 11-cis and the all-
trans isomer have been found in visual pigments. Some vertebrate
rhodopsins contain 3-dehydro-retinal (retinal$_2$) which carries an
additional double bond in the cyclohexene ring. The polypeptide
chain is specific for any species, as proteins usually are. The
molecular weight of most rhodopsins is in the range of 40,000 (1).

The chemical nature of bacteriorhodopsin (BR), which acts as
a light-driven proton pump in Halobacterium (2) and most probably
also controls its step-down photophobic response (3), is similar
to that of visual rhodopsin. The chromophore is 13-cis or all-trans
retinal, respectively. The polypeptide chain contains 245 amino
acids whose sequence is known (4). The molecular weight of BR is
26,500.

2.2 Chromophore-Protein Interaction

Isolated retinal has a maximal absorption at about 380 nm,
depending on the solvent (5), but attached to the polypeptide
(opsin) it shows a remarkable shift of the absorption band towards
longer wavelengths. The absorption peak of visual rhodopsin lies

in most cases around 500 nm (6). BR absorbs maximally at about
570 nm (2). The reason for this red shift is not completely understood. It is generally assumed that the protonated Schiff base and
secondary interactions between the chromophore and the polypeptide
environment both contribute to the bathochromic shift (1). Biochemical studies have led to the assumption that retinal is located
in a "hydrophobic pocket" which is made up by the surrounding amino
acids (7). Reconstitution experiments with retinal analogues revealed that the aldehyde group is essential to fit the chromophore
into the protein, and a high specificity with respect to C=C stereo-
isomers was found. For example, bacterio-opsin (BO) reacts only with
13-cis or all-trans retinal. Retinal with the methyl groups shifted
along the hydrocarbon chain cannot from BR, but analogues which
differ in the chain length, e.g. the C_{22}-compound, can well combine with BO. Also the cyclohexene ring may be slightly modified
without complications (2,8).

The photopigment which controls the step-up photophobic response of Halobacterium (3) could neither be isolated not spectro-
scopically detected so far. However, we have now strong evidences
that in this case also a retinal protein complex is involved, although this absorbs only in the blue and UV and hence differs
probably from BR with respect to chromophore-protein interaction.
Both photosystems of Halobacterium, PS 565 and PS 370, can be re-
constituted, after blocking retinal synthesis through nicotine, by
adding either trans retinal$_1$ (9) or trans retinal$_2$ to the bacterial suspension. In the latter case an expected red shift of about
15 nm of the main peak was found in both photosystems (Schimz and
Sperling, unpublished). The maximum at 370 nm in the action spectrum of the step-up response could well reflect the absorption of
the retinal Schiff base. The large band around 280 nm indicates
that energy transfer from neighbouring aromatic amino acids takes
place.

2.3 Location of Photopigments

In both vertebrate and invertebrate photoreceptor cells
rhodopsin is located in specialized areas of densely packed membranes, which permit a high probability of light absorption. In
invertebrates these consist of a large number of tubolar processes
of the plasma membrane, the microvilli, which, oriented in parallel, build up the highly ordered rhabdomeric structure of the ar-

thropod retinula cell (6). More primitive photoreceptor cells, as
for example the ventral photoreceptor of <u>Limulus</u>, contain irregular
groups of microvilli (10). In contrast, the outer segment of vertebrate rods contains about 1000 flat membraneous vesicles, the
disks (11). Disk membranes are made up of 40 % rhodopsin (dry weight)
besides some other proteins and 50 % lipids (12,13). The outer membrane of the rod outer segment (ROS) contains rhodopsin as well.

Bacteriorhodopsin forms distinct patches of about 0.5 µm diameter, the so-called purple membrane, which cover up to 70 % of the
surface membrane of <u>Halobacterium</u> (14). It seems likely that all
parts can equally well contribute to the step-down response, although stimulation of one end of the cell is sufficient. We assume
that the 370 pigment, together with carotenoids, is located in
those parts of the plasma membrane which are not formed by crystalline BR (9).

2.4 Orientation and Mobility of Pigment Molecules in the Membrane

Rhodopsins,including BR, are integral constituents of the photoreceptive membrane. It is well established that the pigment molecules are embedded in a bilayer lipid phase, but for visual rhodopsins their actual shape and orientation is still controversial.
Recent studies with bovine disks show that rhodopsin spans the whole
thickness of the lipid bilayer (15,16). The tertiary structure of
rhodopsin is stabilized by phospholipid molecules of the environment
(17). While rhodopsin can freely rotate and laterally diffuse in
the lipid phase of vertebrate disk membranes (18,19), it is probably
immobile in the microvillus membrane of invertebrates (6).

The orientation of BR in the purple membrane of <u>Halobacterium</u>
is well known. BR molecules are arranged in trimers which form a
rigid two-dimensional hexagonal crystalline structure (14).
Brownian motion of individual molecules is strongly restricted (20).
The polypeptide chain has a high amount of α-helix structure and
traverses the membrane 7-times (21,22). BR represents the only protein of the purple membrane. Exciton interactions between neighbouring BR molecules could be demontrated by means of circular
dichroism measurements (23,24).

The exact location of the chromophore in rhodopsin and BR is
still obscure. However, a distinct dichroism has been found in photoreceptor structures of arthropods, and several insect species
show high sensitivity to polarized light. Some preferential align-

ment of chromophores along the axis of microvilli has been proposed (25). In vertebrates the chromophores are randomly oriented in a plane which is almost parallel to the plane of the disk membrane. In this way the probability of light absorption is increased about 35 % as compared with a completely random orientation (6). Isolated purple membrane patches show slightly dichroic properties. The absorption transition moment indicates an orientation of 19-27° of the chromophore plane with respect to the plane of the membrane (26,27).

2.5 Accessory Pigments

Substances other than the effective receptor pigment may participate in photosensory processes. A well known example is the participation of carotenoids in both photosynthesis and photobehavior of Rhodospirillum rubrum (28). A photostable pigment, showing maximal absorption at 360 nm, apparently acts as a sensitizer for the visual pigment in certain retinula cells of the fly (29).

It could be demonstrated that carotenoids, probably mainly α-bacterioruberine, act as accessory pigments to the UV-absorbing retinylidene protein (PS 370) in Halobacterium (9). This example is insofar unique as the energy has to be transferred to a compound of shorter wavelength absorption. Such an energy transfer, which may be accomplished by an exciton state of the effective photopigment (Song, personal communication) implies an intimate spatial relation of both substances.

3. PHOTOPIGMENT REACTIONS AND THEIR POSSIBLE FUNCTION IN SENSORY TRANSDUCTION

Following photon capture rhodopsin undergoes a sequence of thermal reactions, and the main problem in photosensory processes is how these events lead to a signal which is transmitted in the cell and which finally activates effector organelles or synaptic endings. In both vertebrate and invertebrate photoreceptor cells the first neurophysiological result of light stimulation is a transient alteration in the ionic conductance of the plasma membrane. This electrically measurable signal occurs some tens to hundreds of milliseconds after the onset of the light stimulus (30-32). In most cases the signal spreads over the membrane in the form of a so-called receptor potential and is transmitted at the

synaptic junction to the subsequent neuron. The receptor potential
is graded depending on the magnitude of the stimulus received. In
unicellular organisms the chain of events following the light stimulus ends with a motor response, which can be observed after a
latent period of some milliseconds up to seconds. For experimental
reasons this is often the only visible sign which indicates that
the system has been activated.

It is generally assumed that rhodopsin after photoexcitation
goes through some conformational transitions one of which couple
photochemical reactions to excitation of the plasma membrane. A
number of short-living metastable intermediates have been detected
in visual rhodopsin as well as in BR using flash photolysis and
low temperature spectroscopy (2,33). These intermediates are characterized by alterations of their absorption spectra, which presumably correspond to conformational changes in the molecule.
Although BR is functionally different from visual rhodopsin one
would expect some similarities at least in the first steps of the
reaction chain.

3.1 Primary Events of the Photochemical Process

Rhodopsin in the dark adapted state contains retinal in the
11-cis configuration. The general effect of light is isomerization
of the chromophore to the all-trans form (1). The situation is
slightly different in BR which, after long-lasting storage in the
dark, contains equal amounts of 13-cis and all-trans retinal. Exposure to light of the proper wavelength shifts this equilibrium
completely towards the trans isomer (2).

For a long time cis-trans isomerization has been considered
as the primary event in the chain of light-induced reactions of
rhodopsin (1). Trans BR, however, exhibits similar absorption
changes as 11-cis rhodopsin after flash illumination. The absorption is first shifted to the red, the so-called batho-intermediate,
in less than 6 ps (34,35) and subsequently back towards the blue
(2,33). Resonance Raman spectroscopy brought additional doubt (36),
and some authors suggest that proton translocation towards the
Schiff base nitrogen (37) and certain charge redistributions at
the isoprenoid chain of retinal (38) may be the first step which
precedes cis-trans isomerization of visual rhodopsin. Recent results
of other authors seem to support the isomerization hypothesis (39,
40).

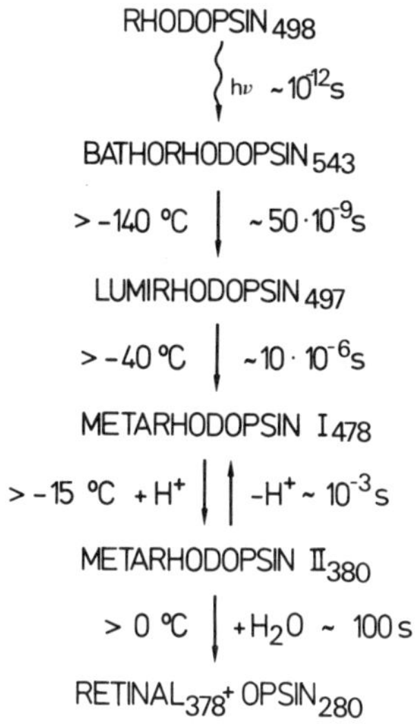

$$\text{RHODOPSIN}_{498}$$

$$\big\} h\nu \sim 10^{12}\text{s}$$

$$\text{BATHORHODOPSIN}_{543}$$

$$> -140 \ °C \ \big| \ \sim 50 \cdot 10^{-9}\text{s}$$

$$\text{LUMIRHODOPSIN}_{497}$$

$$> -40 \ °C \ \big| \ \sim 10 \cdot 10^{-6}\text{s}$$

$$\text{METARHODOPSIN I}_{478}$$

$$> -15 \ °C \ +H^+ \ \big| \big| \ -H^+ \sim 10^{-3}\text{s}$$

$$\text{METARHODOPSIN II}_{380}$$

$$> 0 \ °C \ \big| \ +H_2O \ \sim \ 100\,\text{s}$$

$$\text{RETINAL}_{378} + \text{OPSIN}_{280}$$

FIGURE 1. Reaction chain of vertebrate rod rhodopsin. Absorption maxima of intermediates are indicated by subscripts. Temperature, above which the reaction takes place, and approximate half-lifetime of intermediates at 37°C are given for each step. Modified from (33).

3.2 Thermal Intermediate Sequence of Vertebrate Rhodopsin

The chain of chemical dark reactions of rhodopsin from verte-brate rods (Fig.1) shows a great similarity between different spe-cies (33). The early intermediates are short-living and only stable at low temperatures. They show only minor conformational changes. The most likely step for the coupling to membrane excitation is the meta I to meta II transition. This reaction is characterized by relatively drastic alterations in the rhodopsin molecule. In ac-cordance with the absorption change it was found that the Schiff base becomes deprotonated, and the retinal-opsin binding site be-comes accessible to sodium borohydride at the meta II stage. UV spectroscopy shows perturbations of some aromatic amino acid res-idues in the neighbourhood of the Schiff base of metarhodopsin as compared to unbleached rhodopsin (41). The exposure of a tryptophane residue in meta II was deduced from the results. In a final reaction the chromophore is split off from the protein part (1). However, this reaction is far too slow to account for the excitation process.

3.3 Invertebrate Rhodopsin Reactions

The rhodopsin reactions of invertebrates are less well understood. Moreover, considerable differences may exist between different species. A common feature seems to be that, unlike in vertebrates, the retinal-opsin complex does not cleave, and the chain of dark reactions ends with a metarhodopsin (Fig.2). This exists in two termostable forms which differ probably with respect to the protonation of their Schiff base (33). It has been shown in some arthropod species that metarhodopsin can be reconverted into the initial rhodopsin by the absorption of another photon of the proper energy (42). This effect apparently takes place under normal physiological conditions, and a photostationary equilibrium of rhodopsin and metarhodopsin under steady illumination will result.

Membrane excitation is probably triggered by some intermediate step before the thermostable state is reached (43). The light-induced meta to rhodopsin transition does not lead to excitation, but seems to play an important role in pigment regeneration (42).

3.4 The Photochemical Cycle of Bacteriorhodopsin

Despite some similarities, which may exist with respect to early photochemical events, the function of BR is entirely different

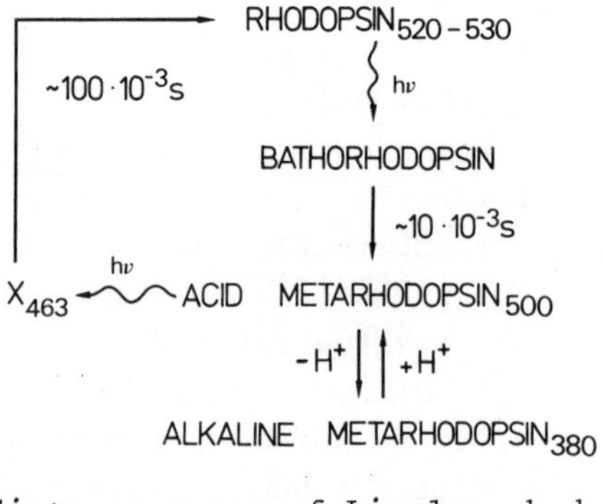

FIGURE 2. Intermediate sequence of _Limulus_ rhodopsin and half-times of intermediate reactions at 25°C. Absorption maxima of intermediates are indicated by subscripts. From (33).

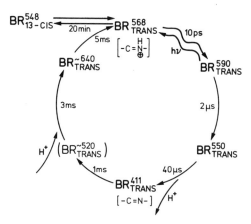

FIGURE 3. Reaction cycle of trans bacteriorhodopsin (BR) and half-
times of intermediate reactions at about 20°C. Absorption maxima
of intermediates are given as superscripts. Reaction cycle of 13-
cis BR not shown. After (2,44).

from that of visual rhodopsins. As already stated, moderate light
causes complete isomerization of all chromophores towards all-trans.
Trans BR shows maximal absorption at 568 nm. After shining light on
it trans BR undergoes a rapid reaction cycle (Fig.3) during which
protons are released and subsequently taken up. At 35°C the cycle
is completed in less than 10 ms (2). Hence, in contrast to rhodopsin,
no noticeable bleaching occurs in BR at room temperature. It is
believed that the release and uptake of protons by the BR molecule
is the basis for the net translocation of protons across the bacterial cell membrane. A number of thermal intermediates have been
detected by means of flash photolysis and low temperature spectroscopy (2). 13-cis BR is subjected to a separate reaction cycle (44)
which is mutually connected by photoreactions with the trans-cycle
(45). However, it seems doubtful whether these reactions have any
physiological significance. It should be emphasized that it is
still controversial whether or not stereoisomerization occurs in
the cycle which starts from trans BR (2,46).

It is presently unknown how BR may control the photophobic
response of Halobacterium. Since the effective stimulus consists
of a temporal decrease in light intensity we can assume that the

reaction cycle is interrupted and the proton translocation rate
decreases. This explanation is consistent with the mechanism proposed for the step-down photophobic response in photosynthetic
procaryotes, which is thought to be triggered by a sudden decrease in
photosynthetic electron flow (47). The possibility that a sudden
light decrease may act via a trans-cis isomerization seems improbable because of the long half time of more than 20 min for this
reaction (48).

4. MECHANISMS OF EXCITATION AND SIGNAL TRANSMISSION

The plasma membrane of visual cells has some fundamental electrochemical properties in common with the nerve membrane. 1) High
permeability for K^+ ions together with a high intracellular concentration of this ion, which is maintained by active transmembrane
transport, give rise to a resting potential of -40 to -70 mV (inside negative), which is close to the K^+ equilibrium potential
(E_K). 2) Na channels which can open or close, depending on the
state of membrane activation, counteract the resting potential allowing Na^+ diffusion down its electrochemical gradient into the
cell (49). But unlike the axon membrane, the receptor cell membrane
is not electrically excitable, i.e. Na channels do not open upon
depolarization.

The basic mechanism of excitation which seems to be valid for
most invertebrate photoreceptor cells is well established (Fig.4).
Na channels are closed in the dark and open transiently upon light
absorption by rhodopsin, thereby causing a depolarizing receptor
potential. Absorption of a single photon generates a small unit
response, the so-called bump. One could think that rhodopsin itself
forms the light-activated Na channel. However, it was found in the
ventral photoreceptor of Limulus that one bump corresponds to a
conductance increase of $5 \cdot 10^4$ pS (50), which is 100-times that associated with the opening of a single Na channel in the axon membrane. This means that at least 100 channels will be opened upon
photoexcitation of one pigment molecule. The basis for this amplification is unknown and, besides others, release of an excitatory
transmitter has been proposed (49).

The situation is further complicated in the vertebrate rod by
the spatial separation of the rhodopsin-containing disks from the
outer plasma membrane. Moreover, a large fraction of Na channels
are open in the dark, and the effect of light is to close some of

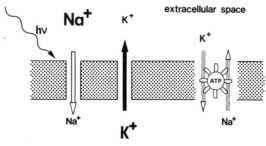

FIGURE 4. Schematic representation of light-induced excitation of invertebrate photoreceptor cell membrane. Active transport is indicated by wavy arrows. Black arrow represents K^+ diffusion in the dark, white arrow represents Na^+ diffusion through light-activated channels. Size of letters give relative concentrations of cations. After (49).

them thereby transiently increasing the membrane potential and generating a hyperpolarized receptor potential (11). Absorption of a single photon causes a decrease in membrane conductance of 1-3 %. This corresponds to an estimated number of 25 to 1000 Na channels (11). Again it is necessary to assume an amplification mechanism. Additionally a chemical messenger, which mediates between the disks and the outer membrane, has to be postulated. A mechanism which could well accomplish this function will be discussed in the next section.

Depolarizing and hyperpolarizing receptor potentials are both electronically transmitted (with decrement) along the plasma membrane and further processed at the synaptic terminal.

Unicellular organisms represent insofar a peculiar system as their receptor structures and motor organelles are located in one cell. Species which are presently best studied with respect to receptor properties and excitation are the ciliates Paramecium caudatum and Stylonychia mytilus (51,53). Although neither of the two is sensitive to light, their mechanoreceptor mechanism will be dealt with because of its eventual key role for understanding receptor-effector coupling. The ionic mechanism is basically similar to that of invertebrates, but differs insofar from the latter as ion channels specifically permeable to Ca^{2+} are opened upon stimulation. The resulting local depolarization spreads over the entire cell membrane and leads to transient opening of another kind of Ca channels which are sensitive to depolarization. These are restricted to the membrane covering the cilia (54). When the membrane becomes excited, Ca^{2+} ions diffuse into the intraciliary space and cause the cilia to reverse their beating direction, which finally leads to the phobic response of the organism.

Unlike ciliates, bacteria are not accessible to intracellular potential recordings. However, various properties could be readily explained by assuming membrane potential changes as the result of sensory activation. 1) Bacteria show spontaneous reversals or tumbling responses. 2) Flagellar activity is coordinated in bipolarly flagellated bacteria such as Spirillum volutans in less than 10 ms over a distance of more than 50 µm (55). 3) The signal is rapidly transmitted to the motor organelle. 4) Signals from different receptor sites are integrated (56). Intracellular diffusion of a transmitter substance could less well accomplish these functions (57). Application of indirect methods revealed that indeed membrane potential changes are associated with behavioral activity of bacteria (58). Although we have no direct proof some results indicate that the membrane potential is involved in photophobic responses of Halobacterium. Triphenyl-methyl-phosphonium ions ($TPMP^+$) inhibit both the step-up and step-down photophobic response at concentrations which also reduce the membrane potential (59). Increasing the K^+ concentration in the medium, so as to reduce the steady membrane potential (60), leads to a significant inhibition of both photoresponses (Hildebrand, unpublished).

According to Mitchell's theory of chemiosmosis, H^+ gradient, electron flow and membrane potential are interconvertible. A temporal decrease of proton pump activity due to the step-down stimulus should consequently lead to some depolarization. Since the motor response of Halobacterium is identical, no matter whether it is triggered by PS 565 or PS 370, and because stimuli simultaneously applied to both receptors are added up (Hildebrand, unpublished), it may be concluded that signals from both photosystems are of depolarizing nature. The most simple explanation for the step-up response is that PS 370 may function as a light-activated ion channel for H^+, Na^+, or Ca^{2+}. The mechanism by which membrane depolarization may trigger the reversal of flagellar rotation cannot be easily explained.

5. SIGNIFICANCE OF CALCIUM

A large number of experiments indicate that ions play an important role in the visual process. Of special interest is the function of Ca^{2+} in the control of membrane conductivity to cations.

A moderate decrease of extracellular Ca^{2+} concentration leads to an enhancement of the photoresponse in the Limulus ventral photoreceptor and other invertebrate photoreceptors (61). However, drastic decrease of extracellular Ca^{2+} has the opposite effect because it causes a drop in the membrane potential so that it comes very close to the Na^+ equilibrium potential, E_{Na}. Simultaneous lowering of the Na^+ concentration antagonizes the effect of Ca^{2+} reduction. These results are consistent with the assumption that Na^+ conductivity varies inversely with the amount of calcium bound to anionic binding sites at the membrane and that Ca^{2+} and Na^+ may compete for these sites (62). Injection of Ca^{2+} into the cell causes a decrease in the receptor potential (63), whereas EGTA-injection facilitates light-induced excitation (64). The ability of Ca^{2+} to control Na^+ conductance at both sides of the membrane is of great importance for the regulation of photoreceptor excitation and adaptation.

A similar action of extra- and intracellular Ca^{2+} on the light-evoked Na^+ conductance change was found in vertebrate photoreceptors (11). However, unlike in invertebrate the electrical signal which in this case is a hyperpolarized receptor potential, increases when Ca^{2+} is increased. Hence, Ca^{2+} mimics the effect of light. Based on this observation Hagins presented an interesting hypothesis which says that Ca^{2+} may be the intracellular transmitter which mediates between light absorption at the disk membrane and the electrical event at the outer membrane of the rod (Fig.5). According to this hypothesis Ca^{2+} ions should be released from the intradisk compartment into the cytoplasmic space, and after diffusion to the outer membrane, will bind to it, thereby decreasing Na^+ influx. Several attempts have been made in order to bring further evidence for this hypothesis, and some results are in favour of it (65). However, as yet there is no direct demonstration of light-induced cytoplasmic Ca^{2+} increase in intact rods. The light-induced Ca^{2+} release measured with isolated disk was maximally one Ca^{2+} per molecule rhodopsin bleached (66), a value which is far too small to account for the amplification mechanism. Only small active calcium uptake by the disk in the dark could be found (67).

Recently some light-dependent enzymatic reactions have been
detected in rod outer segment preparations (68). For example, a
light-activated phosphodiesterase was found which can hydrolyze 10^2
to 10^3 molecules cyclic GMP per bleached rhodopsin per second (69).
Cyclic GMP, on the other hand, antagonizes the effect of Ca^{2+} in
the rod (70) and could thus regulate the action of Ca^{2+} at the
plasma membrane and, moreover, account for the high amplification.
At present there is no candidate for an intracellular transmitter
in invertebrate photoreceptors. Apparently it is not Ca^{2+} which
can rather be regarded as an inhibitory substance as mentioned
above. It is also not known whether cyclic nucleotides play a role
in invertebrate photoreceptors.

As already mentioned in the previous section, Ca^{2+} mediates
ciliary reversal in Paramecium and other ciliates through a transmembrane influx. It seems that in general Ca^{2+} has a regulatory
function in motor responses of at least all flagellated and ciliated
eucaryotic organisms. Moreover, the tumbling response of flagellated
bacteria depends most probably on transmembrane Ca^{2+} influx as has
been shown with E.coli (71). We found that also the reversal response
of Halobacterium depends on the presence of external Ca^{2+} and one
could speculate that light-induced membrane depolarization may

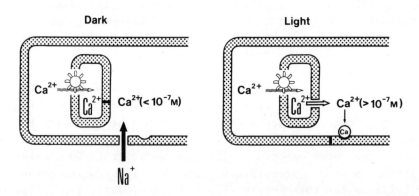

FIGURE 5. Calcium hypothesis of light-induced Na^+ conductance decrease at the plasma membrane of the vertebrate rod outer segment
(ROS). Wavy arrows represent active Ca^{2+} uptake by disk vesicles.
Black arrow represents Na^+ diffusion into the cytoplasmic space
in the dark. White arrow indicates light-induced Ca^{2+} release from
disk vesicles. After (11).

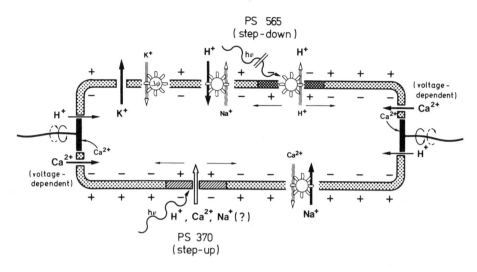

FIGURE 6. Model illustrating hypotetical mechanisms of photosensory
function of <u>Halobacterium halobium</u>. Wavy arrows represent active
cation transport driven by light energy, electrochemical potential,
or cation diffusion (black arrows). Decrease in light intensity of
565 nm wavelength leads via decreasing H^+ pump activity in the
purple membrane to membrane depolarization. A similar effect may
be caused by an UV light stimulus of 370 nm which opens specific
channels which permit cation diffusion into the cell (white arrow).
Depolarizing signals spread over the cell membrane and open electrically excitable Ca^{2+} channels near flagella. Ca^{2+} influx causes
some conformational change at the flagellar basal body which results
in a directional change of proton-driven flagellar rotation.
Modified from (72).

transiently open voltage-dependent Ca gates near the flagella and
that transient cytoplasmic Ca^{2+} increase may cause some conformational changes at the spoke-like substructure of the basal flagellar
ring (Fig.6), which has been postulated in order to explain the
directional change of flagellar rotation (73). The original intracellular Ca^{2+} concentration of probably less than 10^{-7} mole/l could
be quickly resumed by means of active Ca^{2+} efflux which is driven
by the Na^+ gradient across the membrane (74).

6. SITES AND MECHANISMS OF ADAPTATION

As a common feature sensory photoreceptors can adapt to a wide range of light intensities so as to keep to response fairly constant if the relative change in light intensity $\Delta I/I_o$ is constant. Adaptation is defined as a change in light sensitivity, which can be either measured as a variation of the response to a constant stimulus or - more exactly - as an altered light intensity needed to elicit a given response. As a result of adaptation the intensity/ response curve is shifted along the intensity axis (75). Sensitivity decrease due to background or previous illumination is often referred to as light adaptation, whereas the recovery process is called dark adaptation. There are at least three steps in the sensory transduction chain which can be involved in light adaptation: 1) the receptor site, 2) the excitation or response mechanism, 3) the mechanism of amplification.

Adaptation at the receptor site plays an important role in vertebrates because rhodopsin undergoes irreversible bleaching. The sensitivity of rods is primarily determined by the amount of unbleached rhodopsin, but there is no linear proportionality, rather a logarithmic one. For example, after 10 % rhodopsin is bleached the sensitivity decreases to about one hundredth of its original value (76). Regeneration of rhodopsin seems to be the rate determining factor in dark adaptation which is accomplished with a half time of about 10 min (77).

This "pigment adaptation" is of minor importance in invertebrates where sensitivity is determined mainly by the previous response to light (78). This means adaptation is rather related to certain changes of membrane properties than to pigment bleaching and is therefore referred as to "membrane adaptation". It is well established that Ca^{2+} ions are involved in membrane adaptation. Light-induced increase of intracellular free Ca^{2+} associated with the receptor potential was found in several invertebrate photoreceptors. However, the origin of calcium seemsto be different. In the barnacle a transmembrane Ca^{2+} influx occurs, whereas in the ventral photoreceptor of Limulus Ca^{2+} is most probably released from intracellular stores by means of Na^+/Ca^{2+} exchange (64,79). Injection of Ca^{2+}, as already mentioned, decreases sensitivity and hence mimics light adaptation. Local injection of Ca^{2+} causes a local sensitivity decrease which spreads with a velocity which is consistent with Ca^{2+} diffusion in the cytoplasm (63). Recently evidence was brought that Ca^{2+} is also involved in light adaptation of vertebrate photoreceptors (80).

Adaptation can be also the result of alterations of the gain of the amplification process, which is not yet understood. This mechanism is probably interrelated with the action of Ca^{2+}. Another mechanism which may be involved in light/dark adaptation is the enzymatic phosphorylation and subsequent dephosphorylation of opsin which has been found as a late result of bleaching of vertebrate rhodopsin (81).

With respect to the importance of Ca^{2+} in both vertebrate and invertebrate photoreceptors it is tempting to think about the relation between both systems. It should be mentioned that also in Paramecium Ca^{2+} influx causes a transient decrease of excitability, i.e. adaptation, due to calcium-mediated inactivation of Ca gates in the membrane (82,83). While intracellular Ca^{2+} has obviously an inhibitory effect in ciliates and in invertebrate photoreceptors, it acts rather as an excitatory agent in vertebrate photoreceptors. If one considers the disk as a part of the plasma membrane which became separated from the latter, the intradisk compartment must be regarded as extracellular. The situation then turns out to be quite similar in both systems and one could speculate that vertebrates may have used the adaptation mechanism of invertebrates to accomplish excitation in their peculiar photoreceptor system.

It would be interesting to know whether Halobacterium also shows some adaptation. Steady background illumination of the effective wavelength decreases the sensitivity of the step-up photophobic response (PS 370), but not that of the step-down response (PS 565) (3). This means that PS 565 always detects the absolute decrease in light energy flow. Experiments with double stimuli revealed that both systems are absolutely insensitive for about 0.3 s after a response. Sensitivity of PS 565 recovers fully within less than 2 s, but PS 370 needs significantly longer to restore its original sensitivity (72). We attribute the transient sensitivity decrease measured with the step down response to a general refractoriness of the effector system or the excitable membrane, which is superimposed in the case of the step-up response by some pigment adaptation. The lack of any pigment adaptation in PS 565 is consistent with the rapid cycling of BR.

7. EFFICIENCY AND RELEVANCE OF PHOTORECEPTOR SYSTEMS

One photon absorbed by the photoreceptive structure is sufficient to cause a detectable electric signal in vertebrate and invertebrate receptor cells (30,84) and absorption of 5 to 8 photons
in the human eye evokes a visual illusion (85). As compared to this
the sensitivity of Halobacterium is rather low: a single cell has
to experience a decrease of an estimated number of 1500 absorbed
photons of 565 nm wavelength per second to exhibit the phobic response. An amplification mechanism which is present in photoreceptors of highly developed animals is apparently absent in Halobacterium.

All sensory photoreceptors are characterized by their ability
to detect temporal changes in light intensity, $\Delta I/\Delta t$. This requires
refractoriness or adaptation. As we have seen the original sensitivity of Halobacterium after a response is accomplished within a
few seconds which is very fast as compared to animal photoreceptors.
However, unlike bacteria, visual cells can adapt over 4 to 5 log
units background intensity so as to keep the magnitude of response
constant at constant relative intensity changes $\Delta I/I_o$. Adaptation
to background illumination is obviously of great biological value
for pattern recognition and orientation of higher organisms. It
may be less important or even inappropriate for halobacteria which
probably use the photophobic response to accumulate under light
conditions optimal for light-energy conversion or to avoid harmful
UV radiation.

8. CONCLUDING REMARKS

It has been shown that similar retinal protein complexes act
as photosensory pigments in organisms of different degree of complexity, among them, however, a great diversity of functional properties can be found. Despite the chemical similarity of the photopigments their occurence in widely separated groups of organisms
indicates that rhodopsins were independently created by nature at
various stages of evolution. It may be concluded from the examples
being discussed that certain differences in the molecular organization of photoreceptive structures account for different chemical
reactions and sensory mechanisms. The crystalline structure of BR,
which does not allow large conformational changes, favours fast
reversibility suitable for energy conversion (38) and detection

of absolute changes in light intensity. The visual process, however, needs irreversibility and adaptation. This is realized in vertebrate rhodopsin by cis-trans isomerization and cleavage of retinal from opsin. Certain circumstances of behavior and metabolism may have favoured photoreversibility in invertebrates.

It seems that photopigment reactions in all cases are finally linked to rather conservative membrane properties, namely conductivity alterations to Na^+ or, as a probably more primordial feature, to Ca^{2+} ions.

ACKNOWLEDGEMENTS

I thank Dr. Angelika Schimz for reading the manuscript. The work on Halobacterium mentioned in this article was partially supported by the Deutsche Forschungsgemeinschaft (SFB 160).

REFERENCES

1. R. Hubbard, Trends Biochem. Sci. 1:154 (1976).
2. W. Stoeckenius, R. H. Lozier & R. A. Bogomolni, Biochim. Biophs. Acta 505:215 (1979).
3. E. Hildebrand & N. Dencher, Nature 257:46 (1975).
4. Yu. A. Ovchinnikov, N. G. Abdulaev, M. Yu. Feigina, A. V. Kiselev, N. A. Lobanov & I. V. Nasimov, Bioorg. Khim. 4:1573 (1978).
5. W. Sperling & Ch. N. Rafferty, Nature 224:591 (1968).
6. T. H. Goldsmith, J. Exp. Zool. 194:89 (1975).
7. T. Schreckenbach, B. Walckhoff & D. Oesterhelt, Eur. J. Biochem. 76:499 (1977).
8. D. Oesterhelt & V. Christoffel, Biochem. Soc. Trans. 4:556 (1976).
9. N. A. Dencher & E. Hildebrand, Z. Naturforsch. 34c(in press)(1979).
10. A. W. Clark, R. Millecchia & A. Mauro, J. Gen. Physiol. 54:289 (1969).
11. W. A. Hagins, Ann. Rev. Biophys. Bioeng. 1:131 (1972).
12. S. L. Bonting, W. J. de Grip, J. P. Rotmans & F. J. M. Daemen, Exp. Eye Res. 18:77 (1974).
13. H. Heitzmann, Nature New Biol. 235:114 (1972).
14. R. Henderson, Ann. Rev. Biophys. Bioeng. 6:87 (1977).
15. E. A. Dratz, G. P. Miljanich, P. P. Nemes, J. E. Gaw & S. Schwartz, Photochem. Photobiol. 29:661 (1979).
16. B. K. -K. Fung & W. L. Hubbell, Biochemistry 17:4403 (1978).

17. B. J. Litman, Photochem. Photobiol. 29:671 (1979).
18. P. A. Liebman & G. Entine, Science 185:457 (1974).
19. M. M. Poo & R. A. Cone, Nature 247:438 (1974).
20. K. Razi Naqvi, J. Gonzales-Rodriguez, R. J. Cherry & D. Chapman, Nature New Biol. 245:249 (1973).
21. S. B. Hayward, D. A. Grano, R. M. Glaeser & K. A. Fisher, Proc. Natl. Acad. Sci., USA 75:4320 (1978).
22. R. Henderson & P. N. T. Unwin, Nature 257:28 (1975).
23. B. Becher & T. G. Ebrey, Biochem. Biophys. Res. Commun. 69:1 (1976).
24. M. P. Heyn, P. -J. Bauer & N. A. Dencher, Biochem. Biophys. Res. Commun. 67:897 (1975).
25. T. H. Goldsmith & R. Wehner, J. Gen. Physiol. 70:453 (1977).
26. M. P. Heyn, R. J. Cherry & U. Mueller, J. Molec. Biol. 117: 607 (1977).
27. R. Korenstein & B. Hess, FEBS Letters 89:15 (1978).
28. R. K. Clayton, Arch. Mikrobiol. 19:107 (1953).
29. K. Kirschfeld, Biophys. Struct. Mechanism 5:117 (1979).
30. D. A. Baylor, T. D. Lamb & K. -W. Yau, J. Physiol. 288:589 (1979).
31. J.E. Lisman & J. A. Strong, J. Gen. Physiol. 73:219 (1979).
32. H. Stieve, in "Light-Induced Charge Separation in Biology and Chemistry", H. Gerischer & J. J. Katz, eds., Verlag Chemie, Weinheim (1979).
33. S. E. Ostroy, Biochim. Biophys. Acta 463:91 (1977).
34. G. E. Busch, M. L. Applebury, A. A. Lamola & P. M. Rentzepis, Proc. Natl. Acad. Sci. USA 69:2802 (1972).
35. K. J. Kaufmann, P. M. Rentzepis, W. Stoeckenius & A. Lewis, Biochem. Biophys. Res. Commun. 68:1109 (1976).
36. M. Sulkes, A. Lewis & M. A. Marcus, Biochemistry 17:4712 (1978).
37. K. Peters, M. L. Applebury & P. M. Rentzepis, Proc. Natl. Acad. Sci. USA 74:3119 (1977).
38. A. Lewis, Proc. Natl. Acad. Sci. USA 75:549 (1978).
39. Ch. R. Goldschmidt, M. Ottolenghi & T. Rosenfeld, Nature 263: 169 (1976).
40. J. B. Hurley, B. Becher & T. G. Ebrey, Nature 272:87 (1978).
41. Ch. N. Rafferty, Photochem. Photobiol. 29:109 (1979).
42. K. Hamdorf, in "Handbook of Sensory Physiology", H. Autrum, R. Jung, W. R. Loewenstein, D. M. MacKay & H. L. Tueber, eds., Springer Verlag, Berlin (1979).
43. Z. Atzmon, P. Hillman & S. Hochstein, Nature 274:74 (1978).

44. N. A. Dencher, Ch. N. Rafferty & W. Sperling, Ber. KFA Juelich 1374:1 (1976).
45. W. Sperling, Ch. N. Rafferty, K. -D. Koln & N. A. Dencher, FEBS Letters 97:129 (1979).
46. M. A. Marcus & A. Lewis, Biochemistry 17:4722 (1978).
47. W. Nultsch & D. -P. Haeder, Photochem. Photobiol. 29:423 (1979).
48. D. Oesterhelt, M. Meentzen & L.Schuhmann, Eur. J. Biochem. 40: 453 (1973).
49. H. Stieve, Verh. Dtsch. Zool. Ges. 1977:1 (1977).
50. R. Millecchia, A. Mauro, J. Gen. Physiol. 54:331 (1969).
51. R. Eckert, Y. Naitoh & H. Machemer, in "Calcium in Biological Systems", C. J. Duncan, ed., Cambridge Univ. Press (1976).
52. H. Machemer & J. de Peyer, Verh. Dtsch. Zool. Ges. 86 (1977).
53. J. de Peyer & H. Machemer, J. Comp. Physiol. 127:255 (1978).
54. A. Ogura & K. Takahashi, Nature 264:170 (1976).
55. N. R. Krieg, J. R. Tomelty & J. S. Wels, J. Bact. 94:1431 (1967).
56. J. Adler & W. W. Tso, Science 184:1292 (1974).
57. H. C. Berg, Ann. Rev. Biophys. Bioeng. 4:119 (1975).
58. J. P. Armitage & M. C. W. Evans, FEBS Letters 102:143 (1979).
59. W. Nultsch & M. Haeder, Ber. Dtsch. Bot. Ges. 91:441 (1978).
60. G. Wagner, R. Hartmann & D. Oesterhelt, Eur. J. Biochem. 89: 169 (1978).
61. H. Stieve, Biophys. Struct. Mechanism 3:145 (1977).
62. H. Stieve & M. Bruns, Z. Naturf. 33 c:574 (1978).
63. A. Fein & J. S. Charlton, J. Gen. Physiol. 70:591 (1977).
64. J. E. Brown & J. R. Blinks, J. Gen. Physiol. 64:643 (1974).
65. J. E. Brown, J. A. Coles & L. H. Pinto, J. Physiol. 269:707 (1977).
66. H. G. Smith, R. S. Fager & B. J. Litman, Biochemistry 16:1399 (1977).
67. U. B. Kaupp, P. P. M. Schnetkamp & W. Junge, Biochim. Biophys. Acta 552:390 (1979).
68. H. Kuehn, Biochemistry 17:4389 (1978).
69. M. L. Woodruff, D. Bownds, S. H. Green, J. L. Morrisey & A. Shedlovsky, J. Gen. Physiol. 69:667 (1977).
70. W. H. Miller & G. D. Nicol, Nature 280:64 (1979).
71. G. W. Ordal, Nature 270:66 (1977).
72. E. Hildebrand, in "Taxis and Behavior", G. Hazelbauer, ed., Chapman & Hall, London (1978).
73. P. Laeuger, Nature 268:360 (1977).
74. J. W. Belliveau & J. K. Lanyi, Arch. Biochem. Biophys. 186: 98 (1978).

75. S. A. Lipton, S. E. Ostroy & J. E. Dowling, J. Gen. Physiol.
    70:747 (1977).
76. W. A. H. Rushton, Biophys. Struct. Mechanism 3:159 (1977).
77. W. A. H. Rushton, in "Handbook of Sensory Physiology", H. Autrum,
    R. Jung, W. R. Loewenstein, D. M. MacKay & H. L. Tueber, eds.,
    Springer-Verlag, Berlin (1972).
78. E. Hildebrand, H. Stieve, G. Hanowski & H. Gaube, Vision Res.
    14:1399 (1974).
79. J. E. Brown, P. K. Brown & L. H. Pinto, J. Physiol. 267:299
    (1976).
80. D. G. Flaming & K. T. Brown, Nature 278:852 (1979).
81. H. Kuehn, Nature, 250:588 (1974).
82. E. Hildebrand, J. Comp. Physiol. 127:39 (1978).
83. P. Brehm & R. Eckert, Science 202:1203 (1978).
84. F. Wong, Nature 276:76 (1978).
85. S. Hecht, S. Shlaer & M. H. Pirenne, J. Gen. Physiol. 25:819
    (1941).

PHOTOSENSORY TRANSDUCTION CHAINS IN EUCARYOTES

Giuliano Colombetti and Francesco Lenci

C.N.R. - Laboratorio Studio Proprietà Fisiche Biomolecole e Cellule, Pisa (Italy)

1. INTRODUCTION

One of the crucial problems to solve in any sensory transduction
process is how the signal perceived by the receptor(s) is processed and transduced to the effector(s) and eventually to the motor
apparatus. In aneural microorganisms the entire stimulus-response
system is within a single cell, without complicated nervous structures. Therefore, photosensory transduction chains in these living
systems can be defined as multicompartment molecular processes functionally linking the photoreceptor to the motor apparatus. The molecular characteristics of such processes are not only of great interest in their own right but could also constitute proper model-
systems for understanding the underlying elementary events which
generate and control signals in neural networks, at least if it is
true, as we think, that basic mechanisms are generally conserved
through the evolution of living organisms.

A photosensory transduction chain can be schematized as follows
(1-3): the light stimulus (the signal) is perceived by the photoreceptive structure(s), where the primary photophysical and photochemical reactions take place. The outcomes of these reactions, in
their turn, will constitute signals for the processor(s) and the
transducer(s) regulating the activation of the effector. The effector will then operate on the motor apparatus, causing the final
motile response. Whilst in procaryotic organisms,like Halobacterium
halobium and Phormidium uncinatum, e.g., the same pigment systems

seem to act both as photoreceptive units and as photopigments for driving metabolic processes (the purple membrane in <u>Halobacterium</u> and the photosynthetic system in <u>Phormidium</u>), in eucaryotic cells specific and specialized structures can exist dedicated to perceive light stimuli and to translate them into a physiological signal. However, only in very few cases have they been characterized with a sufficient degree of confidence. More or less the same holds true for the other components (with the exception of motor apparatus) of the photosensory chain and for the connections among them. Actually these components and connections have not all yet been identified respectively with specific subcellular organelle   or structures and functional molecular processes.

Rather than reviewing all possible hypotheses and even speculative suggestions about these problems, in this lecture we will discuss some of the most interesting examples of eucaryotic microorganisms in which the suggested models and mechanisms for photosensory transduction are experimentally supported, with particular attention to those processes which can possess a comprehensive and unitary character.

As we will see in more detail in what follows, several lines of experimental approach and a close integration of the results thus obtained have been necessary in most cases in order to achieve sound hypotheses or reliable models. The characterization of the receptor-motor chain, which finally is the central problem of sensory photophysiology, is actually remarkably complex for several reasons. First of all many microorganisms are capable of showing various kinds of photomotile reactions and their final photobehavior can result from the superimposition of elementary photoresponses of different types (non-directional photophobic response, true oriented movements with respect to the light source, photokinetic effects). This ambiguity can constitute a real problem when the different elementary photoresponses are elicited through distinct sensory channels. Consider, as a trivial example, the case in which cell accumulation (or dispersal) in a light trap is due to the concomitance of a photophobic response and of a photokinetic effect, two different mechanisms of light energy utilization (photosensing and photocoupling respectively), generally through two different photopigment systems, for two different photomotile reactions. It is therefore strictly necessary to discriminate among the different possible photoresponses and to find out and analyze the "pure" ones. Potential confusion may also originate from possible interactions among responses to different simultaneous stimuli (4). Just to mention an example, in Euglena the chemoresponse toward oxygen can temporarily be suppressed by blue-light stimulation (5). This blue-light-induced inhibition of chemotransduction can indicate a complex link between the physiological process responsible for chemoaccumulation and the photosensory transduction chain. The importance of an unambiguous identification of photoreceptor pigments and of satisfactory reaction schemes for the elementary photophysical and photochemical events has already been deeply discussed in previous lectures. Also the experimental techniques most commonly used are not devoid of conceptual risk: the use of metabolic inhibitors, uncouplers, ionophores and other drugs, for example, requires much attention to avoid misleading conclusions. For instance, as Bodo Diehn has illustrated in his lecture, the chemical composition of the growth and of the suspension media can affect the motile properties and the photodependent behavior of the cells. Moreover when trying to interpret the results it is not always easy to recognize the drug site of action in a multicompartment system like an eucaryotic cell.

As a general rule, the analysis of the photobehavior of cell populations has to be integrated step-by-step with the study of photomotile reactions of individual cells, properly choosing the behavioral parameters to control. This task can be made even more intricate by an alteration of the general viability of the cells (e.g. following the addition of drugs, application of external electric fields, manipulation of the suspension medium composition). The main trouble in research on sensory transduction consists in the very fact that most measurements are, and have to be, performed on the behavioral response, which is the very final step of a manifold process and which is sometimes very difficult to quantify. As we will quickly see in this lecture, much light on these questions can be shed by studies other than behavioristic, such as flagellar apparatus isolation and electrophysiological techniques.

2. CHLAMYDOMONAS

In the unicellular biflagellate photosynthetic Chlamydomonas reinhardti, Stavis and Hirschberg (6,7) have reported a specific suppression of the photodependent behavior, measured as photoaccumulation in their phototaxigraph, by a drug, sodium azide (NaN_3), often used as an inhibitor of electron transport. This complete

inhibition of the photomotile responses occurred at drug concentrations which had no effect on cell motility and viability, and
was not related to changes in oxygen uptake or in cellular ATP
concentration. Moreover NaN_3 did not seem to block the excitation
of the photopigment by formation of a reversible chromophore-azide
complex. A possible dynamic quenching effect of azide on the excited
states of the photoreceptor seems highly improbable, as very high
NaN_3 concentrations are necessary to slightly quench the excited
states of flavin-type pigments (Colombetti and Lenci, unpublished
data). Of course this is only an indication, as the available data
on the photopigments of Chlamydomonas do not point to their flavinic
nature. A possible mechanism for explaining azide inhibition of
photomotile responses in Chlamydomonas might be suggested by the
depolarizing effect of this drug: ion permeability changes, analogous to those occurring in neurons and in Paramecium (8,9 and references therein), might actually be induced by NaN_3.

Furthermore, it has been shown that an externally applied
electric field of about $10\div20$ V/cm reversibly suppressed the phototactic response of Chlamydomonas (10). According to the authors,
this result supports the hypothesis that the phototactic response
in Chlamydomonas involves the transmission of an electric signal.
Here we only want to mention that in our opinion the application
of external electric fields to test whether an action potential
is involved in light-induced motile responses is a technique to
be used with great care, because of the difficulty in evaluating
the actual charge variation on the cell membrane of microorganisms
swimming between the two electrodes.

The hypothesis that in Chlamydomonas membrane phenomena mediated by certain ions can be responsible for processing and transducing the photic stimulus is supported by the experiments of
Schmidt and Eckert (11) on a CW92 wall-less mutant. In such a cell
the light-induced backward locomotion, that is directly related
to the light-induced motor response, was found to be progressively
suppressed by reducing extracellular Ca^{++} concentration, whilst no
significant Ca^{++} concentration effect was observed on normal swimming. Photo-elicited backward swimming was also clearly reduced
by agents known to block Ca^{++} influx in excitable membranes. Ba
(10^{-2} M) and Mg (10^{-2} M) had a fully reversible effect, whereas
the Ca-blocking α-isopropyl-$\alpha\alpha$-{(N-methyl-N-homoveratryl)-γ-amino-
propyl}-3,4,5-trimetoxyphenylacetonitrile (D-600) at concentrations
higher than 10^{-4} M irreversibly inhibited the photostimulated re-

versal beating response. Also procaine, which reversibly blocks membrane excitability (12) and competes with Ca^{++} in lobster axon (13), reversibly suppressed the photomotile response. On the basis of these findings, Schmidt and Eckert concluded that in Chlamydomonas photostimulation results in an influx of Ca^{++} through the cell membrane, causing a transient increase in the intracellular Ca^{++} concentration.

The findings of Schmidt and Eckert may seem in contrast with those of Nichols and Rikmenspoel (14), who, by means of a microinjection technique, found an inhibitory effect of intracellular Ca^{++} concentration on the flagellar activity of Chlamydomonas. However, according to Nichols and Rikmenspoel themselves, the two effects might be by a different mechanism. Actually Schmidt and Eckert refer to Ca^{++} effect on photostimulated flagellar reversal, whilst Nichols and Rikmenspoel refer to flagellar motility. In any case to two processes are in some way linked, as an inhibitory action on flagellar activity should also be reflected in an inhibition of flagellar response to illumination. Of course also the drastic difference between the two experimental procedures and, in particular, the disrupting character of the microinjection technique, have to be taken into account when comparing the two sets of data. After all, as we will see in what follows, there are no discrepancies, but rather differences due to the different Ca ion concentrations.

The association of Ca^{++} regulation of flagellar activity with the photomotile response of Chlamydomonas seems however established in an elegant experiment of Hyams and Borisy (15). They were in fact able to induce, in isolated flagellar apparatus, transitions from forward to backward motion in vitro by varying the Ca^{++} concentration. The two kinds of movement in isolated flagellar apparatus were directly comparable to the motile responses of living cells. In the isolated flagellar apparatus forward swimming was observed for Ca^{++} concentrations lower than 10^{-6} M, backward motion was reversibly induced by Ca^{++} concentrations above 10^{-6} M. Flagellar motility was inhibited for Ca^{++} concentrations higher than 10^{-3} M. The threshold for Ca^{++} -induced motion reversal was lowered to 10^{-7} M when the flagellar membranes were solubilized, thus pointing at involvement of flagellar membranes in the process of Ca^{++} regulation within the axoneme.

The hypothesis that in Chlamydomonas reinhardti flagellar response and photic stimuli are coupled through Ca ions and Ca^{++} -me-

diated membrane phenomena, has recently been forwarded also by
Nultsch (16) (who has very kindly made available to us some of his
in-press and unpublished results). A specific stimulating effect
of Ca^{++} has been found on the phototactic response of Chlamydomonas
for concentrations up to 10^{-2} M, (significantly enhanced by 10^{-3} M
K^{+}, which, on the other hand, does not influence phototaxis if added alone). Moreover phototaxis was found to be completely inhibited by the Ca^{++} antagonist La^{+++}, but once again fully restored
after Ca^{++} addition. In conclusion, also according to Nultsch (16)
any decrease in Ca^{++} concentration, either by growing the cells in
a Ca^{++} -deficient medium, or by blocking Ca^{++} uptake by means of
La^{+++} or by removing Ca^{++} from the medium, causes a decrease in
the phototactic activity of Chlamydomonas.

There is a point which ought to be briefly discussed: in
Nultsch's experiments (16) the Ca^{++} concentration above which movement inhibition occurs is 10^{-2} M, whilst Schmidt and Eckert (11)
observe such an inhibitory effect for Ca^{++} concentrations above
10^{-3} M. This discrepancy might be only apparent as, beside the different experimental techniques (cell population and single cell)
the wall-less mutant used by Schmidt and Eckert might be more sensitive simply because "more permeable" to ions.

It is rather interesting to remark that a completely similar
Ca^{++} concentration effect is observed on different photomotile responses: actually Stavis and Hirschberg (6,7) and Nultsch (16)
measure respectively photoaccumulation and phototaxis of a cell
population, whilst Schmidt and Eckert (11) measure a step-up photophobic response in individual cells.

That photostimulation of motile reactions results in an influx
of Ca^{++} through the cell membrane has recently been suggested also
for Volvox, a colonial flagellate, by Sagakuchi et al. (17) who
have reported that EGTA concentrations higher than the Ca^{++} concentration in the medium suppressed the photophobic response of the
microorganisms. Incidentally, these authors briefly mention an effect of H^{+}, K^{+}, Valinomycin and CCCP on the photophobic response
of Volvox, which might have been rather interesting to discuss also
in connection with what observed in the case of Euglena gracilis
(see below). Unfortunately all these data are presented in an ab-
stract and therefore are definitely not sufficient for an even approximate critical analysis.

A different approach, which seemed quite promising, to the
problem of characterizing the photosensory transduction chain in

Chlamydomonas,consisted in the isolation of several mutants with
altered phototactic response (18) and of three strains with reversed sign of response (19). Unluckily these mutants have been
studied mainly from a behavioristic point of view, so that their
biochemical and structural characterization is still lacking. A
genetic analysis, possibly in some way analogous to that applied
to Phycomyces (see V. Russo's lecture) might have helped very much
in elucidating the molecular and cellular basis of photomotile responses in Chlamydomonas.

Finally we want to draw the attention to the fact that we can
advance the hypothesis that Ca^{++} transport and Ca^{++} transport-
mediated membrane phenomena are components of the photosensory
chain in Chlamydomonas with a quite good degree of confidence
because of the consistency and integration of different experimental
results obtained by means of a variety of experimental techniques
and approaches such as: use of specific drugs and analysis of their
effects on different physiological parameters (6,7,11), application
of external electric fields (10), controlled variation of the ionic
composition of growth and resuspension media (6,7,11,16), isolation and analysis of flagellar apparatus (15). Even if some important points have been clarified, several key questions are still
unanswered in photoregulated behavior of Chlamydomonas reinhardti;
among them of primary importance seems to us the identification
and characterization of the photoreceptive unit(s) and pigment(s),
a prerequisite for formulating reasonable hypotheses on the early
molecular events occurring in the photoreceptor. Actually Ca^{++}
influx through the cell membrane and the consequent increase in
intracellular Ca^{++} concentration seem to be the result of photostimulation in Chlamydomonas, but nothing is reported about the
primary molecular photoreactions which would initiate this process.
The photoreceptive pigments serving as a light-activated gate for
the passage of ions, especially calcium ions, across the cellular
membrane sounds a fascinating hint, mainly for its resemblance with
the calcium hypothesis of visual transduction. In the light of this
Ca^{++} hypothesis, some decisive questions can be asked about
Chlamydomonas sensory photophysiology. Where should the photoreceptor be located to accomplish this function of triggering a
gating mechanism ? It seems reasonable to think the pigment molecules implanted in the membrane, may be in the region nearby the
flagella, but no evidence is available about their precise localization nor about their supramolecular structural organization.

Would it be conceivable for a light-induced modification of the photopigments to be responsible for conformational variation of some protein, finally leading to the calcium transport process ? Of course these are only a couple of the many non-trivial open questions, to answer which much work is necessary in more related fields such as ultrastructural analysis, spectroscopy, and photo-chemistry.

3. ELECTROPHYSIOLOGICAL APPROACHES

Similarly to what happens in excitable cells, membrane hyper-polarization or depolarization can play a basic role in photosensory transduction processes in aneural organisms. The direct electrophysiological study of photodependent membrane potentials, using microelectrodes to insert into the microorganism or to suck it, can therefore be a very powerful tool to get information on the relevance and characteristics of photopotentials. However it should be kept in mind that impalement with microelectrodes can severely disturb the cells under investigation, and both methods are almost impossible to be used with microorganisms of very small dimensions.

The electrical responses from an intravacuolar electrode to photic stimuli have been investigated in Stentor coeruleus, a blue green ciliate, by Wood (20) and, more recently, by Song, Poff and Häder (21,22) who present and discuss their elegant experiments in this meeting.

Photoinduced extracellular potential differences (PPD) have been measured by Litvin et al. (23) in the biflagellate alga Haematococcus pluvialis, by sucking the cell with a micropipette. This method has the advantage that the analyzed system is only weakly perturbed. In fact the cells are still alive and motile and even exhibit photomotile reactions after 2-3 hour experiments. The kinetics of PPD between the two parts of the cell (inside and outside the micropipette) has been determined; it has moreover been suggested that the photoreceptor for PPD is located in the anterior part of the cell, has rather small dimensions and generates, under light stimulation, a positive potential inside the cell with respect to the outer space. The action spectrum for the initial rate of PPD rise has been determined and found to be very similar to the action spectrum for phototactic cell accumulation in the lighted region. However, the action spectrum for PPD seems to be fairly well resolved, whilst the resolution in the phototactic action

spectrum seems to us to be not so good. Particularly no detailed
informations are given on the characteristics of Haematococcus
photodependent behavior. PPD would be, according to the authors,
a photoreceptor potential responsible for light-induced behavior
of Haematococcus. Interestingly PPD has been found to be reversibly
inhibited by 3mM NaN$_3$. Surprisingly the effect of this drug has not
been tested by the authors on the photobehavior of the cell. The
inhibitory effect of NaN$_3$ on PPD reminds the reversible suppression of photomotile reactions in Chlamydomonas by the same drug
(6,7,16) and such a "coincidence" might represent a welcome unitary
aspect of photosensory transduction in the two organisms. Also according to Litvin et al., however, further experiments are required
to clarify the mechanism of PPD generation.

Intracellular photodependent potentials have been recorded in
Haematococcus by Ascoli and Ristori (24) and it will be interesting to compare intra- and extra-cellular measurements when sufficient data will be available.

4. EUGLENA

Even though the photobehavior of the photosynthetic flagellate Euglena gracilis has been widely investigated (2,25-28 and
references therein) it is still largely unknown how the signal
from the photoreceptor is processed and transduced to the effector.
However, on the contrary to the case of Chlamydomonas, in Euglena
the photoreceptor proper and the screening organelle, as well as
the pigments they contain, have been identified and to some extent
characterized (1,29-33) as we have seen in the lecture on photopigment identification. Concerning the molecular functional pathways, we have shown (34) that the step-down and the step-up photophobic responses of Euglena are not affected by metabolic drugs
such as 3-(3',4' -dichlorophenyl)-1,1-dimethyl urea (DCMU), 2,4-
dinitrophenol (DNP) and NaN$_3$, even at concentrations which severely
impair cell motility and viability and induce serious morphological
alterations. The lack of any effect on the photophobic response
of both DCMU and NaN$_3$, which on the contrary completely inhibit
light-induced oxygen evolution and significantly alter the chloroplast morphology, clearly shows that photosynthesis and photosensory transduction in Euglena are not related. Such a conclusion
is in full agreement with our findings that dark-bleached, non-
photosynthetic cells clearly show step-down photophobic responses

(Colombetti and Lenci, unpublished results) as well as photoaccumulation with an action spectrum identical to that of photosynthetic
green cells (31). The absence of any effect of NaN_3 on Euglena
photobehavior may also suggest that membrane photopotentials, of
the type described for Haematococcus (23) and assumable for
Chlamydomonas (6,7,10,11,16), are not functional pathways operating
in Euglena photobehavior. The inefficiency of the uncoupler DNP,
moreover, might indicate that photo-depending pH variations or proton gradients do not play any significant role in photosensory
physiology of this eucaryote. In conclusion neither the photosynthetic process, in particular non-cyclic photophosphorylation and
electron transport, nor oxidative phosphorylation seem to be connected with photosensory transduction in Euglena.

Ion transport and ion transport-dependent membrane phenomena
might however be involved in Euglena photosensory transduction,
similarly to what happens for other microorganisms (see e.g., the
already discussed case of Chlamydomonas and Stentor coeruleus,
Halobacterium halobium and Phormidium uncinatum in some of the
other lectures (Song,Häder,Nultsch)). Recently we have observed
(Mikolajczyk, Colombetti and Lenci, unpublished data) that drugs
which are able to affect directly or indirectly proton transport
processes, such as Gramicidin D and Carbonyl-cyanide-m-chloro-
phenyl-idrazone (CCCP) on one hand and, on the other hand, 2-n-
heptyl-4-hydroxyquinoline-N-oxide (HOQNO) and Antimycin A, do not
have any effect on the step-down and on the step-up photophobic
responses of Euglena. This suggests that there is no direct linkage
between proton transport processes (at least as far as ΔpH in concerned) and motile responses to photic stimuli, confirming our previous results with DNP.

A support to the ionic nature of photosensory transduction in
Euglena has recently been yielded by Doughty and Diehn (35,36),
who reported that the photobehavioral response of the cells was
unaffected by drugs blocking cation conductance across excitable
membranes, but was strongly influenced by monovalent cations, oubain,
anions. A deeper and more complete critical discussion of the relevance of ionic processes in Euglena photosensory transduction will
be, in our opinion, possible when more data on the subject will be
available both about the very initial steps of the process which
would involve light-induced alterations in coupled monovalent cation
and anion transport across the photoreceptor membrane and about the
final stages, which would consist in Ca^{++} dependent alteration in
the flagellar mechanochemical cycle (36).

Of particular interest, in the case of Euglena gracilis, is
also the discussion about the two-photoreceptor/one-sensory-chain
one-photoreceptor/two-sensory-chains hypotheses. Actually, according
to Mikolajczyk and Diehn (37,38), two photoreceptor systems can be
operating in Euglena: one for the step-down photophobic responses
and one for the step-up (for the sake of clearness we recall that
a step-down photophobic response occurs following a decrease in
light intensity and a step-up response is caused by an increase in
light intensity above a threshold value (39)). This suggestion is based
on the observed inhibition of the step-down response by the ionic detergent cetyltrimethylammonium bromide (CTAB) and by the quencher
of flavin excited states I^- (KI was used in those experiments)
which, on the contrary, do not suppress the step-up response. The
existence of two distinct photoreceptors and/or of two distinct
photopigments for the two photophobic responses, seems to be denied
by our results on action spectra of green and streptomycin-treated
cells, which suggest that the same pigment (a flavin-type chromophore) is responsible for photoaccumulation and photodispersal and
hence for the two photophobic reactions (31). An alternative possibility, suggested by Diehn himself (27), is that CTAB and KI could
have different effects on the step-down and on the step-up responses
because affecting differently two distinct, at least in part, transduction chains, one operating for the step-down and the other for
the step-up response. It would be rather interesting to know the
details of the experiments recently accomplished with monovalent
cations and anions (36) to see if the inhibitory effect of KI on
the photobehavioral responses of Euglena is actually to be entirely
attributed to a specific quenching effect of I^- on the excited states
of flavins (37) or if also a possible effect of K^+ on the transduction process is to be taken into account. Actually in the light
of the one-photoreceptor/two-sensory-chains hypothesis, the I^- induced quenching of flavins excited states observable only for the
step-down response seems not easily sustainable.

As afore mentioned, not only the molecular processes acting
as functional links among components, but even some of the components
(= subcellular organelles) of the transduction chain are yet uni-
denfied in several cases. So it is for the effector structure in
Euglena photobehaviour and for its functional mechanism. It has been
suggested (1) that the role of the effector structure could be
played by the longitudinal fibrils, assumed to be of actomyosinic
nature, surrounding the canal, of the cell. These fibrils would con-

tract modifying the flagellar inclination with respect to the cell
body with consequent variation of the cell trajectory. We have
recently tried to check this hypothesis (40), using cytochalasins
B and D, two drugs which are known to damage actomyosinic systems.
The absence of any effect of both cytochalasins on the step-down
and the step-up responses can be interpreted in two different ways.
One possibility is that the fibrils are really the effector; in
this case they should be of a type not sensitive to the action of
cytochalasins B and D. Alternatively, one could think that the
cytochalasins have no effect on the photophobic responses of Euglena
simply because the fibrils surrounding the canal are not involved
in the sensory transduction process. At present we cannot say which
of the two hypotheses is correct, and work is in progress to try to
clarify this aspect of the problem. It should also be kept in mind
that the paraflagellar rod, the highly ordered quasi-crystalline
struture which extends along the flagellum and in which ATPase activity has been localized (41), has also been suggested as a possible effector (42).

     Unfortunately, it seems that in the case of Euglena most infor-
mations come mainly from behavioristic studies, whose limits have
already been mentioned. Some of these experiments are really accurate and elegant, but some of their conclusions should be checked
with different experimental approaches. Electrophysiological
measurements with intracellular microelectrodes can be very difficult to perform with these cells because the pellicle around the
microelectrode does not adhere to it, thus constituting a short
circuit path to the external medium. Similarly, a short circuit
path is left while sucking the cell into a micropipette because
of the striae into the pellicle. Interesting suggestions can be
put forward by Bodo Diehn from experiments on the flagella with
their paraflagellar body they have recently successfully isolated.
Microspectroscopy possibly coupled with microphotochemistry will
hopefully be used by us in intact cells to try to study the primary molecular events following light-stimulation of the photoreceptor.

     Finally a thorough investigation of the streptomycin-treated
mutants, which is also in our schedule, might help in elucidating
the molecular machinery through which step-up responses are elicited. Actually these mutant Euglena seem to exhibit only step-up
responses, even at extremely low light intensities.

REFERENCES

1. E. Piccinni & P. Omodeo, Boll. Zool. 42:79 (1975).
2. F. Lenci & G. Colombetti, Ann. Rev. Biophys. Bioeng. 7:341 (1978).
3. W. Nultsch & D. P. Häder, Photochem. Photobiol. 29:423 (1979).
4. M. E. Feinleib, Photochem. Photobiol. 27:849 (1978).
5. G. Colombetti & B. Diehn, J. Protozool. 25:211 (1978).
6. R. L. Stavis & R. Hirschberg, J. Cell. Biol. 59:367 (1973).
7. R. L. Stavis, Proc. Natl. Acad. Sci. USA 71:1824 (1974).
8. P. Horowicz, C. Caputo & J. A. Robeson, Fed. Proc. 27:702 (1962).
9. R. Eckert & P. Brehm, Ann. Rev. Biophys. Bioeng. 8:353 (1979).
10. I. Marbach & A. M. Mayer, Isr. J. Bot. 20:96 (1971).
11. J. A. Schmidt & R. Eckert, Nature 262:713 (1976).
12. P. Seeman, Pharm. Rev. 14:583 (1972).
13. M. P. Blaustein & D. E. Goldman, J. Gen. Physiol. 49:1043 (1966).
14. K. M. Nichols & R. Rikmenspoel, J. Cell. Sci. 29:233 (1978).
15. J. S. Hyams & G. G. Borisy, J. Cell. Sci. 33:235 (1978).
16. W. Nultsch, Arch. Microbiol. 123:93 (1979).
17. H. Sagakuchi, S. Saito & K. Iwasa, Proc. VI Intern. Congr. Biophys. IUPAB, Kyoto, Sept. 3-9 (1978).
18. R. Hirschberg & S. Stavis, J. Bacteriol. 129:803 (1977).
19. G. A. Hudock & M. O. Hudock, J. Protozool. 20:139 (1973).
20. D. C. Wood, Photochem. Photobiol. 24:261 (1976).
21. P. S. Song, K. L. Poff & D. P. Häder (in press) (1979).
22. E. B. Walker, T. Y. Lee & P. S. Song, Biochim. Biophys. Acta 587:129 (1979).
23. F. F. Litvin, O. A. Sineschchekov & V. A. Sineschchekov, Nature 271:476 (1978).
24. A. Ascoli & T. Ristori, Proceed Natl. Congr. Cybern. Biophys. Pisa, Apr. 9-11 (in press) (1979).
25. G. Colombetti, R. Ferrara & F. Lenci, Le Scienze (Sci. Am. Ital. Ed.) 94:90 (1976).
26. A. Checcucci, Naturwissensch. 63:412 (1976).
27. B. Diehn, in "Handbook of Sensory Physiology", M. Astrum, ed., Springer Verlag, Berlin (1979).
28. D. P. Häder, in "Encyclopedia of Plant Physiology", W. Haupt M. E. Feinleib, eds., Springer Verlag, Berlin (1979).
29. P. A. Kivic & M. Vesk, Planta 105:1 (1972).
30. P. A. Benedetti & F. Lenci, Photochem. Photobiol. 26:315

31. A. Checcucci, G. Colombetti, R. Ferrara & F. Lenci, Photochem. Photobiol. 23:51 (1976).

32. C. Barghigiani, G. Colombetti, B. Franchini & F. Lenci, Photochem. Photobiol. 29:1015 (1979).

33. P. A. Benedetti, G. Bianchini, A. Checcucci, R. Ferrara, S. Grassi & D. Percival, Arch. Microbiol. 111:73 (1976).

34. C. Barghigiani, G. Colombetti, F. Lenci, R. Banchetti M. P. Bizzaro, Arch. Microbiol. 120:239 (1979).

35. M. J. Doughty & B. Diehn, Biochim. Biophys. Acta (in press) (1979).

36. M. J. Doughty & B. Diehn, Microsens. Newslett. 1:7 (1979).

37. E. Mikolajczyk & B. Diehn, Photochem. Photobiol. 22:269 (1975).

38. E. Mikolajczyk & B. Diehn, J. Protozool. 25:461 (1978).

39. B. Diehn, M. E. Feinleib, W. Haupt, E. Hildebrand, F. Lenci W. Nultsch, Photochem. Photobiol. 26:559 (1977).

40. O. Coppellotti, E. Piccinni, G. Colombetti & F. Lenci, Boll. Zool. 46:71 (1979).

41. E. Piccinni, V. Albergoni & O. Coppellotti, J.Protozool. 22: 331 (1975).

42. A. Checcucci, G. Colombetti, R. Ferrara & F. Lenci, Proc. III Natl. Congr. Cybern. Biophys. Lito Felici, Pisa (1974).

PHOTOSENSORY TRANSDUCTION CHAINS IN PROCARYOTES

Donat-P. Häder

Fachbereich Biologie-Botanik
Lahnberge, D-3550 Marburg
Federal Republic of Germany

1. INTRODUCTION

Due to their simpler structural organisation procaryotes might provide an easier access to the analysis of transduction phenomena than eukaryotes. Photosynthetic prokaryotes, both bacteria and blue-green algae, possess thylakoids not enclosed in an additional chloroplast membrane. The photoreceptor pigments for photomovement phenomena seem to be identical with the pigments used for the light energy conversion in all cases studied up to now. Procaryotes have not developed specialized photoreceptor structures as some eukaryotes (e.g. Euglenophyta, Dinophyta).

Nevertheless the phenomenon of sensory transduction has been studied only in a few photomotile organisms. Therefore this discussion will be confined to a few model cases. Furthermore only parts of the complete reaction chain have been analyzed and in no case the complete process of sensory transduction is known.

1.1 TERMINOLOGICAL CONSIDERATIONS

The terminus sensory transduction has been used by neurophysiologists to describe the conductance of an electrical impuls along a specialized cellular structure. Since in microorganisms the complete signal processing apparatus may be confined to a single cell, the term needs redefinition.

It has been suggested (1,2) to subdivide the reaction chain of light-induced motor responses into three steps: (1) The primary

355

photoperception comprises the absorption of quanta which produces
an excited state of the photoreceptor molecule. (2) The following
step includes the conversion of energy into a transportable form and
the signal transmission to the motor apparatus. (3) This causes a
response which can be detected in a (microscopically) visible reaction. Thus, sensory transduction is the connecting link between photoreceptor and motor apparatus.

Other authors (3) have suggested to confine the term sensory
transduction to the conversion of the stimulus energy absorbed by
the photoreceptor into a different form, while the information
transport is called signal transmission. Since there might be a
multiple conversion from one form of energy to another with interspersed signal transport phases (4) the following scheme will be
used in this chapter.

Sensory transduction links the primary photoperception(absorption of quanta and generation of an excited state) with the motor
response (change in the activity of the motor apparatus). It consists of:

1. Stimulus transformation, which describes one or more conversions
   of one energy form to another and:
2. Signal transmission which comprises all steps in the reaction
   chain which cause a signal transport.

2. BACTERIA

The absorption of a quantum leaves the photoreceptor molecule
in an excited state. This energy can be used to control the motor
apparatus only if the distance between photoreceptor and motor is
negligible, since a transport of this energy involves a resonance
transfer or a similar linkage between molecules. Most likely a stimulus transformation is necessary which converts the absorbed energy
into a transportable form of energy. Some possible mechanisms are
listed in Table 1.

In the case of prokaryotes, where photosensory processes are
governed by the photosynthetic apparatus the generation of an electrochemical gradient seems to be the next step in the reaction
chain following the absorption of a quantum. Further energy conversions may follow.

TABLE 1

Possible mechanisms of stimulus transformation

1. Generation of an electrochemical gradient
2. Subsequent conversion into an ionic gradient
    by antiport or symport mechanisms
3. Production of an electrical potential
4. Generation of a pH gradient
5. Formation of a chemical transmitter

2.1 GENERATION OF AN ELECTROCHEMICAL GRADIENT IN PHOTOSYNTHETIC BACTERIA

The photosynthetic purple bacterium Rhodospirillum rubrum has been shown to perform both photokinesis and photophobic responses (5,6). Since in both cases the action spectra correspond with the absorption spectra of the photosynthetic pigments and the action spectrum for photosynthesis, the involvement of the photosynthetic apparatus in the photosensory responses seems likely. Recently the photosynthetic electron transport chain has been characterized (7,8). The cyclic electron transport includes bacteriochlorophyll, b- and c-type cytochromes and a non-haem iron (Fig.1). Irradiation of chromatophores causes a vectorial proton transport across the membrane. This transmembrane electrochemical potential $\Delta\bar{\mu}\,H^+$ consists of two portions, the electrical potential difference $\Delta\psi$ and the pH gradient (9)

$$\Delta\bar{\mu}\,H^+ = \Delta\psi + \Delta\,pH$$

Quantitative determination of the electrochemical proton gradient can be achieved by a number of techniques (Table 2) even in those cases in which a direct measurement with external or internal electrodes is impossible due to the small dimensions of the cells. Endogenous pigments (e.g. carotenoids) can undergo absorbance changes in response to electrical membrane potentials (10). This electrochromic effect can be calibrated for the membrane potential. This absorbance change is even more pronounced in artificially ap-

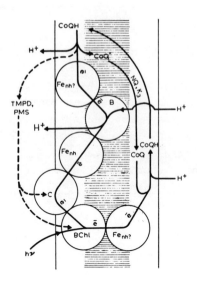

FIGURE 1. Schematic diagram of the cyclic photosynthetic electron
transport in <u>Rhodospirillum</u>. BChl, bacteriochlorophyll; Fe_{nh}, non-
haem iron, B and C, cytochromes; NQ, 1,4-naphtoquinone;
K_3, vitamin K-3; TMPD, N,N,N',N', -tetramethyl-p-phenylenediamine;
PMS, phenazine methosulphate; CoQ, coenzyme Q; after (7)

TABLE 2

<u>Measurement of light-induced potential changes</u>

1. Absorbance changes of endogenous pigments due to
    the electrochromic effect (carotenoids)
2. Absorbance changes of artificially applied
    optical probes (oxonols)
3. Fluorescence changes of fluorescent dyes (9-amino
    acridine)
4. Uptake of charged particles (Nernst potential)
    a) spectrophotometrical estimation of dyes
       (neutral red)
    b) isotope distribution measurement (^{3}H TPMP,
       ^{14}C thiocyanate)

---

plied dyes as in the oxonols (11) which can be incorporated into chromatophore membranes of photosynthetic bacteria. Another way of estimating the electrochemical proton gradient is by use of fluorescent probes like 9-aminoacridine which change their fluorescence properties as a function of the membrane potential (8). The transport and distribution of membrane penetrating ions can be used as well as indicators for electrochemical gradients. The amount of accumulated ions can be detected either spectrophotometrically in the case of coloured dyes (e.g. neutral red (12)) or by labeled substances (8).

Calculations for Rhodospirillum and Spirillum (13,14) have shown that the transport of even a small chemical transmitter is too slow to account for the short response times in phobic responses of purple bacteria (in the msec range). Thus, it is more likely to assume that the electrical potential changes, caused by a sudden change in light intensity, trigger the motor response immediately.

This has been confirmed by Harayama and Iino (15) who studied the correlation between photophobic responses and light-induced potential changes, measured by use of the electrochromic effect. The amplitude of light-induced absorbance changes of a carotenoid absorbing at 525 nm was drastically reduced by the addition of antimycin (20 µM), an inhibitor of the photosynthetic electron transfer (Fig.2). Simultaneously the photophobic response was inhibited to 5% of the uninhibited control.

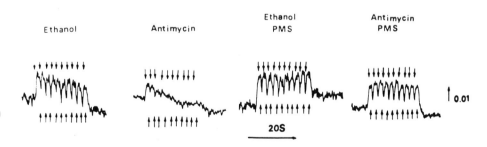

FIGURE 2. Effect of ethanol (0.2%), antimycin (20 µM) and PMS (0.2 mM) on light-induced potential changes in Rhodospirillum rubrum. Up and down arrows indicate light off and on, respectively; after (15).

The effect of antimycin can be overcome by the addition of phenazine methosulfate (PMS) which acts as a bypass around the inhibition site. Valinomycin, an ionophore, had the same effect provided that the penetration barrier for hydrophobic antibiotics, characteristic for gram-negative bacteria (16) was removed by EDTA.

2.2 PRODUCTION OF A CHEMICAL TRANSMITTER

The effect of inhibitors and uncouplers indicate that the photophobic response is directly linked to the electron transport, while photokinesis requires the formation of a high energy inter-mediate (6,17). This is done by means of a protonmotive ATPase which is located in the chromatophore membrane (18,19). In Rhodo-spirillum rubrum photokinetically active light can be substituted by externally applied ATP. Though there is a considerable lag phase which might be caused by a delayed uptake of ATP by the cell, the speed of movement is increased by about 200% (6).

According to recent investigations (20) it is unlikely that ATP is the immediate energy source for movement, but ATP can be converted into a high energy intermediate which drives the bacterial flagellar motor (21,22). At least in some bacteria the motility seems to be linked to the electrochemical gradient $\Delta\bar{\mu}$ H^+. This gradient can be impaired by the protonophorous uncoupler 3,4 dichlo-rocarbonylcyanidephenylhydrazone (CCCP) which abolishes motility (Fig.3) in Rhodospirillum (23) and in E.coli (24). Similar effect

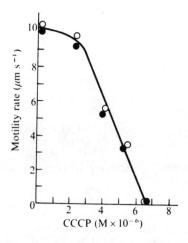

FIGURE 3. Effect of CCCP on motility of Rhodospirillum rubrum; after (23)

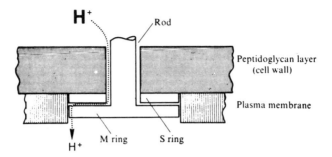

FIGURE 4. Model of the flagellar motor of
Gram-positive bacteria. Protons are assumed
to flow down the electrochemical gradient
driving the M-ring with the attached rod;
after (25)

were obtained by the combined application of valinomycin, which decreases $\Delta\psi$ due to its K^+ conductivity, and acetate which penetrates
the membrane and impairs the pH gradient. These findings resulted
in a model for the flagellar motor of bacteria (Fig.4) according
to which the $\Delta\bar{\mu}\ H^+$ drives the rotation of the M-ring with the attached flagellar rod (25).

2.3 PHOTOPHOBIC RESPONSES IN HALOBACTERIUM

The extreme halophylic bacterium Halobacterium shows both
step-up and step-down photophobic responses (26-29). The photoreceptor for the step-down response seems to be bacteriorhodopsin
which is located in distinct areas, the purple membrane.   The
chromophore, retinal (Fig.5) might undergo an isomeric conformational change upon absorption of a light quantum (30). The chromophore is bound to the protein via a Schiff base. This protein
consists of seven α-helices which span the entire bilipid membrane (31).

Upon illumination bacteriorhodopsin undergoes a photochemical
cycle (Fig.5) during which a proton is taken up from the inside,
transported through the membrane and finally released on the outside of the membrane (32). Thus, a change in the light intensity
could modulate the activity of the motor either directly by a
change in the proton gradient across the cytoplasmic membrane or
by a change in the membrane potential (3). It has been speculated

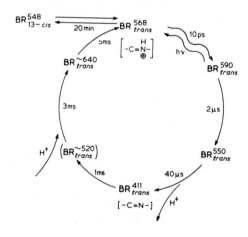

FIGURE 5. Isomeric conformations of retinal and
photochemical cycle of bacteriorhodopsin; after (3)

that voltage sensitive Ca^{2+} gates in the neighborhood of the flagella are triggered by a sudden change in the membrane potential
(caused by a decrease in light intensity) (3). The subsequent
influx of Ca^{2+} might cause a conformational change in the flagellar
base, which results in a reversal of rotation.

3. BLUE-GREEN ALGAE

Unlike the flagellated bacteria blue-green algae move by an

entirely different, though not yet understood, mechanism. Therefore
at least the final steps of stimulus transformation and signal
transmission must necessarily be of a different nature. The primary steps on the other hand seem to be similar and to involve the
generation of light-induced potential changes.

3.1 ELECTRON POOL HYPOTHESIS

In blue-green algae there is a photosynthetic electron transport
which involves two photosystems as in higher plants. Since it has
been shown that the photophobic response is linked to the photosynthetic electron transport chain (33-36), a double irradiation
technique was used to elucidate the role of the two photosystems
(37,38). The organisms were evenly irradiated from above with one
wavelength, while a light field with another wavelength was illuminated from below. Depending on the wavelengths, accumulations
of organisms in the light trap or dispersal from the field was observed.

With a constant trap wavelength of 560 nm, which is absorbed
by C-phycoerythrin and thus activates PS II, phobic reactions occur when the organisms leave the light trap. Variation of the background wavelength results in an action spectrum, which indicates
the activity of PS I.

When, on the other hand, PS I is activated by irradiation of
700 nm (which is absorbed by P_{700}) phobic responses may occur upon
entering or leaving the light trap depending on the background
wavelength. The action spectrum obtained by a variation of the background wavelength resembles a photosynthetic action spectrum of photosystem II.

Studies with photosynthetic inhibitors indicated that there
is an electron pool in the non-cyclic photosynthetic electron transport chain which is filled via photosystem II, while electrons are
drained out of the pool by photosystem I (38,39).

Application of inhibitors of the photosynthetic electron transport chain indicated that the pool is located between the inhibition
sites of 3- 3,4-dichlorophenyl -1,1-dimethylurea (DCMU) and dibromo-
thymoquinone (DBMIB) (40-42). These results suggest that plastoquinone is the biochemical correlate of the pool.

In order to confirm this hypothesis the redox state of plastoquinone was monitored by means of a dual-wavelengths-UV-spectropho-
tometer under conditions which induce photophobic responses in

Phormidium (Fig.6). Alternating activation of photosystems I and
II caused a synchronized change between the oxidized and the reduced forms. The absorption change was observed after averaging
with a minicomputer. The action spectrum (Fig.7) indicates the
involvement of C-phycoerythrin, C-phycocyanin and chlorophyll a,
and resembles the action spectrum of the photophobic response (33).

3.2 SIGNAL TRANSMISSION BY MEANS OF ELECTRICAL POTENTIAL CHANGES

For a number of organisms the signal transmission by means
of electrical potential changes has been suggested or demonstrated
(43-45). The generation of light-induced potential changes can be
explained by Mitchell's chemiosmotic hypothesis (46-47).

The accessory pigments of cyanophyceae are structurally arranged in phycobilisomes (48,49). Absorbed light energy is fun-
neled along the sequence:

C-phycoerythrin $\longrightarrow$ C-phycocyanin $\longrightarrow$ allophycocyanin

to chlorophyll a which is believed to be embedded in the thylakoid
membrane. Absorption of light results in an electron flux through
the electron transport chain. Linked to this process there is a
vectorial transport of protons into the inner space of the thylakoid
(50). The thylakoids have been shown to have a direct connection
to the plasmalemma (51,52). Thus, a superimposition of light-
induced potential changes on the resting potential of the plasma
membrane is reasonable.

3.3 MEASUREMENT OF LIGHT-INDUCED POTENTIAL CHANGES IN PHORMIDIUM

Light-dark cycles which cause repetitive photophobic responses
in blue-green algae cause light-induced potential changes which can
be measured by external electrodes (53) or by internal microelectrodes (54). These potential changes follow the light-dark rhythms
with a lag phase of about 10 s (Fig.8), which corresponds to the
latent period of phobic responses upon a step-down in light intensity. The action spectra of both externally and internally
measured light-induced potential changes resemble the action spectrum of the photophobic response (33,36).

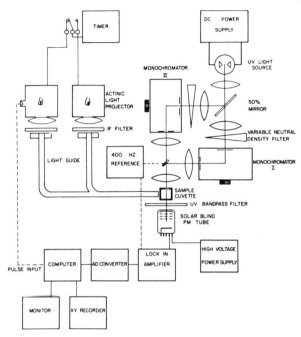

FIGURE 6. Schematic diagram of a double-beam-UV-spectrophotometer.
The UV beam is split into two portions which are passed through
monochromators (260 and 295 nm, respectively). These beams hit
the sample and the photomultiplier after chopping by a tuning fork.
The signal is picked up by a lock-in amplifier and averaged in a
minicomputer. The alternating actinic light is produced by two
slide projectors and passed through light guides (Häder and Poff,
unpublished).

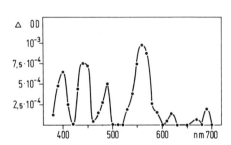

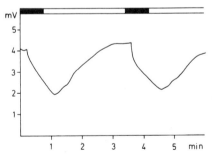

FIGURE 7. Action spectrum of
plastoquinone absorbance
changes induced by 700 nm
and a variable wavelength
(Häder and Poff,unpublished)

FIGURE 8. Internally measured
potential changes induced by a
light-dark rhythm in Phormidium;
after (54)

3.4 INHIBITION OF PHOBIC RESPONSES BY AFFECTING LIGHT-INDUCED POTENTIAL CHANGES

If actually sensory transduction in photophobic responses of blue-green algae is brought about by light-induced potential changes, inhibition of these potential changes should impair photophobic responses. There are various ways of inhibiting specifically resting membrane potentials or light-induced potentials or both.

3.4.1 - Inhibitors of the photosynthetic electron transport chain have been demonstrated to affect photophobic responses. Likewise these substances impair light-induced potential changes which can be measured by means of internal micro-glass-electrodes. DCMU inhibits the non-cyclic photosynthetic electron transport chain close to the primary electron acceptor of photosystem II, thus blocking before the plastoquinone pool. DBMIB impairs both the cyclic and the non-cyclic electron transport by preventing the reoxidation of plastoquinone by the following redox components (55). Both substances specifically reduced the amplitude of light-induced potential changes while the resting membrane potential of the cell is not affected (56)

3.4.2 - Uncouplers with protonophorous properties inhibit phobic responses by a different mechanism. Triphenylmethylphos-phonium bromide (TPMP) penetrates both energy conserving membranes and the cytoplasmic membrane. Thereby it disturbs the generation of a proton gradient. TPMP has been shown to inhibit photophobic responses, impair light-induced potential changes and to reduce the resting potential across the cytoplasmic membrane (56). Similar effects have been demonstrated for the uncoupler CCCP in Rhodospirillum.

3.4.3 - Inhibition of phobic responses by ionophores. Gramicidin A is a linear oligopeptide which can be incorporated into a biological membrane (57). Two molecules in either parallel or end-to-end dimerization form a channel which specifically conducts monovalent cations (58). Since gramicidin inhibits photophobic responses in the blue-green alga Phormidium (Fig.9) it can be concluded that an ionic gradient exists across the membrane which is cancelled by the ionophore.

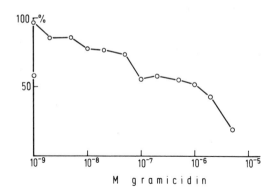

FIGURE 9. Effect of gramicidin on photophobic
responses   in Phormidium (Häder, unpublished)

3.4.4 - Inhibition by external electrical potentials.
Motile filaments of blue-green algae show a temporal polarity
which might be due to an electrical potential difference between
front and rear end. Since this potential is a prerequisite for
the phobic reversal of movement, it is possible to inhibit photophobic responses by the application of external electrical potentials. The potential induced into a single trichome, necessary to
inhibit a phobic response, has the same magnitude as a light-
induced potential change which can be detected with external (59)
electrodes between front and rear end of a trichome. External potentials and inhibitors of the electron transport chain affect the
same basic mechanism. Underoptimal external electrical fields,
which have not effect on the photophobic response, enhance the effectiveness of DCMU by a factor of 100 (56).

3.5 MODEL FOR SENSORY TRANSDUCTION IN THE PHOTOPHOBIC RESPONSE OF
    PHORMIDIUM

The model detailed below is highly hypothetical and only intended to demonstrate the possibility for sensory transduction in
a multicellular prokaryote. The electrical properties of a trichome
can be described by an electrical diagram (Fig.10). Each cell generated a resting potential ψ_R (inside negative) by means of ionic
pumps which are in balance with the opposite ion leakage(symbolized
by the resistor). The mebrane separates electrical charges which

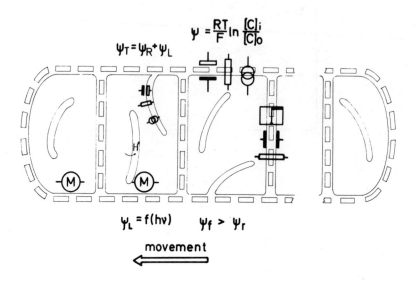

FIGURE 10. Schematic diagram explaining a possible
mechanism for photophobic responses in blue-green
algae (Häder, unpublished)

is demonstrated by the capacitor. In addition to the resting potential each irradiated cell generates a light-induced potential
ψ_L by a vectorial proton transport into the thylakoids by means
of the photosynthetic electron transport. The physiological polarity of a trichome which determines the direction of movement is
manifested in a potential difference in each cell (Fig.11). This
potential difference between front and rear end ($\psi_f < \psi_r$) governs
(not energizes) the movement of the cellular motor apparatus. A
sudden decrease in light intensity reduces the potential of a cell
by ψ_L . This reduces the membrane potential in the rear end of the
cell below a critical threshold (which is not reached in the more
hyperpolarized front membrane). This process might open voltage
sensitive gates which allow a vectorial Ca^{2+} flux into the adjacent
cell (60) where the influx of positively charged ions reduces the
negative potential. Since this triggers a similar process in the
next cell the reversal of polarization is transported from cell
to cell not unlike a chain of connected bistable multivibrators.

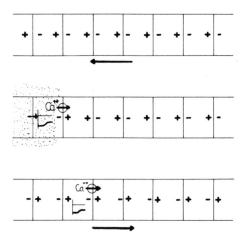

FIGURE 11. Schematic diagram for the reversal of
polarity in a trichome during a photophobic response

The microplasmodesmata found in the septum walls between adjacent
cells of Anabaena might facilitate the ion fluxes (61). The time
consumed by the vectorial Ca^{2+} transport could account for the
observed latent period in phobic responses. The flip-flop mechanism for the reversal of cellular polarity can be initiated by a
sudden decrease in light intensity either only of the front end
or the whole trichome. A shadow over the rear end might reverse
the polarity of the affected cells temporarily but the stimulus
would not be relayed through the trichome. This behaviour is in
good accordance with the observation that a phobic response can
be elicited by casting a shadow over the whole organism or the
front end, but not by shadowing the rear end. Movement in filamentous blue-green algae has been speculated to be brought about
by a parallel array of 5 nm fibrils close to the cytoplasmic membrane. The fibrils are oriented at an angle of 30° to the long
axis of the filament which coincides with the pitch of spirally
rotation during forward movement. It is possible to visualize a
molecular conformational change of the fibrils due to a change
in the internal polarity of the cell or in the local Ca^{2+} concentration which reverses the direction of movement.

REFERENCES

1. D. -P. Häder, in "Research in Photobiology", A. Castellani, ed., Plenum Press, New York, London (1977).

2. D. -P. Häder, in "Encyclopedia of Plant Physiology", W. Haupt & M. E. Feinleib, eds., Springer Verlag, Berlin, Heidelberg, New York (1979).

3. E. Hildebrand, in "Taxes and Behavior", G. L. Hazelbauer, ed., Chapman, London (1978).

4. M. F. Goy, M. S. Springer & J.Adler, Proc. Natl. Acad. Sci. 74:4964 (1977).

5. R. K. Clayton, in, "Photophysiology", A. C. Giese, ed., Acad. Press, New York, London (1964).

6. G. Throm, Arch.Protostenk. 110:313 (1968).

7. L. A. Drachev, V. N. Prolov, A. D. Kaulen, A. A. Kondrashin, V. D. Samuilov, A. Yu. Semenov & V.P. Skulachev, Biochim. Biophys. Acta 440:637 (1976).

8. Z. Gromet-Elhanan & M.Leiser, Arch.Biochem. Biophys. 159:583 (1978).

9. S. Schuldinger, E. Padan, H. Rottenberg, Z. Gromet-Elhanan & M. Avron, FEBS Letters 49:174 (1974).

10. M. Symons, C. Swysen & C. Sybesma, Biochim. Biophys. Acta 462:706 (1977).

11. C. L. Bashford, M. Baltescheffsky & R. C. Prince, FBES Letters 97:55 (1979).

12. A. T. Quintanilha & R.J. Mehlhorn, FBES Letters 91:104 (1978).

13. N. R. Krieg, J. P. Tomelty & J. S. Wels, J. Bact. 94:1431 (1967).

14. A. G. Lee & J. T. R. Fitzsimons, J. Gen. Microbiol. 93:346 (1976).

15. S. Harayama & T. Tino, J.Bacteriol. 131:34 (1977).

16. L. Leive, Biochem. Biophys. Res. Comm. 21:290 (1965).

17. W. Nultsch, "Abhandlungen der Marburger Gelehrten Jg.", W. Fink Verlag, München Nr.2:143 (1974).

18. P. Mitchell, B. Soc. Trans. 4:399 (1976).

19. P. Mitchell, TIBS, 3:N58 (1978).

20. J. Adler, J. Supramol. Struct. 4:305 (1976).

21. J. B. Miller & D. E. Koshland jr., Proc. Natl. Acad. Sci. USA 74:4752 (1976).

22. S. Matsuura, J. Shioi & Y. Imae, FEBS Letters 82:187 (1977).

23. A. N. Glagolev & V. P. Skulachev, Nature 272:280 (1978).

24. S. H. Larsen, J. Adler, J. J. Gargus & r. W. Hogg, Proc. Natl. Acad. Sci. 71:1239 (1974)

25. P. Lauger, Nature 268:360 (1977).

26. E. Hildebrand & N. Dencher, Ber. Dtsch. Bot. Ges. 87:93 (1974).
27. E. Hildebrand & N. Dencher, Nature, 257:46 (1975).
28. E. Hildebrand, Biophys. Struct. Mechanism. 3:69 (1977).
29. W. Nultsch & M. Häder, Ber. Dtsch. Bot. Ges. 91:441 (1978).
30. K. Schulten & P.Tavan, Nature, 272:85 (1978).
31. R. Henderson, P. N. T. Unwin, Nature 257:28 (1975).
32. S. F. Bayley & R. A. Morton, Critical Rev. Microbiol. 6:151 (1978).
33. W. Nultsch, Planta 58:647 (1962).
34. D. -P. Häder & W. Nultsch, Schweiz. Z. Hydrol. 33:566 (1971).
35. W. Nultsch, Arch.Mikrobiol. 55:187 (1966).
36. D. -P. Häder, Diss. Mbg. (1973).
37. D. -P. Häder & W. Nultsch, Photochem. Photobiol. 18:311 (1973).
38. D. -P. Häder, Arch.Microbiol. 96:255 (1974).
39. W. Nultsch & D. -P. Häder, Ber. Dtsch. Bot. Ges. 87:83 (1974).
40. D. -P. Häder, Arch. Microbiol. 103:169 (1975).
41. D. -P. Häder, in "Biophysics of Photoreceptors and Photobehaviour of Microroganisms", G. Colombetti, ed., Lito Felici, Pisa (1975).
42. D. -P. Häder, Arch. Microbiol. 110:301 (1976).
43. R. N. Doetsch, J. Theor. Biol. 35:55 (1972).
44. H. Machemer, J. Exp. Biol. 75:427 (1976).
45. S. Marayana & T. Tino, Photochem. Photobiol. 25:571 (1977).
46. P. Mitchell, Adv. Enzymol. 29:33 (1967).
47. F. M. Harold, Ann. Rev. Microbiol. 31:181 (1977).
48. J. C. Cosner, J. Bacteriol. 135:1137 (1978).
49. D. J. Chapman, in "The Biology of Blue-Green Algae", N. G. Carr & B. A. Whitton, ed., Blackwell Scientific Publications, Oxford, London, Edinburg, Melbourne, Botanical Monographs (1973).
50. R. J. Poole, Ann. Rev. Plant. Physiol. 29:437 (1978).
51. M. M. Allen, J. Bact. 96:842 (1968).
52. J. R. Golecki, Arch. Microbiol. 120:125 (1979).
53. D. -P. Häder, Arch. Microbiol. 118:115 (1978).
54. D. -P. Häder, Arch. Microbiol. 119:75 (1978).
55. U. Siggel, Bioelectrochem. Bioenerg. 3:302 (1976).
56. D. -P. Häder, Arch. Microbiol. 120:57 (1979).
57. G. Szabo & D. W. Urry, Science 203:55 (1979).
58. E. Bamberg, H. -J. Apell, H. Alpes, E. Gross, J. L. Morell, J. F. Harbaugh, K. Janko & P. Läuger, Fed. Proc. 37:2633 (1978).

59. D. -P. Häder, Arch. Microbiol. 114:83 (1977).
60. B. Hille, Biophys. J. 22:283 (1978).
61. T. H. Giddings jr. & L. A. Staehelin, Cytobiologie 16:235
    (1978).

SENSORY TRANSDUCTION IN PHOTOTROPISM: GENETIC AND PHYSIOLOGICAL

ANALYSIS IN PHYCOMYCES

Vincenzo A. Russo

Max-Planck-Institut für Molekulare Genetik

D-1000 Berlin 33   (Germany)

ASEXUAL LIFE CYCLE

Phycomyces blaskesleeanus is a lower fungus of the class Phycomycetes, order Mucorales, family Mucoraceae. Its asexual life cycle is shown schematically in Fig. 1. From the spore a mycelium is formed. After about two days many sporangiophores are produced and the top of these sporangiophores bulges out to form a sporangium in which about 10^5 spores are made. The complete cycle takes about four days under normal laboratory conditions (21°C, humidity above 60%, well aerated cultures). The mycelium is almost completely non-septate and the sporangiophore is completely non-septate (coenocytic). The sporangiophore without sporangium (called stage I), is about 0.1 mm thick and grows at a rate of 1 mm/hr. When the sporangium and the spores are formed the sporangiophore is in stage IV; this stage grows at a rate of 3 mm/hr and can reach a height of more than 10 cm. The growing zone of the stage IV sporangiophore is about 2 mm below the sporangium. A particularity of stage IV is the rotation of the sporangium at an angular velocity of two revolutions per hr (720°/hr). The wall of the growing zone rotates two revolutions per hr at its top and zero at its bottom. This phenomenon is probably important for the phototropic response. This gigantic single-celled sporangiophore that grows at such a rate is unique among the fungi. For this reason and because of the many stimuli to which it reacts, Phycomyces has been relatively well studied for about 100 years. For more information see Bergmann (1), Russo and Galland (2).

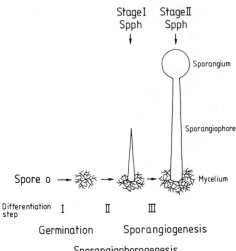

FIGURE 1. Schematic representation of the asexual life cycle of
Phycomyces with the three major differentiation steps. Germination,
Sporangiophorogenesis and Sporangiogenesis.

DESCRIPTION OF THE SENSORY RESPONSE

       There are several stimuli to which Phycomyces responds; both
the mycelium and the sporangiophore can be stimulated. The same
stimulus, blue light, can cause several responses. All these responses are interrelated as can be seen later with the genetical
analysis. The possibility of stimulating the same structure, the
sporangiophore,with different stimuli can be very useful in order
to make a physiological analysis of the sensory transduction of
light, which is the main topic of this chapter, as can be seen
later. For this reason, I will give a brief description of all of
the sensory responses known in Phycomyces.

Light- and Dark-Growth Responses

The growth rate of a stage IV sporangiophore is completely
independent of light intensities ranging from dark to 1 W/m^2.
However, a change of light intensity in this range will induce a
transient change in the growth rate. The light regime can either
be a pulse-up (or pulse-down) or a step-up (or step-down) change.
In Fig. 2 the four basic illumination programs and their outputs
are shown. The important features are: i) with a pulse-up of light
there is first an increase and then a decrease of growth rate
before reaching the base-line level (the opposite happens with
a pulse-down); ii) with a step-up (or step-down) there is only a
transient increase (or decrease) in growth rate. In order to explain
this "adaptation" of the growth response Delbrück and Reichardt (3)
introduced the concept of level of adaptation A. The level of adaptation A is postulated to be a biophysical or biochemical status
of the cell and its magnitude is somehow proportional to the intensity of light used in the pretreatment. They assumed that after a
long pretreatment at constant intensity the adaptation reaches an
equilibrium with intensity, formally $A = I_o$.
    Furthermore, after a change of light intensity, A will be
described by a differential equation:

$$\frac{dA}{dt} = \frac{I - A}{b}$$

where b is the time constant of the system. Finally, the growth is
a function of the subjective light intensity $i = I/A$ at any given
time. With a step-up of light the results are true only within a
small range of light intensities, at both extremes the situation
is more complicated as will be discussed in Fig. 8.

Phototropism

    If the growing zone of a sporangiophore is illuminated from
one side, the sporangiophore will bend toward the light. The
measurable parameters of the phototropic response are the delay
before bending starts and the rate of bending. The delay of bending
can be very different if the light stimulus has an intensity different from the intensity of the pretreatment light. In the case
where the light stimulus has the same intensity as the light at
which the sporangiophore was pretreated for several hours, the

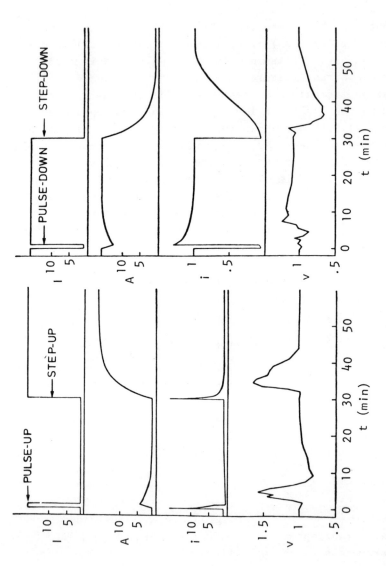

FIGURE 2. Four basic illumination programs and their outputs. The top gives the program, intensity I (linear scale) versus time t. The bottom row gives the growth output, velocity (relative to average velocity) versus time. The second and third row give the level of adaptation A, and the subjective intensity, i = I/A, calculated according to the theory developed. Note that the scale used to plot i (t) is twenty times larger for the "down" than for the "up" programs (3).

delay in phototropism is 6 ± 2 min and it is independent of the
light intensity of the stimulus in the range from 6 to 10^{-6} W/m^2.
The bending rate is of the order of a few degrees/min and is slightly
dependent on the absolute light intensity in the same range (Fig. 3).
A noteworthy phenomenon is that the phototropic response can last
for hours if the geometrical relationship between the growing zone
and light source is kept constant (4). The growth response, however,
lasts only 20 to 30 minutes. Another paradox is that in the light
growth response there is more growth when there is more light
(although only transiently) while in phototropism there is more
growth in the "shadow" part of the sporangiophore, for geometrical
reasons (5), see Fig. 4a. This last paradox can be explained   by
the lens effect. The growing zone is an almost transparent cylinder,
therefore the light is focused on the distal side (or "shadow" side)
of the sporangiophore. Shropshire (6) has shown that the lens effect
is essential for a phototropic response. The first paradox could be
explained by the rotation model of Dennison and Forster (7). They
calculated the actual light intensity which is on the cell wall of
a sporangiophore taking into account the structure of the growing

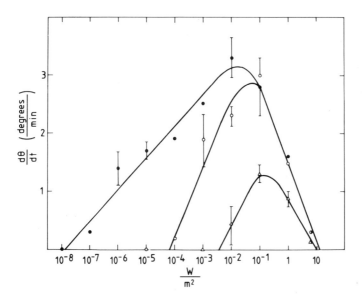

FIGURE 3. Bending rate in degrees/min as function of light intensity. o -- o WT, o -- o madA, Δ -- Δ madB. (Unpublished results).

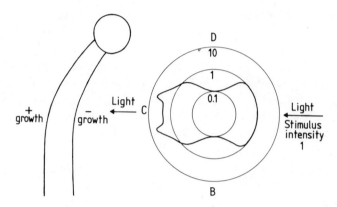

FIGURE 4. (a): During a phototropic response there is more growth
on the distal side than on the proximal side (5).
(b) Theoretical polar logarithmic plots of the light intensity
adjacent to the inner surface of the wall in a 0.01-μm-thick pigment
layer when the cell is illuminated by unpolarized light incident
perpendicular to its axis. The circles are separated by a factor
of 10 in intensity. The circle marker "1" is normalized to one
for an incident intensity of one. All reflections have been taken
into account, but no correction for small particle scattering has
been made. Assumed radial dimensions are 50 μm to outer surface,
49,4 μm to inner surface of wall, and 23 μm to tonoplast. Refractive
indices are assumed to be 1.50 for the wall, 1.38 for the cytoplasm,
and 1.36 for the vacuole (wavelength: 450 nm). Absorption coefficients are assumed to be 0.23 cm^{-1} in the wall, 27.6 cm^{-1} in the
cytoplasm, and 0.23 cm^{-1} in the vacuole. The illumination is
assumed to be a beam of intensity 1.00 from the right. (Redrawn
from Dennison and Forster (7)).

zone (cytoplasmic membrane, cytoplasm, tonoplast, vacuole, all
with a different refractive index). The result of those calculations is shown schematically in Fig.4(b).The rotational model
assumed that the light receptor and growth machinery rotates clockwise as does the wall. Therefore, any element which is in the region
B of Fig. 4b, will reach the region C in about 15 min (360°/hr is
the rotation rate in the middle of the growing zone) and will sense

a step-up of light, i.e. it will grow faster; when it reaches the
position D it will sense a step-down and will grow slower, and so
on. Although the light stimulus is constant in space, any part of
the light-growth machinery will undergo this continuous light and
dark program. That is the reason why the phototropic response is
steady while the light induced growth is transient. Dennison and
Forster (7) bring some evidence in favor of their model. Evidence
against this model is the fact that stage I Spph is phototropic
but does not rotate. The action-spectrum of both light-growth
response and phototropism is the typical action-spectrum of blue
light (8).

Photocarotenogenesis

For many years it has been known that in Phycomyces, as in
many other microorganisms, blue light can stimulate the production
of β-carotene (up to 20 times more (9)).

Photophorogenesis

There are conditions which uncouple the growth of the mycelium
from the formation of the sporangiophore (differentiation II of
Fig. 1). Blue light can induce the formation of the sporangiophore
(10,11). This phenomenon is called photophorogenesis.

Photosporangiogenesis

There are other conditions which uncouple the growth of sporangiophore stage I from the formation of the sporangium (differentiation step III of Fig. 1). Blue light can induce the formation of
the sporangium (12). This phenomenon is called photosporangiogenesis.

Autochemotropism, Rheotropism

A barrier, placed in the vicinity of the growing zone will
induce a reaction of avoidance such that the sporangiophore will
grow away from the barrier. This phenomenon is called avoidance
or autochemotropism. A double barrier will induce a transient
growth response. Both responses are very fast and similar in
magnitude and kinetics to the phototropic and light-growth response.
Johnson and Gamow (13), suggested that the sporangiophore emits
an unknown gas; the barrier somehow will increase the amount of

this gas at the growing zone near to the barrier; the gas is thought
to be a growth promoter. Russo (14-15), suggested that this gas
could be ethylene.

A sporangiophore which is placed in a wind tunnel will grow
toward the wind (16).

Geotropism

The sporangiophore of Phycomyces can sense gravity. If it is
placed horizontally it will grow upwards. The delay of the response
is variable between 20 and 180 min. The bending rate is weak, about
0.3°/min.

WHY A GENETIC ANALYSIS OF PHOTOTROPISM?

How can light induce a uniform growth response, if given
bilaterally, or a differential growth response (phototropism) if
given unilaterally? For sure there must be a molecule which absorbs
the light. We call this molecule the photoreceptor. The photoreceptor
excited by the light will make a certain chemical reaction X
(either transport of ions, or oxydation or something else), this
reaction X will induce a reaction Y, which will induce a reaction
Z and so on until there is more growth of the sporangiophore. I
will call this sequence of events from the light reception to the
growth of the wall the blue light sensory transduction chain in
phototropism. How many steps are there in the sensory transduction
chain? What kind of reactions are they? With a solely physiological
analysis it would be impossible to answer these questions. A purely
biochemical analysis would be very difficult; it will be very
hard for one to be convinced that a certain biochemical change
induced by blue light is relevant to the growth response and not
just a side effect. I will give an example. We have seen that light
induces a biochemical change in the mycelium, more β-carotene production, as well as a morphogenetical change, photophorogenesis.
Does that mean that the production of β-carotene by light is an
intermediary step in the sensory transduction chain of photophorogenesis? With the help of genetics we can answer: no. The experiment is very simple: it is possible to obtain photophorogenesis
in mutants which have no β-carotene and where the light cannot
induce β-carotene (11). That is a simple example where a genetical
analysis helps to decide if a biochemical change is essential for
the sensory transduction of a certain stimulus.

Another example is the following. It is a general belief that
the blue light photoreceptor in Phycomyces is a flavoprotein;
this belief is based on the action-spectrum and on the results of
Delbrück et al. (17). But which flavoprotein? In any living
organism there are many flavoproteins with very different functions
(18). If it were possible to obtain a mutant which is in the photoreceptor (this could be decided by its physiological behaviour as
will be discussed later), it would be of great help for the biochemical identification of the photoreceptor. The flavoprotein
which is absent in the mutant must be the photoreceptor.

The power of genetics lies in the theoretical possibility to
eliminate (by mutation) one and only one enzyme of the cell's
biochemical machinery. Physiological studies will say if this
enzyme is essential in the sensory transduction chain. There are,
however, limitations to the genetic approach. The first limitation
is that some mutations may be lethal. If any enzyme of the sensory
transduction chain of phototropism is essential also for the
growth of Phycomyces, a mutation in this gene will never allow
a viable organism (the mutation is lethal). The second limitation
is due to the fact that with genetical means it is impossible to
decide the function of a certain enzyme which is eliminated by a
certain mutation. Even with those limitations, I think it is very
useful to add a genetical analysis to the physiological and biochemical analysis of a certain sensory phenomenon. The rest of the
chapter will show how useful the genetic analysis of phototropism
in Phycomyces has been.

Mutants abnormal in phototropism were isolated by Bergman et
al. (9). Those mutants were shown to be complementable by Ootaki
et al. (19). Until now seven genes have been known to be important
in the phototropic response. That means that at least seven
enzymes are in the sensory transduction chain of phototropism.
This is a minimal estimation for two reasons: i) the lethal mutations will never be found; ii) there is no certainty that all the
genes which make an enzyme involved in the sensory transduction
chain have been mutated. The seven genes are named madA, madB,
madC, madD, madE, madF, madG. A physiological analysis of these
mutants was made with respect to all the other sensory responses
known. It turns out that the mutants are all pleiotropic except
for those in gene madC. Namely, the mutants in the genes madD,
E, F, G, also block the geotropic and the autochemotropic response
(9), the rheotropic response (Jan, personal communication), the
ethylene induced growth (15). The mutants in the genes madA and

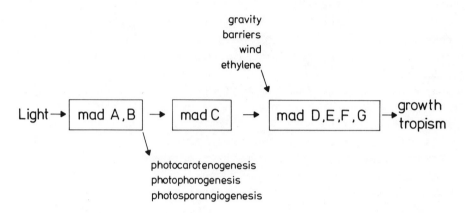

FIGURE 5. Pleiotropism of Phycomyces strains selected as phototropic defective mutants. Mutants madA and madB are defective in
all photoresponses, mutants madD, E, F, G, only in tropic and
growth responses.

madB have a higher light threshold for photocarotenogenesis (9,20),
for photophorogenesis (11), and for photosporangiogenesis (12).
All are abnormal for the light induced growth response (21), (Fig.
8). This pleiotropism, shown schematically in Fig. 5, indicates that
all the blue light effects need the two genes madA and madB and
all the responses where a growth of the sporangiophore is implied
need the genes madD, E, F, G. That implies that probably all the
blue light effects are mediated by the same photoreceptor. Another
logical implication is that along the sensory transduction chain
there is, in time, first the action of madA and madB coded enzymes,
then the action of madC coded enzyme then the action of the other
enzymes. Therefore, the sequence of Fig. 5 also indicates the time
sequence from the light reception to growth stimulation. For enzymes
coded by genes in the same block it is still impossible to decide
the time sequence.

DARK- AND LIGHT-ADAPTATION IN PHOTOTROPISM

     If a sporangiophore is illuminated with a bilateral light beam
of intensity I_o for many hours and then is stimulated unilaterally

with a beam of intensity I, there will be a phototropic response
as long as I is above the threshold. The delay, however, will be
very different depending if a) $I = 2I_0$, b) $I \gg 2I_0$, c) $I \ll 2I_0$.
In the case a) the delay is 6 ± 2 min and is independent of the
light intensity of I in the range from 6 to 10^{-6} W/m^2. In the case
b) the delay can be more than 90 min and depends on the ratio $I/2I_0$.
We call this phenomenon dark-adaptation in phototropism. In case
c) the delay can be 20 - 30 min and depends again on the ratio
$I/2I_0$. This phenomenon is called light-adaptation in phototropism.

The kinetics of dark adaptation for two different values of
I_0 are shown in Fig. 6. In the top curve of Fig. 6C there are a
series of points for wt which are obtained in the following way.
Each point is the average of at least four experiments. In each
experiment the sporangiophore was pretreated with a bilateral
illumination $I_0 = 3$ W/m^2 for 60 min. From time zero, the bending
angle θ was measured every two min with a horizontal microscope
equipped with a goniometric device. At time 10 min the bilateral
illumination was stopped and a unilateral illumination with intensity I was begun. The angle θ was measured until the sporangiophore
bent at least $30°$. In Fig. 6A and Fig. 6B are shown the curves θ
vs. time of two different types of experiments: in Fig. 6A: $I =$
$= 2I_0$, in Fig. 6B: $I = 10^{-5} \times 2I_0$.

From a series of experiments we can now plot the delay as a
function of I as shown in Fig. 6C. The top curve is produced from
a series of experiments where $I_0 = 3$ W/m^2 the bottom with $I_0 =$
$= 6.10^{-5}$ W/m^2. In Fig. 6C it is really plotted I vs. delay because
we want to know how changes the sensitivity of the system of the
sporangiophore, or the level of adaptation A, according to Delbrück
and Reichardt (3). The level of adaptation A is proportional to the
light intensity I_0 in a steady state, $A(t,I_0) = kI_0$. After a
change to a light intensity I, A will change to the new level A
$(t, I) = kI$ in a finite time. We assume that there is a phototropic
response 6 min after is true the equation $A(t,I) = kI$. With these
assumptions we can interpret the wt curves in Fig. 6C as the kinetics of dark decay of the level of adaptation A. From the top curve
it is possible to see that it takes $90-6 = 84$ min before the level
of adaptation drops from its value $A(I_0)$ at $2I_0 = 6$ W/m^2 to the new
value $A(I)$ for $I = 6 \ 10^{-5}$ W/m^2. The shape of the curve indicates
that the kinetics is biphasic. The wt curve in the bottom part of
Fig. 6C is very similar to the top curve, which indicates that
the kinetics of dark adaptation of A is independent of the light
intensity I_0 at which the sporangiophore was pretreated. The

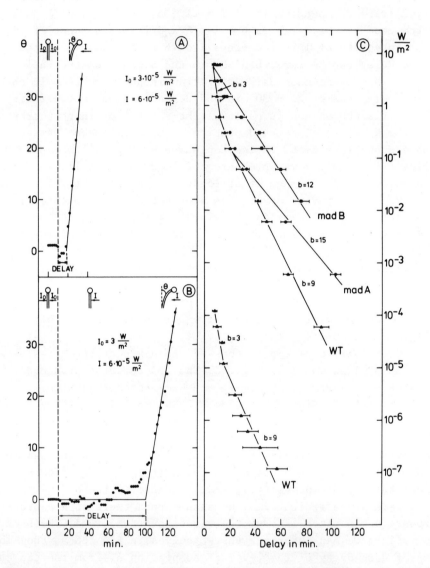

FIGURE 6. (A) Phototropic response at constant light intensity. The
sporangiophores were pretreated for 120 min at a total light intensity of $2I_0 = 6 \times 10^{-5}$ W/m^2. At times 10 min the illumination was
shifted from bilateral to unilateral keeping the same total light

(continue legend to Fig. 6)
intensity $I = 6 \times 10^{-5}$ W/m^2. The delay is obtained by extrapolating
back from the steady state bending.
(B) Phototropic response after a step-down of light intensity.
The sporangiophores were pretreated for 60 min at a total bilateral illumination of $2I_0 = 6$ W/m^2. At time 10 min the illumination
was made unilateral with an intensity of 6×10^{-5} W/m^2. The delay
is obtained in the way as in Fig. 8 (A).
(C) Delay of the phototropic response after a step-down of light
intensity as a function of the light intensity of the phototropic
stimulus I. In the upper part of the figure are given the curves
for wt, madA, madB, all pretreated for 60 min at $2I_0 = 6$ W/m^2. In
the lower part of the figure there is a curve for wt, the pretreatment light intensity is $2I_0 = 1.2 \times 10^{-4}$ W/m^2. The points are the
average of experiments on 4 different sporangiophores. The error
of the mean is given (unpublished results).

behaviour of mutants in the genes madA and madB is also shown in
Fig. 6C. madA is normal in the first part of the curve but slower
in the slower part. madB lacks completely the fast part of the
curve. In a similar way it is possible to study the kinetics of
light adaptation and the results are shown in Fig. 7, curve a.
These results are more difficult to interpret, because after the
step-up of light there is also a growth response. There could be
the possibility that the delay of 20 min for $I = 2 \times 10^{-1}$ W/m^2 is
due to the fact that the growth response is saturated so that a
differential growth response is impossible. To exclude this possibility an experiment with two simultaneous stimuli was carried
out. At time 10 min, when the light illumination was shifted from
bilateral at total intensity $2I_0$ to unilateral at intensity I, the
sporangiophore was presented with a glass barrier between the sporangiophore and the light source. The bending was away from the
barrier and from the light source. The delay as a function of
light intensity is shown in Fig. 7b. Let us consider the step-up
with $I = 2 \times 10^{-1}$ W/m^2: the delay in the case of autochemotropism
is 10 min instead of 20 min for the phototropic response at the
same light intensity. This result indicates that at 10 min it is
possible to obtain a differential growth response if the stimulus
is a barrier but not if the stimulus is light. The simplest interpretation of these data is the following: a) During a light adaptation experiment some processes in the early sensory transduction

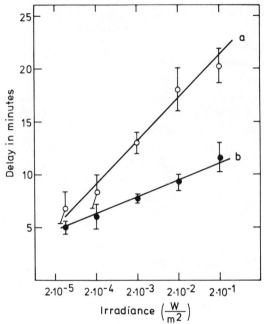

FIGURE 7. (a) Delay of phototropism after a step-up light versus
the light intensity of the stimulus. The sporangiophores were pretreated at a total intensity $2I_o$ = 2 x 10^{-5} W/m^2 for 90 min.
The experiments are similar to the one of Fig. 8 (B) with the difference that at the time of shift from bilateral to unilateral
light the intensity of light stimulus was increased.
(b) Delay of an autochemotropic response after a step-up of light.
The sporangiophores were pretreated for 90 min at $2I_o$ = 2 x 10^{-5}
W/m^2. At time 10 min the total light intensity was increased from
2 x 10^{-5} W/m^2 to the values written on the abscissa and a barrier
was put in the vicinity of the sporangiophore. For this experiment
it is irrelevant whether the illumination after the step-up intensity is bilateral or unilateral (unpublished results).

chain, before the action of genes madD to madC products, is saturated for 15 min and then there is "adaptation". b) In the late
part of the chain, from madD on, there is also a saturation but it
lasts only 5 min. (With a step-up of bilateral light and a double
barrier given at the same time it is possible to show directly
a saturation in this part of the chain because the two responses
are not additive). In the light-adaptation madA behaves like wt
at a light intensity at which it is not blind, while madB has a

curve like wt but it is shifted at high light intensities (unpublished results).

LIGHT- AND DARK-GROWTH RESPONSES

Let us pretreat a sporangiophore with a bilateral light beam
of intensity I_0 for many hours and then make a step-up in intensity
of both beams by a factor of 100. What is the growth response? The
answer is shown in Fig. 8A. The growth response is very dependent
on the light intensity I_0. At a very low intensity there is almost
only a prolonged inhibition of growth of about 10%. At an intermediate intensity there is first a stimulation and then an inhibition.
At medium-high intensities there is only a stimulation which
disappears after a while. At a very high intensity the growth stimulation stays for a long time (this last phenomenon was already
known by Bergman et al. (1), who reported that the stimulation can
be maintained for hours). We can interpret these results by saying
that light can make both an inhibitor and a promoter of growth.
The light threshold for the production of inhibitor is lower than
the threshold for the production of promoter. Forster and Lipson
(21), assumed that at very high intensities the growth rate does
not go back to the basal level because: a) the photoreceptor is an
inhibitor of growth and b) at high intensities the photoreceptors
are converted into non-functional photoproducts, therefore the
inhibitory control of the growth rate is removed. Interestingly
enough the mutants in madA and madB, at the lowest intensity at
which they react, do not behave like wt at its lowest intensity
(Fig. 8B and C). At the highest intensity madA is like wt while
madB is not. We can conclude that these "night blind" mutants do
not behave like grey filters in the sensory transduction chain.

Let us make a step-down of light during a bilateral illumination I_0 to I and let us ask what is the behaviour of the growth
response as a function of the ratio I/I_0. In Fig. 9A is shown a
typical growth curve at $I_0 = 1.5 \times 10^{-1}$ W/m^2; in Fig. 9B the bilateral illumination was changed from $I_0 = 1.5 \times 10^{-1}$ W/m2 to I = 3
x 10^{-6} W/m^2. The growth rate diminishes by 20% and then returns
to the basal level in this case after a time T = 24 min. We then
ask how is T as function of I. The answer is in Fig. 9C. After a
step-down by a factor 50, T = 24 min and stays constant up to a
step-down of $I_0/I = 5 \times 10^4$. Therefore, the "adaptation" of the
growth rate to a step-down of light is very different from the
"adaptation" of phototropism to a step-down of light (see Fig. 6).

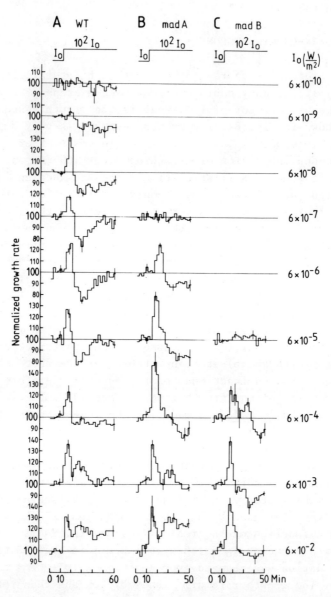

FIGURE 8. Normalized growth response after a step-up from I_0 to
$100\ I_0$ for different value of I_0. Each curve is the average of four
experiments except for the <u>wt</u> curve at $I_0 = 6 \times 10^{-9}$ W/m^2 where eight
experiments were made. A) <u>wt</u> the pretreatment time was three hours

(continues legend to Fig. 8)
for $I_o \leq 6 \times 10^{-8}$ W/m^2 at two hours for $I_o \geq 6 \times 10^{-7}$ W/m^2. B) C21
(<u>madA</u>) the pretreatment time was three hours for $I_o \leq 6 \times 10^{-5}$ W/m^2
and two hours for the other intensities. C) C109 (<u>madB</u>); the pretreatment time was three hours for $I_o \leq 6 \times 10^{-4}$ W/m^2 and two hours for the
other intensities (umpublished results).

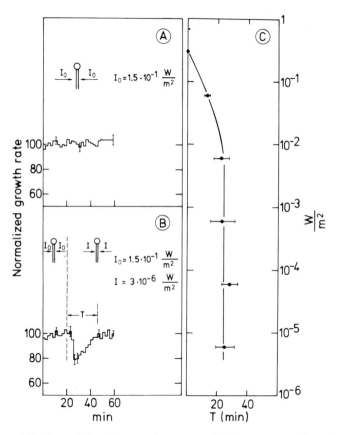

FIGURE 9.   (A) Normalized growth rate of the sporangiopphore at
constant light intensity $I_o = 1.5 \times 10^{-1}$ W/m^2. (B) Dark-growth
response of the sporangiophore after a step-down of light (at 20
min). T is the time at which the growth rate returns to the baseline
level.
(C) Time T as a function of the light intensity after a step-down
of light. The values are obtained from a series of experiments of

the type presented in Fig. 7(b). The sporangiophores were always pretreated at $I_o = 1.5 \times 10^{-1}$ W/m^2 and the step-down was made with different light intensities. The ordinate shows the total light intensity I after a step-down. All experiments were repeated on 4 different sporangiophores. The error of the mean is given. Sporangiophores were pretreated for 60 min at a total light intensity of $2I_o = 3 \times 10^{-1}$ W/m^2 (unpublished results).

We will discuss this phenomenon further.

Temporary Inversion of Tropic Responses

Reichardt and Varjü (22) discovered that a step-up in light intensity during unilateral irradiation causes a temporary inversion of the steady state phototropic reaction. Castle (5) has demonstrated that a saturating pulse of light given from any direction can cause the inversion and Dennison (4) has shown that a step-down will cause an inversion (Fig. 10). Removal of the light during a phototropic response also results in an inversion (unpublished results). At an intensity of 10^{-2} W/m^2 sporangiophores bend at a rate of 2.5 degrees/min; after the light is turned off reversed bending at a rate of 1.5 degrees/min takes place. Bergman et al. (1), reported that an autochemotropic response or a geotropic response can also show an inversion if the sporangiophore is stimulated by a saturating pulse of light. The results of Bergman et al. (1) therefore, indicate that the difference in growth reactivity between distal and proximal sides can be attributed to a late part of the sensory transduction chain. Our light-off experiments suggest that a gradient of an "inhibitor" of growth may be inside the cell and that the distal side has more "inhibitor" than the proximal side. Similar inversion is observed by removal of the barrier during autochemotropism.

BIOCHEMISTRY OF LIGHT-GROWTH RESPONSE

The only information available at the biochemical level is that the level of cAMP drops to 50% of its normal value, after one minute, after a step-up of light (23). This biochemical change is probably a part of the sensory transduction chain for the following reasons: a) externally supplied cAMP induces a transient inhibition of growth (23); b) a change of cAMP level does not occur with a mutant in madD (24).

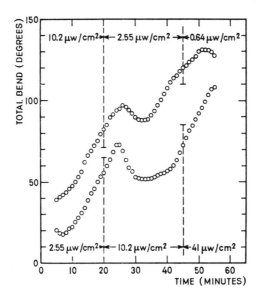

FIGURE 10. Phototropic inversions following an increase in intensity (lower curve) and a decrease of intensity (upper curve). (From Dennison (4)).

TWO TYPES OF ADAPTATION

I would like to present a formal model which may qualitatively explain many of the phenomena presented here. The model is represented schematically in Fig. 11A. It hypothesizes the following transduction chain: Blue light induces the production of a promoter (P1) and of an inhibitor (A) of enzyme E1. The enzyme E1 produces a promoter (E$_2$) and an inhibitor (i) of enzyme E$_2$ which is somehow responsible for growth. The characteristics of these six parameters are as follows: The level of P1 changes very fast with a change of light while the level of A changes slowly (see Fig. 2). The activity of E1 is between a minimum value for P1/A $\ll$ k and a maximum value for P1/A $\gg$ k (k is arbitrary). P$_2$ has a property similar to P1 and $\underline{i}$ a similar property to A but with faster kinetics than A. The activity of enzyme E$_2$ is between a maximal and a minimal value, depending if P2/$\underline{i}$ $\gg$ K of P2/$\underline{i}$ $\ll$ K (K is again arbitrary). The last hypothesis is that adaptation II is in the sensory transduction chain in common with the autochemotropic and geotropic transduction chain. Let us see how this model can explain qualitatively the different phenomena. Adaptation I is responsible for the level of sensitivity and adaptation II is responsible for "adaptation" in growth. Since $\underline{i}$ is faster than A, this explains the difference in dark adaptation after a step-down of a factor 10^5 in phototropism (90 min in Fig. 6) and in growth response (24 min in Fig. 9).

In a phototropic response after a step-up of light (or a step-down of light) there is a delay because E1 activity is maximal (or minimal) on both the distal and the proximal side of the sporangiophore, therefore there is no differential information. The saturation of E_1 activity after a step-up of light produces a saturation of E2; but, since i is faster than A, it is possible to obtain an autochemotropic response after 10 min, instead of 20 min (Fig. 7).

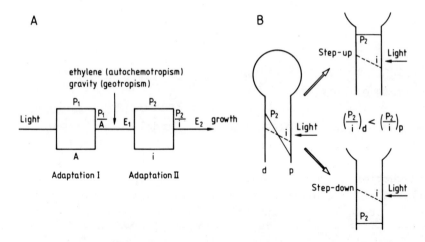

FIGURE 11. A Schematical representation of the two types of adaptation in the sensory transduction chain of phototropism in Phycomyces.
B Hypothesized gradient of P_2 and i during a phototropic response (left) or after a step-up of light during a phototropic response (above right) or after a step-down of light during a phototropic response (below right).

The inversion of the tropic response after turning off the
light during a phototropic response (or removing the barrier during
autochemotropism) indicates that in the sporangiophore there is a
gradient of i as shown in the left part of Fig. 11B. The inversion phenomenon of phototropism (or autochemotropism) after a step-up of
light or a step-down of light (Fig. 10) is due to the saturation
of E_1 activity and to the relatively slow kinetics of i so that
the situation in the sporangiophore is like that shown in the right
part of Fig. 11B. Both, after a step-up or a step-down of light:

$$(\frac{P2}{i})_{distal} < (\frac{P2}{i})_{proximal} \quad therefore \quad (growth)_{distal} < (growth)_{proximal}$$

proximal. Although the model can explain several phenomena, only a
biochemical analysis can say if and to what extent it is correct.
It is a good working hypothesis for planning new genetic and physiological experiments.

CRITERIA FOR THE IDENTIFICATION OF PHOTORECEPTOR MUTANTS

We have seen that it is very important to have criteria for
the identification of mutants in the photoreceptor. On the basis
of physiological studies we have until now two criteria.
    Criterion 1): Forster and Lipson (21) suggested that a mutant
which bends at very high intensities, where wt is non-reactive
(phototropic indifferent, see Fig. 3), and is night blind, is most
probably a mutant where the absorption cross section of photoreceptor is smaller than the cross section of wt. This is based on the
reasonable assumption that at very high intensities the ratio:

$$P_{bl}/P_{tot} = F (\sigma,k,I) = \sigma I/ (\sigma I + k)$$

(where P_{bl} is the number of photoreceptors bleached, P_{tot} the
amount of photoreceptor, σ the absorption cross section (25), k
is the regeneration time and I the light intensity) is near to
unity so that the absorption of light is the same on the distal
and the proximal side of the sporangiophore. A mutant with smaller
correction ($\sigma_{mut} < \sigma_{wt}$) will reach the point of phototropic indif-
ference at a higher light intensity. This criterion is sufficient
but not necessary. No such mutants have been found until now.
    Criterion 2) If we believe that a light response in the low
intensity range, is a function of bleached photoreceptor R = f
(P_{bl}) = f $(P_{tot} \sigma I)$, then any mutant that has less photoreceptor

$(P_{tot})_{mut} < (P_{tot})_{wt}$ should have exactly the same behaviour as wt but at higher light intensity. The mutation would be a kind of grey filter. This criterion is necessary but not sufficient.

Are any of the known mutants in the photoreceptor? According to the analysis presented in Fig. 5, only madA and madB could be candidates among the seven genes known. The results of Fig. 3 show that neither madA nor madB mutants satisfy criterion 1. The results of Fig. 8 and Fig. 6, show that they do not satisfy criterion 2. Therefore, there are no mutants, until now, which are in the photoreceptor.

REFERENCES

1. K. Bergman, P. V. Burge, E. Cerdà-Olmedo, C. N. David, M. Delbrück, K. W. Forster, E. W. Goodell, M. Heisenberg, G. Meissner, M. Zalokar, D. S. Dennison, & W. Shropshire, Jr., Bact. Rev. 33:99 (1969).

2. V. E. A. Russo, & P. Galland, in "Sensory Physiology and Structures of Molecules", P. Hemmerich, ed., Springer, Berlin (1969).

3. M. Delbrück, & W. Reichardt, in "Cellular Mechanism in Differentiation and Growth", Princeton University Press, (1956).

4. D. S. Dennison, J. Gen. Physiol.48:393 (1965).

5. E. S. Castle, Science 133:1424 (1961).

6. W. Shropshire Jr., J. Gen. Physiol. 45:949 (1962).

7. D. S. Dennison, & K. W. Forster, Biophys. J. 18:103 (1977).

8. M. Delbrück, & W. Shropshire, Plant. Physiol. 35:194 (1960).

9. K. Bergman, A. P. Eslava, & E. Cerdà-Olmedo, Mol. Gen. Genet. 123:1 (1973).

10. V. E. A. Russo, Plant. Sci. Let. 10:373 (1977b).

11. P. Galland, & V. E. A. Russo, Photochem. Photobiol. 29:1009 (1979).

12. V. E. A. Russo, P. Galland, M. Toselli, & L. Volpi, in "Proceedings International Conference on the Effect of Blue Light in Plants and Microorganisms", H. Senger, ed. Springer, Berlin-Heidelberg-New York (1979) (in press).

13. D. L. Johnson, & R. I. Gamow, J. Gen. Physiol. 57:41 (1971).

14. V. E. A. Russo, J. Bact. 130:548 (1977a).

15. V. E. A. Russo, B. Halloran, & E. Gallori, Planta 134:61 (1977).

16. R. J. Cohen, Y. N. Jan, J. Matricon, & M. Delbrück, J. Gen. Physiol. 66:67 (1975).

17. M. Delbrück, A. Katzir, & D. Presti, Proc. Natl. Acad. Sci. USA 73:1969 (1976).

18. H. R. Mahler, & E. H. Cordes, "Biological Chemistry" Harper and Row Publishers, New York (1971).

19. T. Ootaki, E. P. Fischer, & P. Lockhard, Molec. Gen. Genet. 131:233 (1974).

20. M. Jayaram, D. Presti, & M. Delbrück, Exptl. Mycology 3:42 (1979).

21. K. W. Forster, & E. D. Lipson, J. Gen. Physiol.62:590 (1973).

22. W. Reichardt, & D. Varjù, Z. für Phys. Chem. Neue Folge 15:297 (1958).

23. R. J. Cohen, Nature 251:144 (1974b).

24. R. J. Cohen, Plant. Sci. Let. 13:315 (1978).

25. E. D. Lipson, Biophys. J. 15:1013 (1975).

Panel Discussion: SENSORY TRANSDUCTION AND PHOTOBEHAVIORS: FINAL

CONSIDERATIONS AND EMERGING THEMES.

      Wolfgang Haupt

      Institute of Botany, University

      8520 Erlangen (W. Germany)

     As has been shown in several lectures during the Institute, there are a few organisms in which the transduction chain connecting stimulus perception and (photo)behavior is so well investigated that at least a superficial understanding seems possible. Nevertheless, even in these cases there are always steps in the transduction sequence which have not yet progressed beyond the hypothetical stage. Summarized below are sensory-transduction schemes for several organisms treated in detail by earlier reports in this Institute, viz. Halobacterium halobium (presented by Hildebrand), Phormidium uncinatum (presented by Häder), Euglena gracilis (presented by Diehn) and Mougeotia spec. (presented by Wagner). In addition, the mechanophobic response of Paramecium is included for comparison, because the knowledge of its transduction chain is relatively complete (presented by Hildebrand, based on the results obtained by Naitoh, Eckert, Kung, Machemer, Ogura and other Authors). In the diagrams (except for Euglena), well-established steps are shown by capitals, highly probable steps by lower-case letters, and hypothetical steps by italics.

     Based on these schemes and on discussion of sensory transduction in these and other systems during the Conference, one can attempt to formulate some general conclusions.

Halobacterium: photophobic response

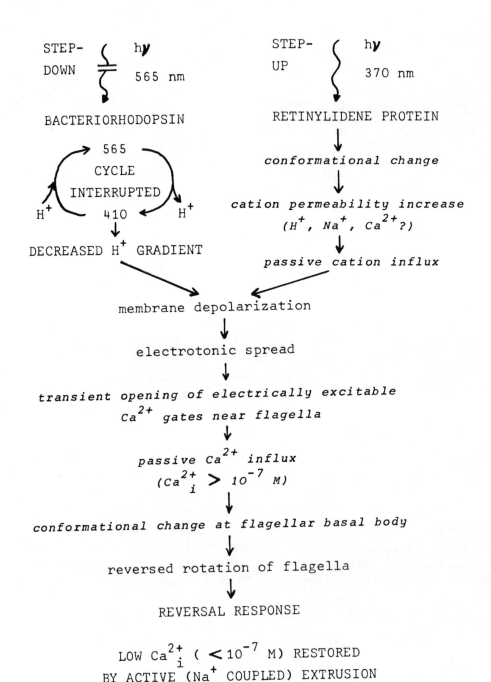

Phormidium: photophobic response

STIMULUS:

$$\frac{dI}{dt}$$

↓

TRANSDUCTION STEPS:

CHANGE IN ELECTRON FLOW RATE

↓

CHANGE IN PLASTOQUINONE

↓

change in proton transport
across thylakoid and
plasma membranes

↓

change in electrical polarity of cell

↓

reversal of anterior-posterior
cell polarity

↓

amplification by
cell-to-cell transfer

↓

cell polarity controls direction
of motor activity

↓

RESPONSE:

REVERSAL OF GLIDING MOVEMENT

Evidence for causal connection of some hypothetical steps: electro-
potential along a trichome 10 mV, light-induced change with 10 s
delay; response inhibited by external electric fields of 10 mV
along a trichome with a minimum pulse length of 10 s; photophobic
response time 10 s.

Euglena gracilis: transduction chain of photophobic
step-down response

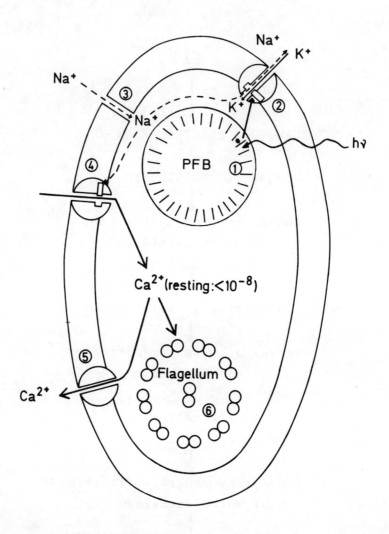

① Flavin chromophores
② Na^+/K^+ pump, light-activated (blocked by ouabain)
③ Na^+ leak or pore
④ Ca^{2+} channel, opened by Na^+ (blocked by La^{3+})
⑤ Ca^{2+} pump, always on
⑥ Flagellum, reorientation upon Ca^{2+} increase (due to
   step-down of the light)

Mougeotia: Chloroplast orientation

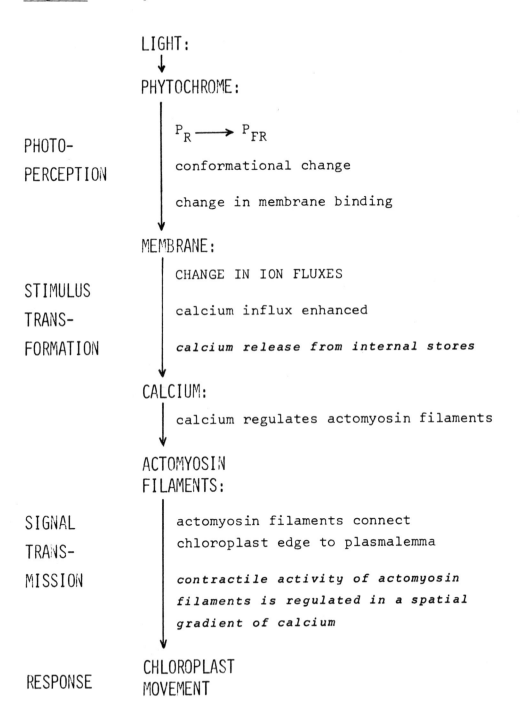

LIGHT:
↓
PHYTOCHROME:

PHOTO-
PERCEPTION

$P_R \longrightarrow P_{FR}$

conformational change

change in membrane binding

MEMBRANE:

STIMULUS
TRANS-
FORMATION

CHANGE IN ION FLUXES

calcium influx enhanced

calcium release from internal stores

CALCIUM:

calcium regulates actomyosin filaments

ACTOMYOSIN
FILAMENTS:

SIGNAL
TRANS-
MISSION

actomyosin filaments connect
chloroplast edge to plasmalemma

*contractile activity of actomyosin
filaments is regulated in a spatial
gradient of calcium*

RESPONSE

CHLOROPLAST
MOVEMENT

Paramecium: mechanophobic response

MECHANICAL (STEP-UP) STIMULUS
↓
ANTERIOR RECEPTOR SITES
AT BODY MEMBRANE
↓
conformational change
of membrane receptor protein
↓
membrane permeability increase for Ca^{2+}
↓
local Ca^{2+} influx
↓
LOCAL MEMBRANE DEPOLARIZATION
↓
ELECTROTONIC SPREAD
↓
TRANSIENT OPENING OF ELECTRICALLY EXCITABLE
Ca GATES AT CILIARY MEMBRANE
↓
REGENERATIVE Ca^{2+} INFLUX ($Ca^{2+}_i > 10^{-7}$M)
↓
activation of ATPase in axoneme
↓
CILIARY REVERSAL
↓
PHOBIC RESPONSE

LOW Ca^{2+}_i ($< 10^{-7}$ M) RESTORED BY
ACTIVE (ATP-DRIVEN) EXTRUSION

1. PERCEPTION LEVEL

On the one hand, many different photoresponses of many different organisms are obviously controlled by identical photoreceptor
pigments. This raises the possibility of a master-reaction at the
beginning of different transduction chains. On the other hand,
there is also an amazing diversity of photoreceptor pigments. Not
only can a given organism make use of different pigments to start
different transduction chains (e.g. Phormidium), but different
organisms can also use different pigments for a given type of
response (e.g. photophobic response). This raises the possibility
of a common early non-pigment-specific transduction step. This concept becomes even more interesting if one is aware of the various
locations of photoreceptor pigments in different organisms. (There
are still well-investigated systems where knowledge of this location is nearly nil!).

Perception of light direction - in case where a photobehavior
is oriented by the light - is understood in principle. Nevertheless
there are still unsolved problems. Even the apparent simple distinction between one-instant and two-instant mechanisms cannot be
established yet in some given organisms, and the function of the
stigma is still obscure in some flagellates.

Understanding the perception of temporal gradients (step-up,
step-down) is barely beyond the speculative stage. Since this perception is probably related to adaptation processes, the latter
urgently need to be analyzed further. It may be that adaptation
processes are not identical in all systems, but can be located at
different sites in the transduction chain. But interestingly, one
general property is suggested by comparison of several systems:
according to Lipson, the adaptive kinetics seem to be asymmetric.
Step-up adaptations tend to proceed with a faster time course than
step-down adaptations. Besides our aneural organisms, this rule
applies to vertebrate vision. Furthermore, it is not restricted to
photoresponses, but has been found also in chemotaxis of E. coli.
Here seems to be an interesting field for further research.

In most photoresponses, there is a qualitative change as a
function of light intensity: positive phototaxis switches to negative, and step-down photophobic response switches to step-up with
increasing light intensity. We have nearly no knowledge about the
nature of "the switch".

2. TRANSDUCTION LEVEL

There is increasing evidence that membranes are intimately involved in transduction steps. Are we always dealing with conformational changes of membrane proteins? And how are such changes brought about by light absorption in the photoreceptor pigment? We are only at the very beginning of an understanding.

Calcium seems to be the most important ion in several well-investigated transduction chains. Calcium pumps are activated or inactivated, calcium gates are opened or closed, giving rise to a localized change in intracellular calcium concentration. This is supposed to result, eventually, in the control of the motor apparatus. And indeed, according to several hypotheses, calcium controls e.g. the beat of flagella or cilia or the activity of microfilaments responsible for intracellular movement. Reasonable as this idea is, we still have to investigate the causal relationship in detail: how is straightening of a flagellum (or cilium) or reversal of its direction of bending controlled by calcium on a molecular level?

Ions other than calcium that are involved in a transduction chain may control the response indirectly by an influence on the calcium pumps or gates. What is the nature of such an influence? Another problem arises with these ions: are we dealing with a change in the leakage properties of the membranes, or is active pumping involved?

It is true, in particular cases, some of these questions can already be answered (and have been answered), but we are still far from a general understanding. It is to be hoped that the next NATO School on Photoreception and Sensory Transduction in Aneural Organisms will deal with progress in this wide field of current and future research.

LIST OF PARTICIPANTS

Members of the Advisory and Organizing Committees are indicated by
an asterisk

AMESZ, J.
State University Leiden
Huygens Laboratory
Biophysics Dept.
Wassenaarseweg 78
2300 RA Leiden
Netherlands

ASCOLI, C.
C.N.R. - Lab. Studio Proprietà
Fisiche di Biomolecole e Cellule
Via F. Buonarroti,9
56100 Pisa
Italy

*BALDOCCHI, M. A.
C.N.R. - Lab. Studio Proprietà
Fisiche di Biomolecole e Cellule
Via F.Buonarroti,9
56100 Pisa
Italy

BARCELO, J. A.
School of Biological Sciences
University of Kentucky
Lexington
Kentucky 40506
U.S.A.

*BENSASSON, R. V.
Lab. Biophysique
Museum National
d'Histoire Naturelle
61 rue Buffon
75005 Paris
France

BIZZARO, M. P.
C.N.R. - Lab. Studio Proprietà
Fisiche di Biomolecole e Cellule
Via F. Buonarroti,9
56100 Pisa
Italy

BLATT, M.
Carnegie Inst. of Washington
290 Panama St.
Stanford
Ca. 94305
U.S.A.

BOZZI, A.
Ist. Chimica Biologica dell'Univ.
Città Universitaria
00185 Roma
Italy

BRIDELLI, M. G.
Istituto di Fisica
Via M. D'Azeglio,85
43100 Parma
Italy

BRUHN, H.
Fritz-Haber Inst. der MPG
Faradayweg 4-6
1 Berlin 33
Federal Republic of Germany

BURR, A. H.
Dept. Biol. Sciences
Simon Fraser University
Burnaby B. C. V5A 1S6
Canada

CARLILE, M. J.
Imperial College
Science and Technology
Dept. Biochemistry
London SW7 2AZ
England

CAUBERGS, R.
Univ. Antwerpen R.U.C.A.
Groenenborgelaan 171
B-2020 Antwerpen
Belgium

CHESSIN, M.
Univ. Montana
Botany Dept.
Missoula
Montana 59812
U.S.A.

*COLOMBETTI, G.
C.N.R. - Lab. Studio Proprietà
Fisiche di Biomolecole e Cellule
Via F. Buonarroti,9
56100 Pisa
Italy

CREUTZ, C.
Univ. Toledo
Biology Dept.
2801 W. Bancroft Street
Toledo
Ohio 43606
U.S.A.

DAHL, D.
C. W. Post Center
Long Island Univ.
C. W. Post College
P.O. Greenvale
N. Y. 11548
U.S.A.

DE LA ROSA AGOSTA, F.
Physiologia   Vegetal
Dept. Bioquimica
Fac. Biologia
Univ. Sevilla
Sevilla
Spain

DENCHER, N.
Inst. Neurobiologie der
Kernforschungsanlage Jülich
Post-Fach 1913
D-Jülich 1
Federal Republic of Germany

DIEHN, B.
Dept. Zoology
Michigan State University
East Lansing
Michigan 48824
U.S.A.

DOHRMANN, U.
Pflanzenphysiologisches Inst.
Universitaet
Untere Karspüle 2
3400 Gottingen
Federal Republic of Germany

FANI, R. E.
Ist. di Genetica
Università di Firenze
Via Romana,19
50143 Firenze
Italy

FEINLEIB, M. E.
Biology Dept.
Tufts University
Medford
Mass. 02155
U.S.A.

FERNANDEZ, H. R. C.
Wayne State University
Dept. Biology
Detroit
Michigan 48203
U.S.A.

FORD, M.
Queen Elizabeth College
London
England

FUKSHANSKY, L.
Albert-Ludwigs-Universität
Biologisches Institut II
D-7800 Freiburg
Federal Republic of Germany

GALLAND, P.
California Inst. of Technology
Division of Biology 156-29
Pasadena
California 91125
U.S.A.

GIRNTH, C.
Carlsberg Lab.
Gamle Carlsbergsvej 10
DK - Copenhagen Valby
Denmark

GHISLA, S.
Fb. Biologie Universität
D-7775 Konstanz
Federal Republic of Germany

GONZALES-RONCERO, I
Dept. of Genetica
Facultated de Biologia
Univ. de Sevilla
Sevilla
Spain

HÄDER, D. P.
Fachbereich Biologie-Botanik der
Philipps Univ. Marburg
Marburg/Lahn.
Federal Republic of Germany

HÄDER, M.
Fachbereich Biologie-Botanik der
Philipps Univ. Marburg
Marburg/Lahn.
Federal Republic of Germany

HAUPT, W.
Inst. für Botanik und
Pharmazeutische Biologie der
Universität
Erlangen
Schlossgarten 4
Federal Republic of Germany

HEELIS, P. F.
The North E Wales Inst.
Kelsterton College
Connah's Quay
Clwyd CH5 4BR
England

HEELIS, D. V.
The North E Wales Inst.
Kelsterton College
Connah's Quay
Clwyd CH5 4BR
England

HEMMERICH, P.
Universität Konstanz
Fachbereich Biologie
Postfach 7733
D-7750 Konstanz
Federal Republic of Germany

*HERTEL, R.
Institut für Biologie III
Universität Freiburg
Schanzlestrasse 1
78-Freiburg i.Br.
Federal Republic of Germany

HILDEBRAND, E.
Inst. fur Neurobiologie der
Kernforschungsanlage Jülich
Postfach 1913
D-Jülich 1
Federal Republic of Germany

HOUBEN, J. L.
C.N.R. - Lab. Studio Proprietà
Fisiche di Biomolecole e Cellule
Via F. Buonarroti,9
56100 Pisa
Italy

HUBER, A.
Botanisches Institut der
Universität München
Menzinger Str. 67
8000 München 19
Federal Republic of Germany

KRAML, M.
Inst. für Botanik und
Pharmazeutische Biologie der
Univ. Erlangen-Nürnberg
Schlossgarten 4
D-8520 Erlangen
Federal Republic of Germany

*LENCI, F.
C.N.R. - Lab. Studio Proprietà
Fisiche di Biomolecole e Cellule
Via F. Buonarroti,9
56100 Pisa
Italy

LERCARI, B.
Ist. Orticultura Floricultura
dell'Università
Viale delle Piagge,23
56100 Pisa
Italy

LEUTWILER, L.
Division of Biology
California Inst. of Technology
Pasadena
California 91125
U.S.A.

LIPSON, E.
Syracuse Univ.
Dept. of Physics
201 Physics Building
Syracuse
New York 13210
U.S.A.

LOPEZ-DIAZ, I.
Departemento de Genetica
Facultad de Biologia
Universidad de Sevilla
Sevilla
Spain

MALDONADO, J. M.
Fisiologia Vegetal
Dept. of Bioquimica
Facultad de Biologia
Univ. de Sevilla
Sevilla
Spain

MASSEY, V.
Dept. of Biological Chemistry
The Univ. Michigan
Medical School
Ann Arbor
Michigan 48109
U.S.A.

MELANDRI, B. A.
Università di Bologna
Ist. e Orto Botanico
Via Irnerio,42
40126 Bologna
Italy

MIGLIORE, L.
Inst. Genetica dell'Università
Cattedra di Ecologia
Città Universitaria
00185 Roma
Italy

MITZKA-SCHNABEL, U.
Botanisches Inst. der
Universität München
Menzinger Str.   19
8000 München 19
Federal Republic of Germany

MONTAGNOLI, G.
C.N.R. - Lab. Studio Proprietà
Fisiche di Biomolecole e Cellule
Via F. Buonarroti,9
56100 Pisa
Italy

MOORE, T. A.
Arizona State University
Dept. of Chemistry
Tempe
Arizona 85281
U.S.A.

MOREL, N.
Tufts University
Dept. of Biology
Medford
Mass. 02155
U.S.A.

*MORELLI, R.
C.N.R. - Lab. Studio Proprietà
Fisiche di Biomolecole e Cellule
Via F. Buonarroti,9
56100 Pisa
Italy

*NULTSCH, W.
Lehrstul für Botanik
Fachbereich Biologie der
Philipps Univ. Marburg
Marburg/Lahn.
Federal Republic of Germany

OMODEO, P.
Università Degli Studi
di Padova
Ist. di Biologia Animale
Via Loredan,10
35100 Padova
Italy

*PACCHINI, C.
Via di Pratale,30
56100 Pisa
Italy

PANAJOTOPULU, E.
Greek Atomic Commission
Nuclear Res. Center Demokritos
Biology Dept.
Aghia Paraskevi
Athens
Greece

*PASSARELLI, V.
C.N.R. - Lab. Studio Proprietà
Fisiche di Biomolecole e Cellule
Via F. Buonarroti,9
56100 Pisa
Italy

PASTERNAK, C.
The Weizman Inst. of Sciences
Dept. Membrane Research
Rehovot
Israel

PFAU, J.
Fachebereich Biologie der
Philipps Universität Botanik
355 Marburg/Lahn.
W. Germany

POFF, K. L.
MSU-DOE Plant Res. Laboratory
Michigan State University
East Lansing
Michigan 48824
U.S.A.

RAUGEI, G.
Ist. Genetica
Università Degli Studi di
Firenze
Via Romana, 19
50126 Firenze
Italy

RISTORI, T.
C.N.R. - Lab. Studio Proprietà
Fisiche di Biomolecole e Cellule
Via F. Buonarroti,9
56100 Pisa
Italy

*RUSSO, V. E. A.
Max-Planck-Inst. für
Molekulare Genetik
Inhestrasse 63-73
1000 Berlin 33
W. Germany

SCHEUERLEIN, R.
Botanical Inst. Erlangen
Schlossgarten 4
D-8520 Erlangen
Federal Republic of Germany

SCHÖNBOHM, E.
Fachbereich Biologie der
Philipps Univ. Botanik
355 Marburg/Lahn.
Federal Republic of Germany

SCHUCHART, H.
Fachbereich Biologie der
Philipps Univ. Botanik
355 Marburg/Lahn.
Federal Republic of Germany

SIEP, E.
Albert-Ludwigs-Universität
Inst. für Biologie III
Schanzlenstrasse
7800 Freiburg i.Br.
Federal Republic of Germany

SONG, P. S.
Dept. of Chemistry
Texas Tech   University
Lubbock
Texas 79409
U.S.A.

SPIKES, J. D.
The University of Utah
Dept. of Biology
Biology Building 201
Salt Lake City
Utah 84112
U.S.A.

STAVENGA, D. G.
Laboratorium voor Algemene
Natuurkunde Rijksuniversiteit
Westersingel 34
9718 CM Groningen
Netherlands

TADDEI, C.
Lab. Cibernetica
Via Toiano,2
80072 Arco Felice
Napoli
Italy

TOSELLI, M.
Max-Planck-Institut für
Molekulare Genetik
Ihnestrasse 63-73
1 Berlin 33
Dahlem
Federal Republic of Germany

VERMUE, N.
Laboratorium voor Algemene
Natuurkunde Rijksunversiteit
Westersingel 34
9718 CM Groningen
Netherlands

VIERSTRA, R.
MSU-DOE Plant Research Lab.
Michigan State University
East Lansing
Michigan   48824
U.S.A.

VOLPI, L.
Istituto di Genetica
dell'Università
Via Celoria,10
20133 Milano
Italy

WAGNER, G.
Inst. für Botanik und
Pharmazeutische Biologie
der Universität
Erlangen-Nurnberg
Schlossgarten 4
D-8520 Erlangen
Federal Republic of Germany

WENDEROTH, K.
Fachbereich Biologie der
Philipps-Universitat Botanik
355 Marburg/Lahn.
Federal Republic of Germany

INDEX